Springer Handbook of Auditory Research

Volume 66

More information about this series at http://www.springer.com/series/2506

The ASA Press

The ASA Press imprint represents a collaboration between the Acoustical Society of America and Springer dedicated to encouraging the publication of important new books in acoustics. Published titles are intended to reflect the full range of research in acoustics. ASA Press books can include all types of books published by Springer and may appear in any appropriate Springer book series.

ASA Press

Hans Slabbekoorn • Robert J. Dooling
Arthur N. Popper • Richard R. Fay
Editors

Effects of Anthropogenic Noise on Animals

ASA Press

Springer

Editors
Hans Slabbekoorn
Faculty of Science
Institute of Biology Leiden (IBL)
Leiden University
Leiden, The Netherlands

Arthur N. Popper
Department of Biology
University of Maryland
Silver Spring, MD, USA

Robert J. Dooling
Department of Psychology
University of Maryland
College Park, MD, USA

Richard R. Fay
Loyola University Chicago
Chicago, IL, USA

ISSN 0947-2657 ISSN 2197-1897 (electronic)
Springer Handbook of Auditory Research
ISBN 978-1-4939-8572-2 ISBN 978-1-4939-8574-6 (eBook)
https://doi.org/10.1007/978-1-4939-8574-6

Library of Congress Control Number: 2018953164

Printed on acid-free paper

This Springer imprint is published by the registered company Springer Science+Business Media, LLC part of Springer Nature.
The registered company address is: 233 Spring Street, New York, NY 10013, U.S.A.

This volume is dedicated to the memory of Professor Douglas B. Webster (1934–2017), a remarkable teacher, mentor, scholar, colleague, and friend. All of the editors know Doug's fascinating work on hearing in desert rodents, but Dick Fay and Art Popper had the

pleasure of working closely with Doug on two very important volumes in which his scholarly expertise was critical to the success of the books. The first was The Evolutionary Biology of Hearing *(Webster, D. B., Fay, R. R., and Popper, A. N., 1992, Springer-Verlag, New York) and the second was volume 1 in the SHAR series in which Doug was the first senior editor (*The Mammalian Auditory Pathways: Neuroanatomy, *Webster, D. B., Popper, A. N., and Fay, R. R., 1992, Springer-Verlag, New York). Art is particularly grateful to Doug because it was in Doug's lab at New York University that Art (then an undergraduate) was first introduced to comparative biology and research on the auditory system.*

Acoustical Society of America

The purpose of the Acoustical Society of America (www.acousticalsociety.org) is to generate, disseminate, and promote the knowledge of acoustics. The Acoustical Society of America (ASA) is recognized as the world's premier international scientific society in acoustics, and counts among its more than 7000 members, professionals in the fields of bioacoustics, engineering, architecture, speech, music, oceanography, signal processing, sound and vibration, and noise control.

Since its first meeting in 1929, the ASA has enjoyed a healthy growth in membership and in stature. The present membership of approximately 7000 includes leaders in acoustics in the United States of America and around the world. The ASA has attracted members from various fields related to sound including engineering, physics, oceanography, life sciences, noise and noise control, architectural acoustics; psychological and physiological acoustics; applied acoustics; music and musical instruments; speech communication; ultrasonics, radiation, and scattering; mechanical vibrations and shock; underwater sound; aeroacoustics; macrosonics; acoustical signal processing; bioacoustics; and many more topics.

To assure adequate attention to these separate fields and to new ones that may develop, the Society establishes technical committees and technical groups charged with keeping abreast of developments and needs of the membership in their specialized fields. This diversity and the opportunity it provides for interchange of knowledge and points of view has become one of the strengths of the Society.

The ASA's publishing program has historically included *The Journal of the Acoustical Society of America, JASA-Express Letters, Proceedings of Meetings on Acoustics*, the magazine *Acoustics Today*, and various books authored by its members across the many topical areas of acoustics. In addition, ASA members are involved in the development of acoustical standards concerned with terminology, measurement procedures, and criteria for determining the effects of noise and vibration.

Series Preface

Springer Handbook of Auditory Research

The following preface is the one that we published in volume 1 of the Springer Handbook of Auditory Research back in 1992. As anyone reading the original preface, or the many users of the series, will note, we have far exceeded our original expectation of eight volumes. Indeed, with books published to date and those in the pipeline, we are now set for over 75 volumes in SHAR, and we are still open to new and exciting ideas for additional books.We are very proud that there seems to be consensus, at least among our friends and colleagues, that SHAR has become an important and influential part of the auditory literature. While we have worked hard to develop and maintain the quality and value of SHAR, the real value of the books is very much because of the numerous authors who have given their time to write outstanding chapters and to our many co-editors who have provided the intellectual leadership to the individual volumes. We have worked with a remarkable and wonderful group of people, many of whom have become great personal friends of both of us. We also continue to work with a spectacular group of editors at Springer. Indeed, several of our past editors have moved on in the publishing world to become senior executives. To our delight, this includes the current president of Springer US, Dr. William Curtis. But the truth is that the series would and could not be possible without the support of our families, and we want to take this opportunity to dedicate all of the SHAR books, past and future, to them. Our wives, Catherine Fay and Helen Popper, and our children, Michelle Popper Levit, Melissa Popper Levinsohn, Christian Fay, and Amanda Fay Sierra, have been immensely patient as we developed and worked on this series. We thank them and state, without doubt, that this series could not have happened without them. We also dedicate the future of SHAR to our next generation of (potential) auditory researchers—our grandchildren—Ethan and Sophie Levinsohn, Emma Levit, and Nathaniel, Evan, and Stella Fay.

Preface 1992

The Springer Handbook of Auditory Research presents a series of comprehensive and synthetic reviews of the fundamental topics in modern auditory research. The volumes are aimed at all individuals with interests in hearing research including advanced graduate students, postdoctoral researchers, and clinical investigators. The volumes are intended to introduce new investigators to important aspects of hearing science and to help established investigators to better understand the fundamental theories and data in fields of hearing that they may not normally follow closely.

Each volume presents a particular topic comprehensively, and each serves as a synthetic overview and guide to the literature. As such, the chapters present neither exhaustive data reviews nor original research that has not yet appeared in peer-reviewed journals. The volumes focus on topics that have developed a solid data and conceptual foundation rather than on those for which a literature is only beginning to develop. New research areas will be covered on a timely basis in the series as they begin to mature.

Each volume in the series consists of a few substantial chapters on a particular topic. In some cases, the topics will be ones of traditional interest for which there is a substantial body of data and theory, such as auditory neuroanatomy (Vol. 1) and neurophysiology (Vol. 2). Other volumes in the series deal with topics that have begun to mature more recently, such as development, plasticity, and computational models of neural processing. In many cases, the series editors are joined by a co-editor having special expertise in the topic of the volume.

Richard R. Fay, Chicago, IL, USA
Arthur N. Popper, College Park, MD, USA

SHAR logo by Mark B. Weinberg, Potomac, Maryland, used with permission.

Volume Preface

Over the past several years, many investigators interested in the effects of man-made sounds on animals have come to realize that there is much to gain from studying the broader literature on hearing and on the effects of sound, as well as from data on the effects of sound on humans. It has also become clear that knowledge of the effects of sound on one group of animals (e.g., birds or frogs) can guide studies on other groups (e.g., marine mammals or fishes) and that a review of all such studies together would be very useful to get a better understanding of the general principles and underlying cochlear and cognitive mechanisms that explain damage, disturbance, and deterrence across taxa.

The purpose of this volume, then, is to provide a comprehensive review of the effects of man-made sounds on animals, with the goal of fulfilling two major needs. First, it was thought to be important to bring together data on sound and bioacoustics that have implications across all taxa (including humans) so that such information is generally available to the community of scholars interested in the effects of sound. This is done in Chaps. 2–5. Second, in Chaps. 6–10, the volume brings together what is known about the effects of sound on diverse vertebrate taxa so that investigators with interests in specific groups can learn from the data and experimental approaches from other species. Put another way, having an overview of the similarities and discrepancies among various animal groups and insight into the "how and why" will benefit the overall conceptual understanding, applications in society, and all future research.

Chapter 1 by Hans Slabbekoorn, Robert J. Dooling, and Arthur N. Popper provides an overview of the topic of the book and an introduction to the chapters.

The next two chapters address hearing complications and perceptual strategies under challenging conditions in terms of noisy and complex acoustic environments. Chapter 2 by Robert J. Dooling and Marjorie R. Leek addresses the phenomenon of masking of biologically relevant sounds while in Chap. 3, Micheal L. Dent and Mark A. Bee go beyond masking as they address the perceptual mechanisms for extracting relevant signals from a background of potentially distracting sounds. This is followed by Chap. 4 in which James C. Saunders and Robert J. Dooling address issues related to sound levels that are very high and result in potential hearing damage. At the end of this part of the book, Chap. 5 by Ole Næsbye Larsen and Craig Radford covers the physical properties of air and water in terms of sound transmission.

In Chap. 6, Anthony D. Hawkins and Arthur N. Popper start the taxonomic reviews by addressing what is known about fishes. In Chap. 7, Andrea Megela Simmons and Peter M. Narins address the literature on the effects of man-made sounds on amphibians, which include not only frogs and toads but also salamanders, newts, and the caecilians (limbless amphibians). Then, in Chap. 8, Wouter Halfwerk, Bernard Lohr, and Hans Slabbekoorn focus on birds, including the relatively homogeneous and well-studied songbirds, about half of all 10,000 species of birds.

The final two chapters turn to mammals. In Chap. 9, Hans Slabbekoorn, JoAnn McGee, and Edward J. Walsh address the wide-ranging hearing abilities as well as the wide-ranging types of investigations on the sound impact for the diverse group of terrestrial mammals. Then, in Chap. 10, Christine Erbe, Rebecca Dunlop, and Sarah Dolman address marine mammals, including cetaceans (whales, dolphins, and porpoises) and sirenians (sea cows) that are fully aquatic, and also several marine carnivores (seals and walruses) that spend time both on land and in water.

Hans Slabbekoorn, Leiden, The Netherlands
Robert J. Dooling, College Park, MD
Arthur N. Popper, College Park, MD
Richard R. Fay, Chicago, IL

Contents

1 **Man-Made Sounds and Animals** 1
Hans Slabbekoorn, Robert J. Dooling, and Arthur N. Popper

2 **Communication Masking by Man-Made Noise** 23
Robert J. Dooling and Marjorie R. Leek

3 **Principles of Auditory Object Formation by Nonhuman Animals** . . . 47
Micheal L. Dent and Mark A. Bee

4 **Characteristics of Temporary and Permanent Threshold Shifts in Vertebrates** 83
James C. Saunders and Robert J. Dooling

5 **Acoustic Conditions Affecting Sound Communication in Air and Underwater** 109
Ole Næsbye Larsen and Craig Radford

6 **Effects of Man-Made Sound on Fishes** 145
Anthony D. Hawkins and Arthur N. Popper

7 **Effects of Anthropogenic Noise on Amphibians and Reptiles** 179
Andrea Megela Simmons and Peter M. Narins

8 **Impact of Man-Made Sound on Birds and Their Songs** 209
Wouter Halfwerk, Bernard Lohr, and Hans Slabbekoorn

9 **Effects of Man-Made Sound on Terrestrial Mammals** 243
Hans Slabbekoorn, JoAnn McGee, and Edward J. Walsh

10 **Effects of Noise on Marine Mammals** 277
Christine Erbe, Rebecca Dunlop, and Sarah Dolman

Contributors

Mark A. Bee Department of Ecology, Evolution, and Behavior, University of Minnesota, St. Paul, MN, USA

Micheal L. Dent Department of Psychology, University at Buffalo, State University of New York (SUNY), Buffalo, NY, USA

Sarah Dolman Whale and Dolphin Conservation, Chippenham, Wiltshire, UK

Robert J. Dooling Department of Psychology, University of Maryland, College Park, MD, USA

Rebecca Dunlop Cetacean Ecology and Acoustics Laboratory, School of Veterinary Science, University of Queensland, Gatton, QLD, Australia

Christine Erbe Centre for Marine Science and Technology, Curtin University, Perth, WA, Australia

Wouter Halfwerk Department of Ecological Sciences, Faculty of Earth and Life Sciences, VU University Amsterdam, Amsterdam, The Netherlands

Anthony D. Hawkins Loughine Ltd, Aberdeen, UK

Ole Næsbye Larsen Department of Biology, University of Southern Denmark, Odense M, Denmark

Marjorie R. Leek VA Loma Linda Healthcare System, Loma Linda, CA, USA

Bernard Lohr University of Maryland, Baltimore County (UMBC), Baltimore, MD, USA

JoAnn McGee Developmental Auditory Physiology Laboratory, Boys Town National Research Hospital, Omaha, NE, USA

Peter M. Narins Department of Integrative Biology and Physiology, University of California, Los Angeles, Los Angeles, CA, USA

Department of Ecology and Evolutionary Biology, University of California, Los Angeles, Los Angeles, CA, USA

Arthur N. Popper Department of Biology, University of Maryland, College Park, MD, USA

Craig Radford Leigh Marine Laboratory, Institute of Marine Science, University of Auckland, Warkworth, New Zealand

James C. Saunders Department of Otorhinolaryngology, Head and Neck Surgery, The Perelman School of Medicine, University of Pennsylvania, Philadelphia, PA, USA

Andrea Megela Simmons Department of Cognitive, Linguistic and Psychological Sciences, Brown University, Providence, RI, USA

Department of Neuroscience, Brown University, Providence, RI, USA

Hans Slabbekoorn Faculty of Science, Institute of Biology Leiden (IBL), Leiden University, Leiden, The Netherlands

Edward J. Walsh Developmental Auditory Physiology Laboratory, Boys Town National Research Hospital, Omaha, NE, USA

Chapter 1
Man-Made Sounds and Animals

Hans Slabbekoorn, Robert J. Dooling, and Arthur N. Popper

Abstract The world is full of sounds of abiotic and biotic origin, and animals may use those sounds to gain information about their surrounding environment. However, it is becoming increasingly clear that the presence of man-made sounds has the potential to undermine the ability of animals to exploit useful environmental sounds. This volume provides an overview of how sounds may affect animals so that those interested in the effects of man-made sounds on animals can better understand the nature and breadth of potential impacts. This chapter provides an introduction to the issues associated with hearing and man-made sound and serves as a guide to the succeeding chapters. Chapters 2, 3, 4 and 5 cover the basic principles of sound and hearing, including an introduction to the acoustic ecology of the modern world in which man-made sounds have become very prominent. They also address how noisy conditions may hinder auditory perception, how hearing adaptations allow coping under acoustically challenging conditions, and how man-made sounds may damage the inner ear. The role of sound propagation in affecting signals and noise levels is treated for both terrestrial and aquatic habitats. This chapter also provides an overview of hearing and the effects of sound on particular taxa, which are the focus of Chaps. 6, 7, 8, 9, and 10. Those chapters address the concepts and insights in five different vertebrate taxa: fishes, amphibians and reptiles, birds, terrestrial mammals, and marine mammals. The overall aim of this volume is to stimulate and guide future investigations to fill in taxonomic and conceptual gaps in the knowledge about how man-made sounds affect animals.

H. Slabbekoorn (✉)
Faculty of Science, Institute of Biology Leiden (IBL), Leiden University, Leiden, The Netherlands
e-mail: H.W.Slabbekoorn@Biology.LeidenUniv.NL

R. J. Dooling
Department of Psychology, University of Maryland, College Park, MD, USA
e-mail: rdooling@umd.edu

A. N. Popper
Department of Biology, University of Maryland, College Park, MD, USA
e-mail: apopper@umd.edu

H. Slabbekoorn et al. (eds.), *Effects of Anthropogenic Noise on Animals*, Springer Handbook of Auditory Research 66,
https://doi.org/10.1007/978-1-4939-8574-6_1

Keywords Acoustic deterrence device · Anthropogenic noise · Comparative review · Experimental design · Man-made sound · Noise impact studies · Vertebrates

1.1 Introduction to the Volume

1.1.1 The Problem

The past decades have seen increased interest in questions concerning the effects of man-made sounds on animals (e.g., Fletcher and Busnel 1978; Popper and Hawkins 2012, 2016). The overall issue, however, is not new, especially with regard to the potential effects of sound on humans. Indeed, a few years ago, the World Health Organization published a report on the topic (World Health Organization 2011), and the issue of potential effects of noise on humans has been the subject of much research and regulation (e.g., Le Prell et al. 2012; Murphy and King 2014). Moreover, it is now quite clear that many of the issues associated with the potential effects of man-made sound on humans (Miedema and Vos 2003; Basner et al. 2014) apply equally to animals (Francis and Barber 2013; Shannon et al. 2016).

The increased concern about the effects of man-made sounds on animals arises from the substantial increase in environmental noise produced by everything from roadway traffic to airplane overflights and from vessel noise to offshore exploration for oil and gas (Andrew et al. 2002; Mennit et al. 2015). The nature of these sounds varies dramatically, from the brief or intermittent high-impact signals produced by destruction or construction activities to the continuously increased background sound levels due to gradually fluctuating amounts of car and vessel traffic (e.g., Singh and Davar 2004; Hildebrand 2009).

The potential effects on animals (as on humans) also vary rather substantially, from immediate death due to overexposure from extremely intense sounds to changes in physiological stress levels that may or may not have long-term consequences. The potential effects may also range from temporary or permanent hearing loss to behavioral changes that result in animals interrupting activities or leaving their normal home range (Kight and Swaddle 2011; Popper and Hawkins 2016).

Additionally, more subtle man-made sounds may make biologically important signals or cues inaudible due to masking or may undermine optimal reception by distraction, which are effects that may have indirect but severe, detrimental consequences (Slabbekoorn et al. 2010; Shannon et al. 2016). Not being able to hear or pay sufficient attention to conspecific communication signals may mean missing important social aggregations or mating opportunities. Failing to recognize acoustic cues from the surrounding habitat may also result in the inability to find shelter or the right migratory route. Not hearing prey may prevent animals finding food. Not detecting a predator may even lead to sudden death.

The distribution and probability of potential effects on free-ranging animals can be viewed from different perspectives as reflected in the diversity in schematic

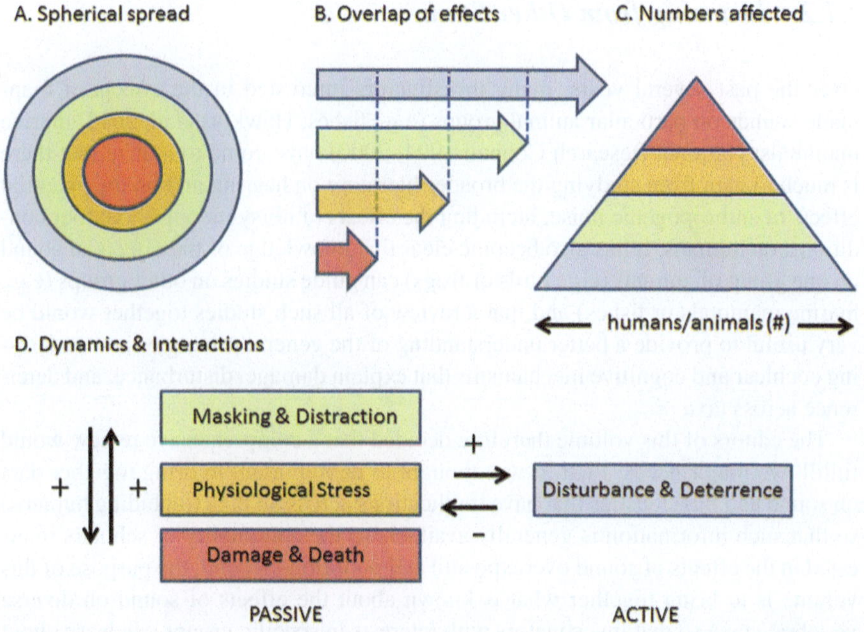

Fig. 1.1 Schematic illustrations providing insight into the nature of potential impact of man made sounds on animals and emphasizing different aspects. **A**: noise impact severity is likely to decrease with distance away from the sound source in all directions due to propagation loss of sound energy via spherical spread (based on Richardson et al. 1995). **B**: the variety of potential effects accumulates with proximity to the sound source because the effects typically do not exclude each other but exhibit zones of overlap (see Hawkins and Popper, Chap. 6). **C**: a pyramid of noise-induced health effects, with the growing severity of the effect toward the top segment and the growing number of individuals that are likely to be affected toward the bottom segment (Babisch 2002). **D**: some potential effects are an inherent and passive consequence of sound exposure, whereas others depend on an active response of the animal itself (from Slabbekoorn 2012). Many potential effects are likely to be positively correlated (*up and down arrows* and *arrow to the right*). If exposure causes severe impact in one direction, it will likely do so in the other. However, a negative correlation may arise by negative feedback (*arrow to the left*) when an active behavioral response makes animals less vulnerable in other ways

representations (see Fig. 1.1). Increasing the distance from the source is almost always correlated with lower exposure levels, less severe effects, and less different overlapping effects (Richardson et al. 1995; Hawkins and Popper 2016). However, as severity goes down at the individual level, the potential effects at the community or even population level may go up as the number of individuals exposed becomes larger (Babisch 2002; Kunc et al. 2016). Some effects are an inherent consequence of passive exposure, whereas other effects only arise after an active decision by the animal itself (Slabbekoorn 2012). Furthermore, different overlapping effects may not only occur together, but they are also likely to covary and may have positive or negative feedback interactions.

1.1.2 Learning from Other Studies

Over the past several years, many investigators interested in the effects of man-made sounds on particular animal groups (e.g., fishes: Hawkins et al. 2015; marine mammals: National Research Council 1994, 2000) have come to realize that there is much to gain from studying the broader literature on hearing and on the potential effects of anthropogenic noise, including the effects of noisy indoor or outdoor conditions, on humans. It has also become clear that knowledge of the effects of sound on one group of animals (e.g., birds or frogs) can guide studies on other groups (e.g., marine mammals or fishes) and that a review of all such studies together would be very useful to provide a better understanding of the general principles and underlying cochlear and cognitive mechanisms that explain damage, disturbance, and deterrence across taxa.

The editors of this volume therefore decided that a comprehensive review would fulfill two major needs. First, it was thought to be important to bring together data on sound and bioacoustics that have implications across all taxa (including humans) so that such information is generally available to the community of scholars interested in the effects of sound overexposure and pollution. Second, the purpose of this volume is to bring together what is known about the effects of sound on diverse vertebrate taxa so that investigators with interests in specific groups can learn about the data from other species as well as about the experimental approaches used to obtain the data. Put another way, having an overview of the similarities and differences among various animal groups and insight into the "how and why" will benefit the overall conceptual understanding, applications in society, and future research.

Accordingly, this volume has two parts. Chapters 2, 3, 4, and 5 discuss the fundamental principles of sound and how noisy conditions may hinder auditory perception and how perceptual abilities, shaped over evolutionary time, can make the best of a potentially bad situation. In addition, damage to the inner ear after exposure to loud sounds for substantial periods is addressed, along with its consequences for hearing and conditions for complete or partial recovery of auditory function. How sounds propagate through the environment and how attenuation and degradation alter the signals from the senders and the cues from abiotic and biotic sources before they end up at the receivers are also covered.

In Chaps. 6, 7, 8, 9, and 10, the ideas and phenomena addressed in Chaps. 2, 3, 4, and 5 are applied to all major vertebrate taxa. To keep the length of this volume reasonable, it was decided to focus on vertebrates. However, it is recognized that many invertebrates detect sound and use their acoustic environment. This includes insects (Hoy et al. 1998; Montealegre-Z et al. 2012), crustaceans (Montgomery et al. 2006; Filiciotto et al. 2016), bivalves (Wilkens et al. 2012; Lillis et al. 2013), cephalopods (e.g., Mooney et al. 2010), and coral larvae (e.g. Vermeij et al. 2010). Several reviews have addressed these groups and what is known about the effects of man-made sounds although, in fact, very little is known about the effects on invertebrates and this is an area of growing interest and concern, particularly in the aquatic environment (Morley et al. 2014; Hawkins and Popper 2016).

Thus, the variety in hearing abilities and functions found among the vertebrate taxa is already impressive and a challenging but still reasonable and appropriate target for the current overview and integration. The variety of investigations addressed in this volume makes it clear that the basic ideas and principles discussed in the first chapters apply to all these vertebrate taxa treated later and to many invertebrate groups as well. This volume will hopefully stimulate and guide future investigations to fill in taxonomic and conceptual gaps in our knowledge.

1.2 Acoustic Ecology of the Modern World

1.2.1 Perceiving the Auditory Scene

Auditory challenges in human-altered environments may be novel selection pressures at an evolutionary timescale, and sound levels may often exceed typical naturally occurring ambient amplitudes. However, so-called competition for acoustic space in itself is not new, and the natural world surrounding animals (and humans) is full of sound (Brumm and Slabbekoorn 2005; Hildebrand 2009). The auditory senses that serve critical functions for survival and reproduction should therefore be regarded as shaped by selection under naturally fluctuating and often noisy conditions (Klump 1996; Wiley 2017). Sounds naturally occurring in the environment include abiotic sounds, generated by wind or rain and by rivers or oceans, and biotic sounds, generated by all members of more or less noisy local animal communities.

Habitat features above and below the water surface determine whether and how sounds originating at one point arrive at receivers and whether and how they may play a role in affecting their behavior (Wiley and Richards 1978; see Larsen and Radford, Chap. 5). Vegetation may attenuate and filter out or resonate and amplify particular frequencies. Sounds may be reflected by the ground below or the surface above, and reverberations may accumulate over distance and with habitat complexity. Industrialization and urbanization have not only added new, diverse sound sources to the modern world (Pine et al. 2016; Shannon et al. 2016) but also dramatic changes in propagation as a result of altered vegetation or novel obstacles and a multitude of reflective surfaces (roads, houses, and buildings; Warren et al. 2006).

In a world full of sounds, there is much to learn about the surrounding environment if these sounds are detected, discriminated, and recognized (see Fig. 1.2) by the appropriate sensory and processing tools (see Dooling and Leek, Chap. 2; Dent and Bee, Chap. 3). Sound is not only highly suitable for eavesdropping on biologically relevant events (e.g., listening for cues of predators or prey) but also for communication among conspecifics (e.g., signals that have evolved by natural selection through the effects on mate attraction or competitor deterrence). For example, sound, unlike visual signals, is not hampered by lack of light or the presence of vegetation and other objects in the environment. Furthermore, although attenuation and degradation during propagation will limit the range of potential use, the resultant

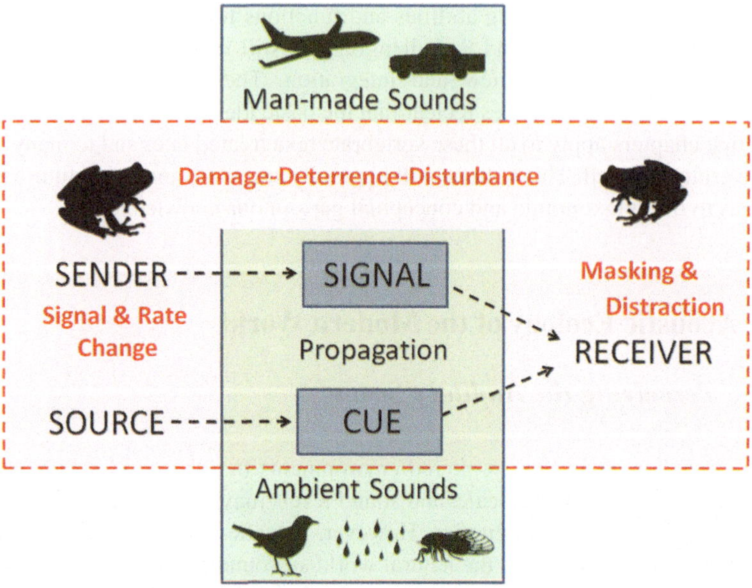

Fig. 1.2 Schematic illustration of the potential effects of man-made sounds in the context of senders and receivers of signals and cues, given the inherent degradation of perceptual potential due to propagation loss through the environment and the presence of natural ambient sounds of biotic and abiotic sources. Note that signals have evolved by natural selection and senders benefit from them being heard by receivers, which is not the case for cues. Man-made noise may cause damage, deterrence, and disturbance to senders and receivers. Senders may alter the acoustic structure of their signal or change the rate of calling or singing under noisy conditions, whereas receivers may be masked or distracted (i.e., informational masking) by the presence of man-made sounds

acoustic changes may also add information about the distance and direction of the sound source (Naguib and Wiley 2001).

Sound is obviously not the only medium by which animals and humans gather information about the world around them (Partan and Marler 1999; Munoz and Blumstein 2012). The visual, chemical, and tactile senses often serve in parallel in affecting auditory perception. Depending on the species, thermal, magnetic, and electrical senses may be added to the multimodal complexity of perceiving the world. Many signals or cues are explicitly multimodal, having, for example, an acoustic and a visual component, which may result in redundancy. This means that relevant information can still be extracted through one channel despite masking problems in the other, and animals have been shown to perceptually shift attention to the sensory information from the channel with the least interference (Pohl et al. 2012; Gomes et al. 2016). However, incoming stimuli in any channel may not only benefit an animal, but it may also interfere with the perception of information in another channel by attentional distraction or by general demands on processing capacity (Chan et al. 2010; Schlittmeier et al. 2015). The study of effects of man-made sounds on signal perception and animal performance is thereby an inherently multimodal discipline (van der Sluijs et al. 2011; Halfwerk and Slabbekoorn 2015).

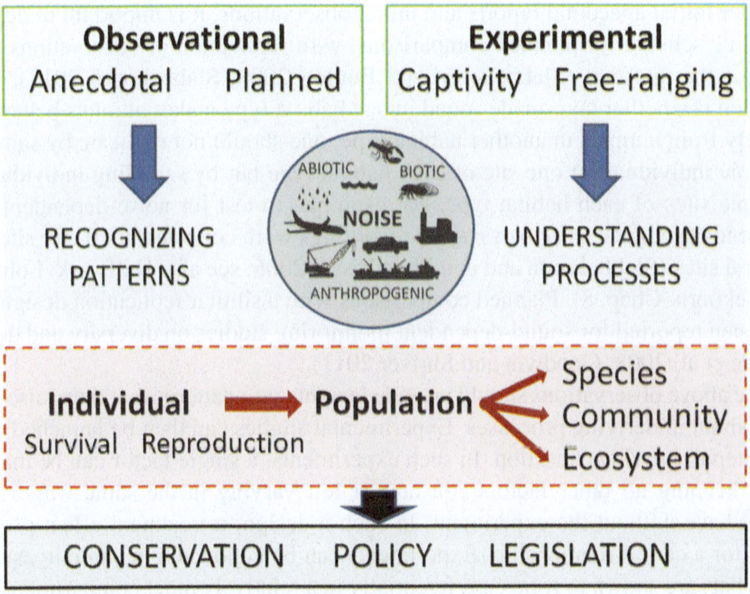

Fig. 1.3 Framework for the nature and contribution of noise impact studies on the understanding and mitigation of potential effects of man-made sounds on animals. Scientific investigations can be observational, ranging from anecdotal reports to planned and well-replicated sampling, or experimental, applying manipulations of sound conditions with replicated sampling and controls, either in captivity or on free-ranging animals in the wild. Observational studies provide correlational data and contribute most to recognizing patterns, whereas experimental data allow interpretations of causation and contribute most to understanding the processes. Both types of data can be combined with theoretical models for extrapolation across space or in time or for evaluation of critical parameters for the effects on survival and reproduction of individual animals. The effects on individuals will accumulate at the population level and potentially have consequences at the level of the species, community, or ecosystem, which can all be critical levels for conservationists, policy makers, and legislators

1.2.2 Studying the Potential Effects of Man-Made Sounds

The investigations into the potential effects of man-made sounds are diverse in terms of the types of studies, the types of effects, and the range of species (see Fig. 1.3). Research interest and awareness of potential problems often start with observational data and reports of anecdotal nature (something seen, somewhere, in some species). As an example, the issue of sonar impact on whale stranding started as anecdotal observations. However, after many observations, it has become clear that sonar use may indeed be associated with stranding whales, but it also has become equally clear that strandings can also happen without any sonar and not all sonar use leads to strandings. Thus, strandings may be caused by but are not inherently linked to sonar use (D'Amico et al. 2009; Zirbel et al. 2011).

After initial anecdotal reports and initial observations, it is important to design a sampling scheme for planned comparisons, with replication of observations at the level of the question (Slabbekoorn and Bouton 2008; Slabbekoorn 2013). If the question is whether man-made sound in one habitat type makes animals behave differently from animals in another habitat type, one should not replicate by sampling multiple individuals at one site of each habitat type but by sampling individuals at multiple sites of each habitat type. For example, to test for noise-dependent song frequency use, great tit (*Parus major*) recordings were collected at 10 city sites and 10 rural sites (Slabbekoorn and den Boer-Visser 2006; see also Halfwerk, Lohr, and Slabbekoorn, Chap. 8). Planned comparisons with a similar replication design have also been reported for sound-dependent monitoring studies on diversity and density (Bayne et al. 2008; Goodwin and Shriver 2011).

The above observations should provide insights into patterns that then raise questions about underlying processes. Experimental studies can then be launched as the next step to proving causation. In such experiments, a single factor can be manipulated, keeping all other factors constant or left varying in the same way as they would have without the experiment. In such a design, a treatment effect provides proof for a causal relationship. Experiments can be conducted in captivity, such as when bats are shown to forage preferentially in a relatively quiet compartment of an aviary instead of an experimentally elevated noisy compartment (Schaub et al. 2008; see also Slabbekoorn, McGee, and Walsh, Chap. 9). They can also be conducted outside at the waterside for frogs or in natural bird territories (Sun and Narins 2005; Halfwerk and Slabbekoorn 2009). Again, if the question requires it, sampling should be replicated on a geographic scale (cf. Mockford and Marshall 2009; LaZerte et al. 2016).

Studies in captivity are often limited in terms of the spatial or social restrictions such that animals may not show a natural behavior or be in a behavioral or physiological state associated with a specific context because it only occurs in free-ranging conditions (Calisi and Bentley 2009; Neo et al. 2016). Studies in captivity also often use a specific subset of test animals raised in captivity or accidentally caught or stranded, which raises uncertainty about generalizability. However, captive studies also have advantages in that the test animals are typically well known (e.g., background, age, size, condition, coping style), available in sufficient numbers for replication, and selected for homogeneity in groups to compare. This type of study is therefore often suitable for investigating processes in the laboratory environment that explain patterns that occur in natural conditions, whereas absolute numbers or the nature of response patterns should not be extrapolated to the natural conditions of free-ranging animals in the wild (Slabbekoorn 2016).

In addition to the limitations mentioned above for studies in captivity, there are also possible difficulties with the acoustic test conditions of captive studies. Reverberant bird cages, speaker limitations in spectrum or level, or sound field conditions in fish tanks can make sound exposure very artificial and unlike anything animals would experience in the wild (Duncan et al. 2016; Rogers et al. 2016; see also Hawkins and Popper, Chap. 6). This obviously makes extrapolation of data problematic. A similar extrapolation problem also applies to tests done at a single natural location in the wild with unique features or with a single, highly specific sound stimulus (Slabbekoorn and Bouton 2008). However, captive and single-location studies can be useful for investigating mechanisms and help answer fundamental questions. Inherently, studies of different types, on either animals spatially and behaviorally restricted in captivity or free ranging in the natural environment of their own choice in the wild, will never yield straightforward answers to the big questions by themselves. They just provide parts of the grand puzzle that requires complementary insights of both types of study (Slabbekoorn 2016).

Eventually, conservationists, policy makers, and legislators are interested in effects, not at the individual level but at the population or community level (New et al. 2014; Kunc et al. 2016). Detrimental effects of any kind can potentially affect an individual and determine its fitness through an impact on survival and reproductive output. The accumulated effects on all exposed individuals will translate into population-level consequences, which are of interest for the conservation of species, communities, or ecosystems. The societal relevance of sound impacts on animals in nature, farms, zoos, and laboratories is not only providing funding opportunities but is also guiding research interests. This sometimes results in unrealistic targets, such as single dose-response curves for a particular taxonomic group expressed in a sound unit unable to cover the necessary acoustic parameters required for impact assessment (Neo et al. 2014; Farcas et al. 2016). However, it is also the responsibility of scientists to identify such issues and comment on whether applications are appropriate and to gather the information and understanding required for societal needs.

1.2.3 Acoustic Deterrence by Man-Made Sounds: Foe or Friend?

A large part of the literature on the potential effects of man-made sounds on animals concerns behavioral changes and spatial responses in particular, such as moving away or acoustic deterrence (see Fig. 1.4). The sound-induced effects on decisions about movement probably involve a trade-off between reasons to stay and reasons to leave. An animal may stay as it exploits local resources related to feeding or breeding when it is familiar with local risks. However, the sense of fear for predation may be elevated by an unfamiliar sound. The decision about exchanging familiar conditions and certain resources for an unfamiliar and uncertain destiny may be detrimental but will vary with species and context. Furthermore, after repeated or continuous exposure to a sound, animals may habituate and respond less to the same stimulus if they do not experience direct negative consequences (Bejder et al. 2006; Neo et al. 2016). Alternatively, if there is some sort of negative reinforcement, animals may also exhibit sensitization and respond more strongly to subsequent exposure to the same stimulus (Götz and Janik 2011).

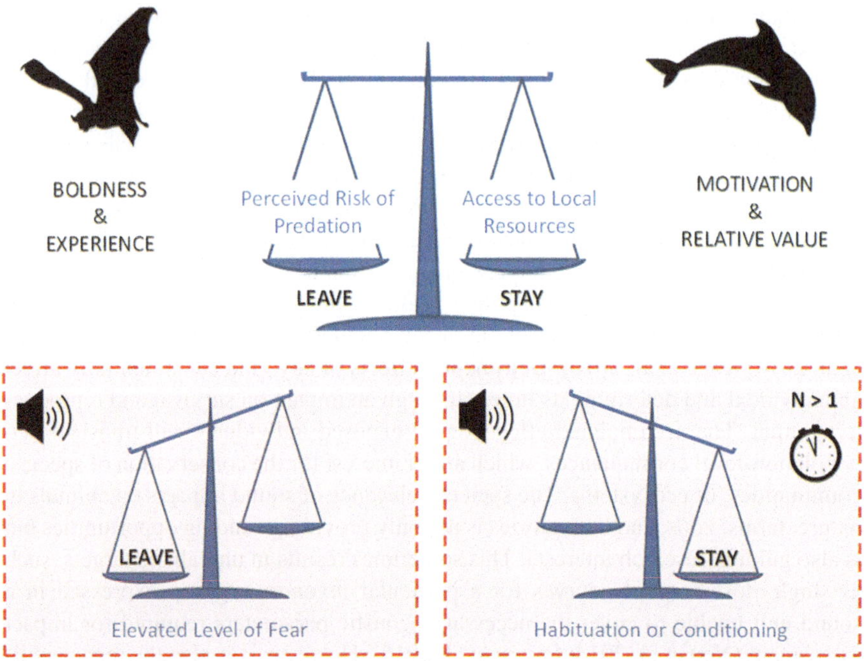

Fig. 1.4 Balance of decision making about spatial behavior in the context of exposure to man-made sounds. Whether an animal leaves or stays in a particular area likely depends on evaluation of the costs and benefits as determined by the perceived risk of predation (Frid and Dill 2002) as well as access to local resources (Sutherland 1983). Hunger or reproductive state may affect the motivation to stay or leave and particularly bold or experienced animals may be less easily frightened than shy and naive animals. The expectations about the relative value of local resources versus those likely to be encountered when leaving will also affect the balance. Exposure to man-made sound may shift the balance to "LEAVE" due to elevated fear or anxiety and make animals move away from the source (*bottom left*). Repeated or long-term exposure to man-made sound (*bottom right*) may cause the balance again to shift back to "STAY" due to habituation. Association between sound and beneficial conditions may result in conditioning and even attract animals toward the sound source (e.g., "dinner bell effect")

It is also important to realize that species vary in relative site fidelity depending, for example, on their degree of territoriality. And some species are bolder than others and perhaps more resistant to noise pollution than others (cf. Møller 2010; Sol et al. 2013). This means that one has to exercise caution in extrapolating from one species to another because relative tolerance or quick habituation to man-made sounds may reflect a lack of threat from noise pollution in one species but not in another (Tablado and Jenni 2015). Furthermore, it is also possible that relatively resistant species are the same everywhere and that effects of man-made sounds at a community level are not simply reflected in local species diversity. Urban bird communities are, for example, often relatively species rich but typically concern the same set of species in cities that can be far apart. The so-called urban homogenization thereby entails a diversity decline across large areas (McKinney 2006; Slabbekoorn 2013).

One potential problem is that man-made sounds could deter animals from staying in a particular place. At the same time, that sounds can deter animals can be an effective tool to move pest animals from places at which they are not wanted. The need to move pest animals from certain places is widespread. For example, sounds may be useful in moving animals from gardens, agricultural fields, aquacultural facilities, and fishing nets. Animals may become threats to aircraft due to collision, spread disease, or induce direct physical harm or even death to domestic animals or humans themselves. In these cases, the potential deterrent effect of sound exposure to animals may or may not be harmful to the animals but often concerns an application to the benefit of humans. However, experience shows that acoustic deterrent devices currently on the market vary dramatically in effectiveness, in part because animals are flexible and adaptable. The two different types of studies on the deterrent effects of man-made sounds on animals (i.e., as a useful tool or a conservation or welfare problem, as a friend or foe) could possibly benefit from some integration and collaboration.

A few examples from the applied literature on deterrent devices in a wide variety of taxa clearly show several aspects that reflect the studies and insights addressed in this volume. Swaddle et al. (2016), for example, reported positive results in keeping starlings (*Sturnus vulgaris*) away from an airfield using a "sonic net" of sound overlapping their species-specific spectral range. This success is not necessarily the case for all attempts of acoustic deterrence because some species habituate quickly and sometimes people just select inappropriate sound stimuli. Jenni-Eiermann et al. (2014), for example, studied feral pigeons (*Columba livia*) as pests in terms of potential damage to buildings and as hazards to public health. The birds did not show any effect of an ultrasonic deterrent device; there were no changes in corticosterone levels of caged pigeons and no deterring effect on free-ranging pigeons. This should not have been a surprise because ultrasonic sounds are inaudible to humans but are also well above the sensitive range for pigeons.

Deterrent efficiency is reported to vary among species or with the spectral match between sound stimulus and hearing range. Domestic cats, for example, were shown to vary dramatically among and within individuals in their response to an audible (for them) ultrasonic device and did not necessarily avoid the area covered acoustically (Mills et al. 2000; Nelson et al. 2006). Moreover, badgers (*Meles meles*) were not frightened by an acoustic deterrent device and were even reported to be attracted by the sound to sites when it was associated with bait (Ward et al. 2008). Apparently, there are factors that vary among (e.g., age, boldness, strength) and within (e.g., over time, across motivational states, or after experience) individuals that help shape decisions about responses to man-made sounds. Natural recordings of predators or conspecific calls of distress are sometimes better deterrents (e.g., Spanier 1980; Ramp et al. 2011), although occasionally the opposite is found (Yokoyama and Nakamura 1993). It is also true that multimodal stimuli may be most effective and delay habituation best (Lecker et al. 2015).

Acoustic deterrence has also been applied in the aquatic environment, mainly to marine predators and fishes, although not without problems and often only with very limited success (Bomford and O'Brien 1990; Popper and Carlson 1998).

Man-made sounds are, for example, used not only in the context of bycatch and depredation problems but also to keep animals away from potentially harmful human activities such as explosions or pile driving. The so-called "pingers" have been shown to work for some, but not all, marine mammal species, although habituation may limit long-term applications (Cox et al. 2001; Rankin et al. 2009). The application has also raised concerns about unwanted side effects such as hearing loss in target species (Brandt et al. 2013; see also Saunders and Dooling, Chap. 4). Furthermore, acoustic deterrent devices may even become a sort of "dinner bell" when animals learn that the sound is not associated with any danger but with an exceptional aggregation of food (Carretta and Barlow 2011; Schakner and Blumstein 2013). Obviously, more studies are often needed to initially design and repeatedly improve such applications. Fundamental insights about the impact of man-made sounds on animals, as addressed in the following chapters, may serve as a guide and inspiration.

1.3 Chapter Contents

1.3.1 Basic Principles for Impacts of Noisy Conditions

Chapters 2, 3, 4, and 5 focus on the basic principles that are applicable to all animal groups, including humans (see Fig. 1.5). Although there is a wide variety of the potential effects of man-made sounds, only directly auditory phenomena (related to hearing and damage to the ear) are treated but not the more indirect consequences of exposure (physiological stress, behavioral deterrence, and disturbance). Many animals use sounds to detect predators and prey and to find partners or deter competitors, all critical matters for survival and reproduction. Thus, it makes sense to consider the impacts of noise pollution in the typical framework used for acoustic communication, with a sender generating a signal (or a source generating a cue) that propagates through the environment before it reaches receivers. Aspects of hearing such as masking and distraction are restricted to the receiver side, whereas damage to the ear can apply to both senders and receivers and thereby affect the production and perception of signals for communication. The fact that signal production can also be affected by noisy conditions without any physical hearing damage will only be addressed in the taxonomically organized chapters that follow after this part of the book.

Chapter 2 by Robert J. Dooling and Marjorie Leek and Chap. 3 by Micheal L. Dent and Mark A. Bee address hearing complications and perceptual strategies under challenging conditions in terms of noisy and complex acoustic environments. As mentioned in Sect. 1.2.1, the natural world is often very noisy so the complications that occur and the strategies used by animals are not novel or special for man-made sounds. Long evolutionary histories and strong selection pressures for hearing particular sounds against a naturally noisy background explain the auditory phenomena

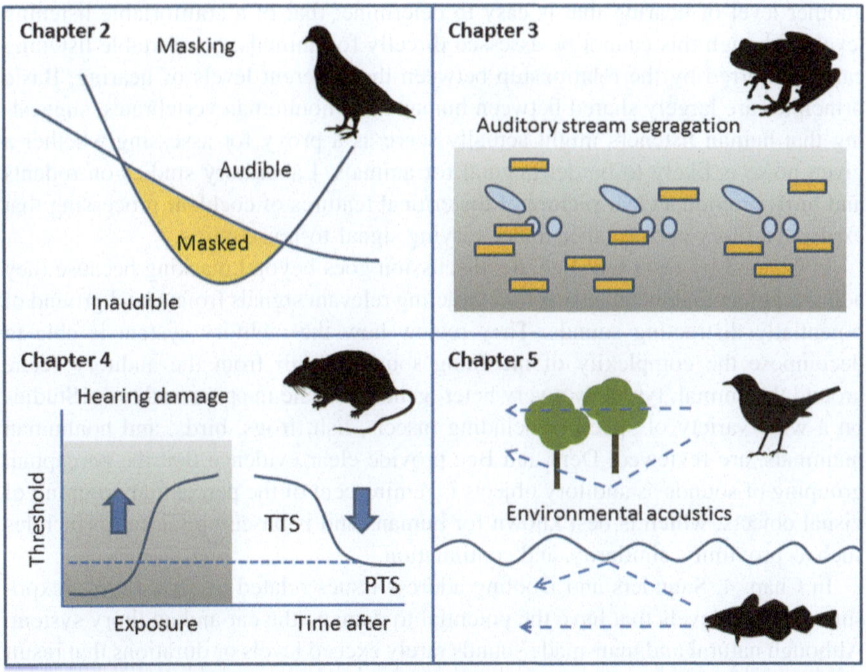

Fig. 1.5 Overview of the core concepts addressed in Chaps. 2, 3, 4, and 5. Chapter 2 by Dooling and Leek: auditory masking, reflected by a hearing curve of detection threshold across frequencies, overlaid by a noise masker with an energy bias toward low frequencies. Chapter 3 by Dent and Bee: auditory stream segregation, depicted by a repetitive, biologically relevant signal amid a scattered background of irrelevant sounds. Chapter 4 by Saunders and Dooling: temporary threshold shifts (TTS) and permanent threshold shifts (PTS) in the auditory detection threshold after exposure for a particular duration. Chapter 5 by Larsen and Radford: environmental acoustics in air and in water, where absorption, scatter, and reflections by objects and surfaces in animal surroundings affect propagation of sounds in a variety of ways. The animal silhouettes reflect potential model species, but the phenomena apply across taxa

reviewed and the wide variety in hearing capacities across animal taxa. In Chap. 4 by James C. Saunders and Robert J. Dooling, acoustic overexposure is addressed, which is obviously less of a natural phenomenon and a matter of physical trauma that is of interest especially in the context of artificial sound impact assessments. Chapter 5 by Ole Næsbye Larsen and Craig Radford is not about hearing or damage but about attenuation and degradation, which are fundamental principles for all sounds propagating through an environment.

Chapter 2 by Dooling and Leek addresses the phenomenon of masking of biologically relevant sounds with a focus on communication sounds. Masking can be defined as the interference of detection of one sound by another. The presence of man-made sounds, at levels above naturally present ambient noise, may result in increased hearing thresholds for detection, discrimination, and recognition of target sounds depending on the overlap in time and frequency. And in humans, there is another level of hearing that is easy to determine: that of a comfortable listening level. Although this cannot be assessed directly for animals, comfortable listening can be inferred by the relationship between the different levels of hearing. Basic principles are largely shared between humans and nonhuman vertebrates, suggesting that human listeners might actually serve as a proxy for assessing whether a given noise is likely to be detrimental for animals. Laboratory studies on rodents and birds provide a clear picture of the critical features of cochlear processing that explain auditory performance under varying signal-to-noise ratios.

In Chap. 3 by Dent and Bee, the discussion goes beyond masking because they address perceptual mechanisms for extracting relevant signals from a background of potentially distracting sounds. They review how the auditory system is able to decompose the complexity of incoming sound stimuli from the auditory scene around the animal, typically a very heterogeneous scene in space and time. Studies on a wide variety of species, including insects, fish, frogs, birds, and nonhuman mammals, are reviewed. Dent and Bee provide clear evidence that the perceptual grouping of sounds as auditory objects is reminiscent of the perceptual grouping of visual objects, which is best known for humans and is based on Gestalt principles such as proximity, similarity, and continuation.

In Chap. 4, Saunders and Dooling address issues related to acoustic overexposure to sound levels that have the potential to damage the ear and auditory system. Although natural and man-made sounds rarely exceed levels or durations that result in physical injuries, exceptional overexposures can occur when animals are near blast explosions or pile driving or when animals remain for an extended period of time in close proximity to a noise source such as runways with jet aircraft or in waters with relatively long-term seismic surveys. Many comprehensive studies on several rodent species highlight the acoustic parameters that are important in causing damage to the inner ear and yielding temporary or permanent hearing loss (temporary [TTS] and permanent [PTS] threshold shifts).

In Chap. 5, Larsen and Radford cover the physical properties of air and water in terms of sound transmission. Sound properties of biologically relevant sounds and potential maskers are critical for acoustic receivers aiming at detecting signals and cues, but the signal-to-noise ratio matters only at the receiver and not at the source. Attenuation and degradation during propagation through the environment from source to receiver may alter perceptual opportunities and are of critical importance to understand the potential for detrimental effects of man-made sounds, which are affected in the same way as any other sound.

1.3.2 Taxon-Specific Insights for Sound Impact on Animals

After discussion of the basic principles applicable to all animals, Chaps. 6, 7, 8, 9, and 10 provide overviews of the concepts addressed and insights gained from five different taxonomic groups: fishes, amphibian and reptiles, birds, terrestrial mammals, and marine mammals. The chapters cover a wide variety of sensitivities,

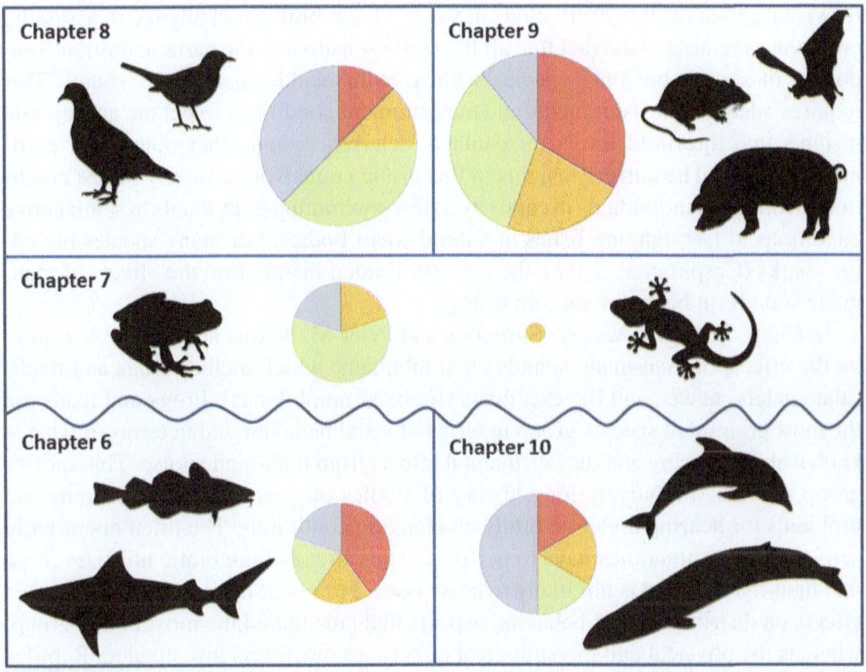

Fig. 1.6 Overview of the taxonomic groups of vertebrates discussed in Chaps. 6, 7, 8, 9, and 10 and an indication of the number of studies published on the effects of man-made sounds in these taxa. Relative size of the pie charts indicates a single or very few to hundreds of studies. The colors in the pie charts reflect the estimated proportion of studies addressing particular effects: *red,* damage and death; *orange,* physiological stress; *yellow,* masking and distraction; *lilac,* disturbance and deterrence. The pie charts are based on the material covered in these chapters: fishes, Chap. 6 by Hawkins and Popper; anurans and reptiles, Chap. 7 by Simmons and Narins; birds, Chap. 8 by Halfwerk, Lohr, and Slabbekoorn; terrestrial mammals, Chap. 9 by Slabbekoorn, McGee, and Walsh; marine mammals, Chap. 10 by Erbe, Dunlop, and Dolman

habitats, contexts, and conditions and include observational as well as experimental studies and investigations of animals in captivity and in the wild, in air and underwater. Despite the variety and number of species addressed, it should be realized that still only a limited subset of species has been investigated and the overall efforts and allocation strategies toward particular effects vary per group (see Fig. 1.6). It also becomes clear in these chapters that the basic principles discussed in Chaps. 2, 3, 4, and 5 are relevant to all species.

In Chap. 6, Anthony D. Hawkins and Arthur N. Popper address what is known about the effects of man-made sound on fishes. This group, which comprises more species than all of the other vertebrate groups combined, is characterized by relative low-frequency hearing abilities, a range that includes the acoustic energy produced by many man-made sound sources such as shipping, pile driving, and seismic exploration. The potential problems of man-made sounds for aquatic animals such as fishes have only recently attracted more attention, which may be due to challenges for visual and acoustic observations in water. One prominent challenge in assessing problems is related to the fact that all fishes are sensitive to the particle motion component of sound, but only species with a swim bladder can detect sound. This requires adequate measurements and measurement conditions to get the appropriate insights into threshold levels for damage, deterrence, and other potentially detrimental effects. The current insights in this group come from a variety of test conditions, from fixed individuals in captivity to free-swimming individuals in semicaptive conditions to free-ranging fishes in natural water bodies. For many species including sharks (Casper et al. 2012), there is still limited insight into the effects of man-made sounds on behavior and physiology.

In Chap. 7, Angela Megela Simmons and Peter M. Narins address the literature on the effects of man-made sounds on amphibians, which include frogs and toads, salamanders, newts, and the caecilians (limbless amphibians). Frogs and toads are the most prominent species group in terms of vocal behavior and in terms of what is known about hearing and the detrimental effects from man-made noise. This species group also has a relatively long history of studies on noisy chorus conditions and problems for hearing under naturally challenging conditions. The often nocturnally active animals sometimes have to perform under already high biotic noise levels so that man-made sound is not likely to be an issue. However, many studies do suggest effects on distribution and behavior, aspects that are studied the most in this group, whereas the physical and physiological effects are much less investigated. Reptiles are also included in this chapter, but there are actually very few studies on noise pollution and sound exposure in turtles, tortoises, crocodilians, snakes, and lizards.

Birds are the subject of Chap. 8 by Wouter Halfwerk, Bernard Lohr, and Hans Slabbekoorn. The chapter discusses the relatively homogeneous and well-studied songbirds that comprise about half of the 10,000 bird species. Birds are particularly interesting because the typical spectral range of sensitivity is very similar to that of humans. Furthermore, the ability to regenerate hair cells after noise-induced inner ear trauma and (temporary) hearing loss has made birds an attractive model for medical investigations. Besides the laboratory work on damage and recovery and various aspects of masking, there is also a lot of work on wild birds. Birds that are territorial and vocally advertising are well suited for monitoring studies and experimental playback studies. Such investigations have yielded much insight into the effects of man-made sounds on distribution, density, and noise level-dependent signal changes and, recently, to some extent on physiological stress levels, behavioral performance, reproductive success, and ecological interactions. Nonsongbirds, including, for example, doves, parrots, grouse, and hearing specialists like owls, are much less well investigated for the effects of man-made sounds.

In Chap. 9, Hans Slabbekoorn, JoAnn McGee, and Edward J. Walsh address the wide-ranging hearing abilities as well as the wide-ranging types of investigations into the sound impact on the diverse group of terrestrial mammals. The species treated include placental mammals and marsupials as well as the primitive monotremes. Laboratory rodents have traditionally been the model for medical investigations into the auditory system of humans and there are many studies into the fundamental understanding of hearing, hearing problems, and inner ear damage. There is also a lot known from this work on the nonauditory effects of loud and long overexposure, including aspects of brain development and physiological performance declines. Besides the lab work, terrestrial mammals have been investigated in farms and zoos, particularly for stress and behavioral changes related to noisy holding conditions and visitors. Furthermore, various studies have investigated the impact of traffic noise and industrial and recreational activities on the distribution and foraging activities of a diverse set of small-to-large mammal species in habitat types ranging from boreal pine forests to tropical rainforests. Bats, with their exceptionally high-frequency hearing ranges and echolocation abilities, form a special group that also has been investigated for the various effects of man-made sounds.

In Chap. 10, Christine Erbe, Rebecca Dunlop, and Sarah Dolman address marine mammals, including cetaceans (whales, dolphins, and porpoises) and sirenians (sea cows), that are fully aquatic and several marine carnivores (seals and walruses) that spend time both on land and in water. Like terrestrial mammals, the hearing ranges vary dramatically among species; they can go high in the ultrasonic range and many species echolocate. Marine mammals inhabit all of the world's oceans, including distant offshore waters where some species dive several kilometers deep and shallow coastal waters where the smaller species roam close to the surface. These habitat preferences will likely affect exposure probability. For example, close to the coast are many human activities and therefore more man-made sounds, whereas propagation of low-frequency sounds is more restricted in shallow waters. Furthermore, it is also important to understand whether species generate low-frequency, long-ranging calls like some baleen whales or high-frequency, short-range calls like those produced by many dolphins and porpoises. The conditions for collecting data are among the most challenging of any taxa, but many modern studies include advanced technology such as passive acoustic monitoring, tagged data loggers, and experimental exposure, complemented with studies on animals in captivity.

1.4 Inspiration and Guideline: Ready to Dive into the Book

The editors hope that the targets of this volume will be met and that both parts of the book and all chapters on the various taxonomic groups will be read with equal interest. The questions and issues range from the cellular to the community level and concern both fundamental and applied studies. Some issues and species groups have a relatively long history of investigation, whereas others have just recently come to the forefront and many others remain untouched. Consequently, although many topics are well explored, there are also many gaps in our understanding of how man-made sounds affect animals. It is therefore clear that the future will be bright and variable for studies of man-made sound impact and it is the hope of the editors that this volume serves as an inspiration and guide to that future.

Compliance with Ethics Requirements
Hans Slabbekoorn declares that he has no conflict of interest.
Robert J. Dooling declares that he has no conflict of interest.
Arthur N. Popper declares that he has no conflict of interest.

References

Andrew, R. K., Howe, B. M., & Mercer, J. A. (2002). Ocean ambient sound: Comparing the 1960s with the 1990s for a receiver off the California coast. *Acoustics Research Letters Online, 3*, 65-70.

Babisch, W. (2002). The noise/stress concept, risk assessment and research needs. *Noise and Health,* 4, 1-11.

Basner, M., Babisch, W., Davis, A., Brink, M., Clark, C., Janssen, S., & Stansfeld, S. (2014). Auditory and non-auditory effects of noise on health. *Lancet*, 383, 1325-1332.

Bayne, E. M., Habib, L., & Boutin, S. (2008). Impacts of chronic anthropogenic noise from energy-sector activity on abundance of songbirds in the boreal forest. *Conservation Biology,* 22, 1186-1193.

Bejder, L., Samuels, A., Whitehead, H., & Gales, N. (2006). Interpreting short-term behavioural responses to disturbance within a longitudinal perspective. *Animal Behaviour,* 72, 1149-1158.

Bomford, M., & O'Brien, P. H. (1990). Sonic deterrents in animal damage control: A review of device tests and effectiveness. *Wildlife Society Bulletin,* 18, 411-422.

Brandt, M. J., Höschle, C., Diederichs, A., Betke, K., Matuschek, R., Witte, S., & Nehls, G. (2013). Far-reaching effects of a seal scarer on harbour porpoises, *Phocoena phocoena. Aquatic Conservation: Marine and Freshwater Ecosystems,* 23, 222-232.

Brumm, H., & Slabbekoorn, H. (2005). Acoustic communication in noise. *Advances in the Study of Behavior*, 35, 151-209.

Calisi, R. M., & Bentley, G. E. (2009). Lab and field experiments: Are they the same animal? *Hormones and Behavior*, 56, 1-10.

Carretta, J. V., & Barlow, J. (2011). Long-term effectiveness, failure rates, and "dinner bell" properties of acoustic pingers in a gillnet fishery. *Marine Technology Society Journal,* 45, 7-19.

Casper, B. M., Halvorsen, M. B., & Popper, A. N. (2012). Are sharks even bothered by a noisy environment? In A. N. Popper & A. Hawkins (Eds.), *The Effects of Noise on Aquatic Life* (pp. 93-97). New York: Springer-Verlag.

Chan, A. A. Y.-H., Giraldo-Perez, P., Smith, S., & Blumstein, D. T. (2010). Anthropogenic noise affects risk assessment and attention: The distracted prey hypothesis. *Biology Letters*, 6, 458-461.

Cox, T. M., Read, A. J., Solow, A., & Tregenza, N. (2001). Will harbour porpoises (*Phocoena phocoena*) habituate to pingers? *Journal of Cetacean Research Management,* 3, 81-86.

D'Amico, A., Gisiner, R. C., Ketten, D. R., Hammock, J. A., Johnson, C., Tyack, P. L., & Mead, J. (2009). Beaked whale strandings and naval exercises. *Aquatic Mammals*, 35, 452-472.

Duncan, A. J., Lucke, K., Erbe, C., & McCauley, R. D. (2016). Issues associated with sound exposure experiments in tanks. *Proceedings of Meetings on Acoustics*, 27, 070008.

Farcas, A., Thompson, P. M., & Merchant, N. D. (2016). Underwater noise modelling for environmental impact assessment. *Environmental Impact Assessment Review,* 57, 114-122.

Filiciotto, F., Vazzana, M., Celi, M., Maccarrone, V., Ceraulo, M., Buffa, G., Arizza, V., de Vincenzi, G., Grammauta, R., Mazzola, S., & Buscaino, G. (2016). Underwater noise from boats: Measurement of its influence on the behaviour and biochemistry of the common prawn (*Palaemon serratus*, Pennant 1777). *Journal of Experimental Marine Biology and Ecology,* 478, 24-33.

Fletcher, J. L., & Busnel, R.-G. (1978). *Effects of Noise on Wildlife*. New York: Academic Press.

Francis, C. D., & Barber, J. R. (2013). A framework for understanding noise impacts on wildlife: An urgent conservation priority. *Frontiers in Ecology and the Environment*, 11, 305-313.

Frid, A., & Dill, L. (2002). Human-caused disturbance stimuli as a form of predation risk. *Conservation Ecology*, 6(1), 11.

Gomes, D. G. E., Page, R. A., Geipel, I., Taylor, R. C., Ryan, M. J., & Halfwerk, W. (2016). Bats perceptually weight prey cues across sensory systems when hunting in noise. *Science*, 353 6305, 1277-1280.

Goodwin, S. E., & Shriver, W. G. (2011). Effects of traffic noise on occupancy patterns of forest birds. *Conservation Biology*, 25, 406-411.

Götz, T., & Janik, V. M. (2011). Repeated elicitation of the acoustic startle reflex leads to sensitisation in subsequent avoidance behaviour and induces fear conditioning. *BMC Neuroscience*, 12, 30.

Halfwerk, W., & Slabbekoorn, H. (2009). A behavioural mechanism explaining noise-dependent frequency use in urban birdsong. *Animal Behaviour*, 78, 1301-1307.

Halfwerk, W., & Slabbekoorn, H. (2015). Pollution going multimodal: The complex impact of the human-altered sensory environment on animal perception and performance. *Biology Letters*, 11, 20141051.

Hawkins, A. D., & Popper, A. N. (2016). A sound approach to assessing the impact of underwater noise on marine fishes and invertebrates. *ICES Journal of Marine Science: Journal du Conseil*, 74(3), 635-671.

Hawkins, A. D., Pembroke, A. E., & Popper, A. N. (2015). Information gaps in understanding the effects of noise on fishes and invertebrates. *Reviews in Fish Biology and Fisheries*, 25, 39-64.

Hildebrand, J. A. (2009). Anthropogenic and natural sources of ambient noise in the ocean. *Marine Ecology Progress Series*, 395, 5-20.

Hoy, R. R., Popper, A. N., & Fay, R. R. (Eds.). (1998). *Comparative Hearing: Insects*. New York: Springer-Verlag.

Jenni-Eiermann, S., Heynen, D., & Schaub, M. (2014). Effect of an ultrasonic device on the behaviour and the stress hormone corticosterone in feral pigeons. *Journal of Pest Science*, 87, 315-322.

Kight, C. R., & Swaddle, J. P. (2011). How and why environmental noise impacts animals: An integrative, mechanistic review. *Ecology Letters*, 14, 1052-1061.

Klump, G. M. (1996). Bird communication in the noisy world. In D. E. Kroodsma & E. H. Miller (Eds.), *Ecology and Evolution of Acoustic Communication in Birds* (pp. 321-338). Ithaca, NY: Cornell University Press.

Kunc, H. P., McLaughlin, K. E., & Schmidt, R. (2016). Aquatic noise pollution: Implications for individuals, populations, and ecosystems. *Proceedings of the Royal Society B: Biological Sciences*, 283, 20160839.

LaZerte, S. E., Slabbekoorn, H., & Otter, K. A. (2016). Learning to cope: Vocal adjustment to urban noise is correlated with prior experience in black-capped chickadees. *Proceedings of the Royal Society B: Biological Sciences*, 283, 20161058.

Lecker, C. A., Parsons, M. H., Lecker, D. R., Sarno, R., & Parsons, F. E. (2015). The temporal multimodal influence of optical and auditory cues on the repellent behaviour of ring-billed gulls (*Larus delewarensis*). *Wildlife Research*, 42, 232-240.

Le Prell, C. G., Henderson, D., Fay, R. R., & Popper, A. N. (Eds.). (2012). *Noise-Induced Hearing Loss: Scientific Advances*. New York: Springer-Verlag.

Lillis, A., Eggleston, D. B., & Bohnenstiehl, D. R. (2013). Oyster larvae settle in response to habitat-associated underwater sounds. *PLoS ONE*, 8, e79337.

McKinney, M. L. (2006). Urbanization as a major cause of biotic homogenization. *Biological Conservation*, 127, 247-260.

Mennit, D. J., Fristrup, K. M., & Nelson, L. (2015). A spatially explicit estimate of environmental noise exposure in the contiguous United States. *The Journal of the Acoustical Society of America*, 137, 2339-2340.

Miedema, H. M. E., & Vos, H. (2003). Noise sensitivity and reactions to noise and other environmental conditions *The Journal of the Acoustical Society of America*, 113, 1492-1504.

Mills, D. S., Bailey, S. L., & Thurstans, R. E. (2000). Evaluation of the welfare implications and efficacy of an ultrasonic 'deterrent' for cats. *The Veterinary Record*, 147, 678-680.

Mockford, E. J., & Marshall, R. C. (2009). Effects of urban noise on song and response behaviour in great tits. *Proceedings of the Royal Society B: Biological Sciences*, 276, 2979-2985.

Møller, A. P. (2010). Interspecific variation in fear responses predicts urbanization in birds. *Behavioral Ecology*, 21, 365-371.

Montealegre-Z F., Jonsson, T., Robson-Brown, K. A., Postles, M., & Robert, D. (2012). Convergent evolution between insect and mammalian audition. *Science*, 338, 968-971.

Montgomery, J. C., Jeffs, A. G., Simpson, S. D., Meekan, M., & Tindle, C. (2006). Sound as an orientation cue for the pelagic larvae of reef fishes and decapod crustaceans. *Advances in Marine Biology*, 51, 143-196.

Mooney, T. A., Hanlon, R. T., Christensen-Dalsgaard, J., Madsen, P. T., Ketten, D. R., & Nachtigall, P. E. (2010). Sound detection by the longfin squid (*Loligo pealeii*) studied with auditory evoked potentials: Sensitivity to low-frequency particle motion and not pressure. *Journal of Experimental Biology*, 213, 3748-3759.

Morley, E. L., Jones, G., & Radford, A. N. (2014). The importance of invertebrates when considering the impacts of anthropogenic noise. *Proceedings of the Royal Society B: Biological Sciences*, 281, 20132683.

Munoz, N. E., & Blumstein, D. T. (2012). Multisensory perception in uncertain environments. *Behavioural Ecology*, 23, 457-462.

Murphy, E., & King, E. A. (2014). *Environmental Noise Pollution*. Burlington, MA: Elsevier.

Naguib, M., & Wiley, R. H. (2001). Estimating the distance to a source of sound: Mechanisms and adaptations for long-range communication. *Animal Behaviour*, 62, 825-837.

National Research Council. (1994). *Low-Frequency Sound and Marine Mammals: Current Knowledge and Research Needs*. Washington, DC: National Academies Press.

National Research Council. (2000). *Marine Mammals and Low-Frequency Sound: Progress Since 1994*. Washington, DC: National Academies Press

Nelson, S. H., Evans, A. D., & Bradbury, R. B. (2006). The efficacy of an ultrasonic cat deterrent. *Applied Animal Behaviour Science*, 96, 83-91.

Neo, Y. Y., Seitz, J., Kastelein, R. A., Winter, H. V., ten Cate, C., & Slabbekoorn, H. (2014). Temporal structure of sound affects behavioural recovery from noise impact in European seabass. *Biological Conservation*, 178, 65-73.

Neo, Y. Y., Hubert, J., Bolle, L., Winter, H. V., ten Cate, C., & Slabbekoorn, H. (2016). Sound exposure changes European seabass behaviour in a large outdoor floating pen: Effects of temporal structure and a ramp-up procedure. *Environmental Pollution*, 214, 26-34.

New, L. F., Clark, J. S., Costa, D. P., Fleishman, E., Hindell, M. A., Klanjšček, T., Lusseau, D., Kraus, S., McMahon, C. R., Robinson, P. W., Schick, R. S., Schwartz, L. K., Simmons, S. E., Thomas, L., Tyack, P., & Harwood, J. (2014). Using short-term measures of behaviour to estimate long-term fitness of southern elephant seals. *Marine Ecology Progress Series*, 496, 99-108.

Partan, S., & Marler, P. (1999). Communication goes multimodal. *Science*, 283, 1272–1273.

Pine, M. K., Jeffs, A. G., Wang, D., & Radford, C. A. (2016). The potential for vessel noise to mask biologically important sounds within ecologically significant embayments. *Ocean & Coastal Management*, 127, 63-73.

Pohl, N. U., Leadbeater, E., Slabbekoorn, H., Klump, G. M., & Langemann, U. (2012). Great tits in urban noise benefit from high frequencies in song detection and discrimination. *Animal Behaviour*, 83, 711-721.

Popper, A. N., & Carlson, T. J. (1998). Application of sound and other stimuli to control fish behavior. *Transactions of the American Fisheries Society*, 127, 673-707.

Popper, A. N., & Hawkins, A. (Eds.). (2012). *The Effects of Noise on Aquatic Life*. New York: Springer-Verlag.

Popper, A. N., & Hawkins, A. (Eds.). (2016). *The Effects of Noise on Aquatic Life II*. New York: Springer-Verlag.

Ramp, D., Foale, C. G., Roger, E., & Croft, D. B. (2011). Suitability of acoustics as non-lethal deterrents for macropodids: The influence of origin, delivery and anti-predator behaviour. *Wildlife Research*, 38, 408-418.

Rankin, C. H., Abrams, T., Barry, R. J., Bhatnagar, S., Clayton, D. F., Colombo, J., Coppola, G., Geyer, M. A., Glanzman, D. L., Marsland, S., McSweeney, F. K., Wilson, D. A., Wu, C.-F., & Thompson, R. F. (2009). Habituation revisited: An updated and revised description of the behavioral characteristics of habituation. *Neurobiology of Learning and Memory*, 92, 135-138.

Richardson, W. J., Greene, C. R., Malme, C. I., & Thomson, D. H. (1995). *Marine Mammals and Noise*. San Diego, CA: Academic Press.

Rogers, P. H., Hawkins, A. D., Popper, A. N., Fay, R. R., & Gray, M. D. (2016). Parvulescu revisited: Small tank acoustics for bioacousticians. In A. N. Popper & A. Hawkins (Eds.), *The Effects of Noise on Aquatic Life II* (pp. 933-941). New York: Springer-Verlag.

Schaub, A., Ostwald, J., & Siemers, B. M. (2008). Foraging bats avoid noise. *Journal of Experimental Biology*, 211, 3174-3180.

Schakner, Z. A., & Blumstein, D. T. (2013). Behavioral biology of marine mammal deterrents: A review and prospectus. *Biological Conservation*, 167, 380-389.

Schlittmeier, S. J., Feil, A., Liebl, A., & Hellbrück, J. (2015). The impact of road traffic noise on cognitive performance in attention-based tasks depends on noise level even within moderate-level ranges. *Noise & Health*, 17, 148-157.

Shannon, G., McKenna, M. F., Angeloni, L. M., Crooks, K. R., Fristrup, K. M., Brown, E., Warner, K. A., Nelson, M. D., White, C., & Briggs, J. (2016). A synthesis of two decades of research documenting the effects of noise on wildlife. *Biological Reviews*, 91, 982-1005.

Singh, N., & Davar, S. C. (2004). Noise pollution-Sources, effects and control. *Journal of Human Ecology*, 16, 181-187.

Slabbekoorn, H. (2012). The complexity of noise impact assessments: From birdsong to fish behavior. In A. N. Popper & A. Hawkins (Eds.), *The Effects of Noise on Aquatic Life* (pp. 497-500). New York: Springer-Verlag.

Slabbekoorn, H. (2013). Songs of the city: Noise-dependent spectral plasticity in the acoustic phenotype of urban birds. *Animal Behaviour*, 85, 1089-1099.

Slabbekoorn, H. (2016). Aiming for progress in understanding underwater noise impact on fish: Complementary need for indoor and outdoor studies. In: A. N. Popper & A. Hawkins (Eds.), *The Effects of Noise on Aquatic Life II* (pp. 1057-1065). New York: Springer-Verlag.

Slabbekoorn, H., & den Boer-Visser, A. (2006). Cities change the songs of birds. *Current Biology*, 16, 2326-2331.

Slabbekoorn, H., & Bouton, N. (2008). Soundscape orientation: A new field in need of sound investigation. *Animal Behaviour*, 76, e5-e8.

Slabbekoorn, H., Bouton, N., van Opzeeland, I., Coers, A., ten Cate, C., & Popper, A. N. (2010). A noisy spring: The impact of globally rising underwater sound levels on fish. *Trends in Ecology & Evolution*, 25, 419-427.

Sol, D., Lapiedra, O., & Gonález-Lagos, C. (2013). Behavioural flexibility for a life in the city. *Animal Behaviour*, 85, 1101-1112.

Spanier, E. (1980). The use of distress calls to repel night herons (*Nycticorax nycticorax*) from fish ponds. *Journal of Applied Ecology*, 17, 287-294.

Sun, J. W. C., & Narins, P. M. (2005). Anthropogenic sounds differentially affect amphibian call rate. *Biological Conservation*, 121, 419-427.

Sutherland, W. J. (1983). Aggregation and the 'ideal free' distribution. *The Journal of Animal Ecology*, 52, 821–828.

Swaddle, J. P., Moseley, D. L., Hinders, M. K., & Smith, E. P. (2016). A sonic net excludes birds from an airfield: Implications for reducing bird strike and crop losses. *Ecological Applications*, 26, 339-345.

Tablado, Z., & Jenni, L. (2015). Determinants of uncertainty in wildlife responses to human disturbance. *Biological Reviews,* 92, 216-233.

van der Sluijs, I., Gray, S. M., Amorim, M. C. P., Barber, I., Candolin, U., Hendry, A. P., Krahe, R., Maan, M. E., Utne-Palm, A. C., & Wagner, H.-J. (2011). Communication in troubled waters: Responses of fish communication systems to changing environments. *Evolutionary Ecology*, 25, 623-640.

Vermeij, M. J. A., Marhaver, K. L., Huijbers, C. M., Nagelkerken, I., & Simpson, S. D. (2010). Coral larvae move toward reef sounds. *PLoS ONE,* 5, e10660.

Ward, A. I., Pietravalle, S., Cowan, D. P., & Delahay, R. J. (2008). Deterrent or dinner bell? Alteration of badger activity and feeding at baited plots using ultrasonic and water jet devices. *Applied Animal Behaviour Science*, 115, 221-232.

Warren, P. S., Katti, M., Ermann, M., & Brazel, A. (2006). Urban bioacoustics: It's not just noise. *Animal Behaviour*, 71, 491-502.

Wiley, R. H. (2017). How noise determines the evolution of communication. *Animal Behaviour,* 124, 307-313.

Wiley, R. H., & Richards, D. G. (1978). Physical constraints on acoustic communication in the atmosphere: Implications for the evolution of animal vocalizations. *Behavioral Ecology and Sociobiology*, 3, 69-94.

Wilkens, S. L., Stanley, J. A., & Jeffs, A. G. (2012). Induction of settlement in mussel (*Perna canaliculus*) larvae by vessel noise. *Biofouling,* **28,** 65-72.

World Health Organization. (2011). *Burden of Disease from Environmental Noise. Quantification of Healthy Life Years Lost in Europe*. Available at www.euro.who.int/en/health-topics/environment-and-health/noise/publications.

Yokoyama, H., & Nakamura, K. (1993). Aversive response of tree sparrows *Passer montanus* to distress call and the sound of paper flag. *Applied Entomology and Zoology*, 28, 359-370.

Zirbel, K., Balint, P., & Parsons, E. C. M. (2011). Public awareness and attitudes towards naval sonar mitigation for cetacean conservation: A preliminary case study in Fairfax County, Virginia (the DC Metro area). *Marine Pollution Bulletin*, 63, 49-55.

Chapter 2
Communication Masking by Man-Made Noise

Robert J. Dooling and Marjorie R. Leek

Abstract Conservationists and regulators are often challenged with determining the masking effects of man-made sound introduced into the environment. A considerable amount is known from laboratory studies of auditory masking of communication signals in birds, so that it is now feasible to develop a functional model for estimating the masking effects of noise on acoustic communication in natural environments not only for birds but for other animals as well. Broadband noise can affect the detection, discrimination, and recognition of sounds and whether acoustic communication is judged comfortable or challenged. Estimates of these effects can be obtained from a simple measure called the critical ratio. Critical ratio data are available in both humans and a wide variety of other animals. Because humans have smaller critical ratios (i.e., hear better in noise) than other animals, human listeners can be used as a crude proxy for estimating the limits of effects on animals. That is, if a human listener can barely hear a signal in noise in the environment, it is unlikely that an animal can hear it. The key to estimating the amount of masking from noise that can occur in animals in their natural habitats is in measuring or estimating the signal and noise levels precisely at the animal's ears in complex environments. Once that is done, a surprising amount of comparative laboratory critical ratio data exists, especially for birds, from which it is possible to predict the effect of noise on acoustic communication. Although best developed for birds, these general principles should hold for all animals.

Keywords Comfortable communication · Critical ratios · Detection · Discrimination · Masking · Recognition · Signal-to-noise ratio

R. J. Dooling (✉)
Department of Psychology, University of Maryland, College Park, MD, USA
e-mail: rdooling@umd.edu

M. R. Leek
VA Loma Linda Healthcare System, Loma Linda, CA, USA
e-mail: Marjorie.Leek@va.gov

H. Slabbekoorn et al. (eds.), *Effects of Anthropogenic Noise on Animals*, Springer Handbook of Auditory Research 66,
https://doi.org/10.1007/978-1-4939-8574-6_2

2.1 Introduction

Man-made (anthropogenic) sound can cause a variety of adverse effects on animals, including behavioral and physiological changes, auditory system damage from acoustic overexposure, and masking of communication and other important biological sounds. A precise understanding of these effects is of interest to conservationists and regulators, but it has nevertheless been difficult to reach a clear consensus on the causal relationships between noise levels and these adverse effects. One reason for this is that surprisingly few animal studies can definitively identify man-made sound alone as the principal source for these effects (Dooling and Popper 2016). Another issue is that animals have different auditory capabilities and sensitivities such as critical ratios, the ratio of signal level to noise level at threshold. Thus, the same spectrum and type and level of noise that is problematic for one species may be less of a problem for another. Not unexpectedly, attempts to reach common ground among conservationists and regulators as well as corporate and military entities often end with a general and unsatisfactory conclusion that more data are needed. Here the case is made that for auditory masking, a surprising amount of laboratory data already exists, and these data, if properly utilized, could bring a new approach and level of precision to the problem of predicting the masking of communication signals by noise in animals.

An essential starting place for clearly understanding masking is to separate the masking effects of man-made noise from other effects. There are generally considered to be four types of man-made noise effects on humans and other animals that are largely dependent on the intensity, duration, and bandwidth of noise. There are several ways one might characterize the various, often times overlapping, effects of noise. These include permanent threshold shift and auditory system damage, temporary threshold shift with potential auditory system damage, masking, and other broad physiological and behavioral effects (Dooling et al. 2009; Dooling and Popper 2016). Masking can occur along with or independent of any of the other effects. In some ways, masking may be the most concerning of these effects for conservationists and regulators because although it seems benign in comparison to other effects, it is at the same time very pervasive. This review focuses only on the principles of masking and its effects on acoustic communication in animals. Nearly all of the available animal research on the masking effects on acoustic communication signals in field studies comes from birds, with a few contributions from other animals. For this reason, the emphasis here will be field and laboratory masking studies of bird acoustic communication. However, there are laboratory masking data available for a wide variety of other animals (e.g., Fay 1988), and the general principles described here for predicting the masking of vocal signals and other biologically important sounds should apply for all vertebrates.

2.1.1 Masking and Levels of Hearing for Communication

Obviously, the masking of communication signals and other important biological sounds (e.g., sounds of an approaching predator) can have significant adverse consequences for individuals, species behavior, and populations. Such consequences may be particularly deleterious in birds because they rely so heavily on acoustic communication for species and individual recognition, mate selection, territorial defense, and parent-offspring communication in addition to the detection of predators and prey. Understanding precisely how masking can affect hearing and communication between individuals is an important first step toward determining the level of impact on them and on the species. It is likely that the general effects and principles of masking apply across all vertebrates including humans. Everyday experiences with the masking of speech, because it is so well studied, might also enrich our understanding of the effects of masking on animal communication (Miller 1947; Bronkhorst 2000).

The most elementary definition of masking is the process whereby there is interference with the detection of one sound by another. The American National Standards Institute (ANSI) defines energetic masking as the amount, in decibels, by which an auditory detection threshold is raised in the presence of a masker (ANSI 2013). More broadly, masking refers to the increase in thresholds for detection, discrimination, or recognition of sounds caused by the presence of another sound. And to add to this, there is also the familiar experience of whether communication in noise feels comfortable or not. For example, two people talking at a comfortable level in a quiet room can easily hear one another because the level of the speech signal arriving at the receiver's ear is sufficiently greater than the ambient background noise. If the level of background noise increases significantly, say, from a vacuum cleaner or dishwasher, speech perception can be adversely affected. This is best illustrated by the common experience of having a conversation in a crowded restaurant where it becomes harder to hear one another because the level of noise from other talkers (i.e., speech) approaches the level of the speech signal from your companion, an example of the masking of speech by speech. Communication in this circumstance is occurring but is no longer judged to be comfortable because it requires greater effort by both the talker and the listener. So, in addition to the changes in hearing thresholds or levels listed above, this fourth level represents another level disruption in acoustic communication by noise. Although no data are available, it would not be hard to imagine that a similar problem of comfortable communication might occur in highly vocal birds that spend a great deal of time in extremely large flocks.

It is clear from decades of scientific studies on masking that, in addition to the four levels of energetic masking above, there is an additional mechanism of masking called informational masking (Kidd and Colburn 2017). Energetic masking is the most familiar and refers to the relative levels of the signal and the noise. Informational masking is less straightforward and is less well studied but typically refers to an extraneous sound that affects a listener's ability to attend to the signal, perhaps because of distraction or because of the acoustic similarity of the two sounds. In the case of speech communication, and probably animal acoustic communication, informational masking can adversely affect a receiver's ability to know which sound is the communication signal. It is clear from the above discussion of the familiar cases of masking of speech communication, both energetic and informational, that masking is ubiquitous. And it is also clear that unless the masking is extreme, it often goes unnoticed. That is because humans, birds, and other animals engage in behaviors, some purposeful and some reflexive, that mitigate the effects of masking noise on acoustic communication (see Sect. 2.2). These should always be taken into account in assessing the effect of noise on acoustic communication.

The energetic masking of speech occurs from all kinds of sounds that have energy in the spectral region of speech, such as household appliances, automobiles, and other man-made devices that are a part of modern living. The same principle holds with animals; it is energy in the spectral region of communication signals that is the most problematic. Fortunately, masking of all kinds of sounds, communication signals included, can easily be measured in the laboratory for both humans and animals, providing confidence that the effects of masking can be predicted under various conditions. Again, using speech as an example, communication involves more than just detecting a sound even though masking is commonly understood as being unable to detect that a sound occurred. It is one thing to say that a speech stream can be heard (i.e., detected), another that a listener can hear that speech has been produced by two different talkers (i.e., discriminated), and quite another to say that a listener is able to understand (i.e., recognize) what is being said. The detection, discrimination, and recognition of vocal signals have been measured in the laboratory with a high degree of precision in both birds and humans and the differences in signal-to-noise ratios (SNRs) required for detection versus discrimination versus recognition are nearly the same in birds and in humans (Dooling et al. 2009; Dooling and Blumenrath 2013). This provides a remarkable opportunity to predict the effects of masking by noise of animal communication signals in a natural environment.

What about comfortable communication? The SNR that represents a comfortable communication level in animals is likely impossible to assess. It has been measured in humans, however, and that SNR is about 15 dB (Franklin et al. 2006; Freyaldenhoven et al. 2006). Because the SNR differences between the hearing levels of detection, discrimination, and recognition are similar for birds and humans, it is possible that a comfortable communication level also exists for birds and that level would be about the same SNR (15 dB) as it is for humans. If so, then it is reasonable to postulate that there are four different levels of hearing in birds that are relevant for communication, each of which require different SNRs (Dooling and Blumenrath 2013). For humans, the distinction between these levels of hearing is intuitive. Field researchers who rely on song playback techniques and monitor behavioral responses of birds to determine whether a song was heard or not are familiar with a similar phenomenon (Brenowitz 1982; Nelson and Marler 1990). Klump (1996) described this issue as just-noticeable differences that may be tested in the laboratory versus just-meaningful differences between stimuli that may be measured in the field.

2.2 Behaviors That Reduce Masking Effects on Communication

Some level of environmental noise is always present. In this sense, it is pervasive, whether it originates from natural environmental sound sources or those created by humans. Because of the pervasive nature of noise, humans and animals evolved effective behaviors that enhance their abilities to use sound for communication in noisy environments and many of these strategies are almost reflexive in that they occur without conscious thought (Brumm 2013). Typically, the goal of such strategies is to create a condition where either the signal level or the masker level changes to improve the SNR. Obvious actions animals and humans might take are to move closer to the signal (so long as the SNR is at least favorable enough to detect that there is a relevant sound present) or away from the masker so that recognition or even comfortable listening might be achieved. Sometimes, as is intuitive for human listeners, simply changing the orientation of the head might already provide some benefit (Dent et al. 1997).

In addition to an improvement in SNR due to shortening the distance between sender and receiver, other movements of both senders and receivers could also result in a more favorable sound transmission path that may mitigate the masking effects (Brumm and Slabbekoorn 2005). For example, a more effective transmission path for a signal without affecting the transmission of habitat noise may be achieved by moving to a higher level relative to the ground (Mathevon et al. 1996, 2005). It is estimated that European blackbirds (*Turdus merula*) and great tits (*Parus major)* could receive an improvement in SNR equivalent to the benefit from closing the interbird distance in half by simply moving upward about 9 meters to a higher perch (Dabelsteen et al. 1993; Blumenrath and Dabelsteen 2004). It is also the case that moving from a lower position to a higher position for the receiver had a greater impact on whether a vocalization was heard than when the receiver stayed in the same position but the sender moved from a lower position to a higher position.

Then there is the well-known Lombard effect in human hearing and vocal production where the speech level is increased in the presence of background noise (Lane and Tranel 1971). This effect has been demonstrated in the field in various bird species (Brumm and Todt 2002; Brumm and Zollinger 2011) and in the laboratory in budgerigars under operant control (Manabe et al. 1998; Osmanski and Dooling 2009). Others have shown that birds in urban areas (high noise) sing more loudly than when they are in a quiet countryside and that they have increased vocal levels during weekdays over weekends when noise levels are usually attenuated (Brumm 2004; see Larsen and Radford, Chap. 5). There is also evidence that birds may shift the frequency region of their vocalizations in noise so the vocalizations occur in a spectral region of a more favorable SNR (Halfwerk and Slabbekoorn 2009; Gross et al. 2010; Verzijden et al. 2010). This might partly or completely explain the geographic patterns of noise level-dependent and habitat-dependent song frequency use (Slabbekoorn and Peet 2003; Slabbekoorn 2013).

Environment noise often fluctuates widely in both temporal and spectral characteristics. Humans and other animals can and do take advantage of periods of quiet or reduced levels in the ongoing noise (Wiley and Richards 1982; Klump 1996). This release from auditory masking in temporally fluctuating noise is termed "dip listening" and reflects a listener's ability to catch a brief "acoustic glimpse" of the signal of interest such as speech when the background noise dips to a relatively low level (Vestergaard et al. 2011; Brumm 2013). Another improvement, called comodulation masking release, occurs when the envelope of the masker fluctuates over time and these fluctuations are correlated across different frequency bands. This demonstrates the ability of the auditory system to perform a highly flexible spectrotemporal pattern analysis (Moore 2003). Comodulation masking release can add up to 10 dB to the detection level in birds (Klump and Langemann 1995; Dent et al. 1997). There is every reason to think that the other levels of hearing (i.e., discrimination, recognition, and comfortable communication) would also show similar degrees of enhancements.

In addition to the strategy of listening in the quieter periods within temporally variable maskers and comodulation masking release, yet another release from masking can occur by spatial separation of the two sources of signal and noise (Arbogast et al. 2002). Benefits of 10-15 dB have been reported in humans (e.g., Saberi et al. 1991) and other animals (Hine et al. 1994; Dent et al. 1997). This suggests that a simple means of achieving greater spatial release from masking would be to reorient the head to increase the acoustic differences of sound arriving at the two ears.

2.2.1 Higher Order Cognitive Strategies

A listener's ability to discriminate and recognize sounds is also likely affected by higher order cognitive processes, which relatively simple masking communication models do not take into account. The pioneering work of Bregman and Campbell (1971) and Bregman (1990) along with other human psychophysical studies have described auditory scene analysis as an important element of the so-called "cocktail party effect" (see Dent and Bee, Chap. 3). This work shows that humans routinely segregate concurrent sounds into separate auditory objects using a variety of cues such as common onset and offset times and common amplitude modulations as well as similar frequency ranges and spatial location (Bregman 1990; Vliegen and Oxenham 1999). The formation of auditory objects is achieved by complex sound processing and cognitive integration of prior sensory experience and other simultaneous sensory input.

The question of whether such an important phenomenon also occurs in nonhuman animals has been addressed some time ago in starlings. Hulse et al. (1997)

and Hulse (2002) showed that starlings could be trained to identify a sample of one species' birdsong presented concurrently with a sample of another species' birdsong. Moreover, these birds could learn to discriminate among many samples of the songs of two individual starlings and could maintain that discrimination when songs of a third starling were digitally added to both song sets and songs from additional starlings were added as further background distracters (Wisniewski and Hulse 1997; MacDougall-Shackleton et al. 1998). These results with starlings and the results from many subsequent animal studies suggest that auditory scene analysis plays an important role in auditory perception in birds (Bee and Klump 2005; Micheyl et al. 2005) and other nonhuman vertebrates (Bee and Micheyl 2008) that must parse the world into auditory objects (Barber et al. 2003; see Larsen and Radford, Chap. 5; Halfwerk, Lohr, and Slabbekoorn, Chap. 8).

These auditory strategies that are available to humans and animals to mitigate the masking effects of noise on acoustic communication are both varied and sophisticated. On one hand, this raises the specter that it may be impossible to predict, with precision, the full effects of any environmental noise, man-made or otherwise, on acoustic communication in the real world. On the other hand, it also shows clearly that animals have a variety of means with which to cope with such noise (e.g., Gil and Brumm 2014). But regardless of the various mitigating strategies that animals and humans use to hear in noisy environments, the fundamental factor in determining whether a signal will be masked is the SNR at the animal's ear, and a huge amount is known in humans and animals, especially in birds, about the required SNR to hear a signal in noise (Dooling et al. 2000). The large number of laboratory masking studies available in birds coupled with the extensive understanding of their acoustic communication systems provides an excellent opportunity to clarify the general principles of masking that affect acoustic communication in nonhuman animals. And because of this wealth of masking data from the laboratory, reaching an understanding of the effects of noise on acoustic communication in the real world is perhaps easier than expected.

2.3 Measurement of Hearing: Absolute Thresholds

In humans, hearing thresholds are typically measured in the quiet of an acoustic test booth that blocks most ambient noise from the environment (ANSI 1999). The detection thresholds for pure tones measured across the frequency range of hearing under these conditions are called absolute thresholds, and taken together, this set of thresholds is commonly referred to as an audiogram, absolute threshold curve, or audibility curve. It is important to note that the quiet conditions within an acoustic test booth where such measurements are taken are rarely, if ever, attained in the real world: Some environmental noise is always present, and if this sound is of a sufficient level, it will interfere with hearing. So, although the audiogram provides a useful measure of the general functioning of the auditory system, it almost never provides an adequate description of what can be heard under natural conditions.

Background noise, even at surprisingly low levels, usually determines whether a sound is audible or not outside the quiet conditions of a test booth. In other words, appeals to the shape or sensitivity of an animal's audiogram in discussions of whether masking might be occurring is not very useful. Instead, it is the level of background noise and the animal's auditory system that determine what is heard.

2.3.1 Critical Bands and Critical Ratios

The simplest kind of masking experiment is to measure the sound detection threshold for a pure tone (the signal) in the presence of a broadband noise. This experiment was performed decades ago in humans by Fletcher (1940), who was perhaps the first to clearly articulate the idea that the auditory periphery might operate as a series of overlapping band-pass filters whose width increased with increasing frequency. More importantly for our present purposes, if a pure tone is just masked by a broad band of noise, gradually decreasing the width of the band of noise without lowering its level at some point allows the tone to become audible. This bandwidth is called the *critical band*. At the test-tone frequency, the ratio of the power in the test tone at threshold level to the power in the critical band of frequencies surrounding the tone is called the *critical ratio*. The noise used in masking experiments such as these is usually spectrally flat and described in terms of a spectrum level (i.e., sound energy per hertz). The signal level in the case of a pure tone is simply the level of the tone in decibels sound pressure level (SPL; i.e., re 20 μPa).

These relationships are illustrated schematically in Fig. 2.1 and support the power spectrum model of auditory masking (Scharf 1980). Figure 2.1 illustrates a critical ratio of 20 dB where a 3-kHz pure tone at 60 dB SPL is just masked by a flat band of noise with a spectrum level (per cycle energy distribution) of 40 dB. Research over the years has shown that the power spectrum model may be overly simplified and relies on a number of assumptions that are not always strictly true (Scharf 1980; Moore 2003). Nevertheless, this model accounts for a consider-

Fig. 2.1 Schematic representation of the critical ratio. For a critical ratio of 20 dB, a 60-dB tone at 3 kHz is just masked by a broadband noise with a spectrum level of 40 dB. The critical ratio is defined as the level of the tone minus the spectrum level of the noise. From Dooling and Popper (2016)

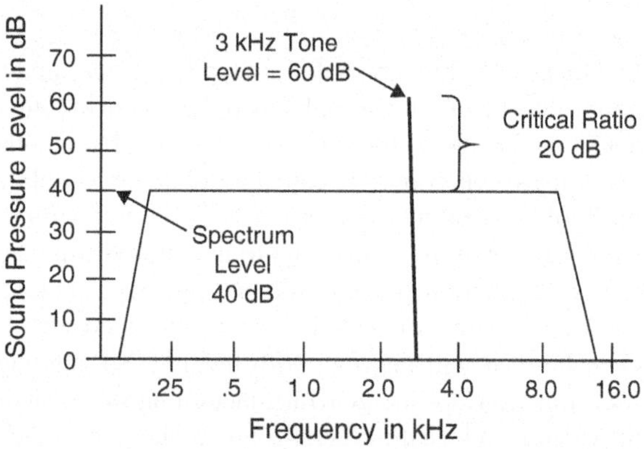

able amount of simple masking behavior in both humans and animals. Especially important is the general finding from masking studies in birds and other animals that, at low to middle SPLs, it is the noise in the frequency region of a signal that is most important in masking the signal, not noise at more distant frequency regions (Dooling et al. 2000). This is the central dogma in creating a simple model for predicting the masking effects on complex vocalizations from natural and man-made noises.

To be sure, at higher noise levels, there is an upward spread of masking that occurs for mammals but not for birds (Dooling et al. 2000), although at such high levels of noise, there are likely to be other behavioral and physiological effects along with masking. Nevertheless, for humans and other mammals, the noise in the environment is often at a higher level in the low frequencies than in the higher frequencies. Mammalian auditory systems experience a spread of masking from high-level low-frequency sounds that can mask or interfere with communication signals with higher frequency energy, resulting in increased masking of the high frequencies in a signal.

2.3.2 The Power Spectrum Model of Masking

Much of the early work on masking focused on filter bandwidths in hearing. It arose from studies in humans focused on understanding how the peripheral auditory system functions in detecting and discriminating sounds and on the relationship between the concepts of the critical ratio and the critical band (Scharf 1980; Yost and Shofner 2009). Although important for theories of hearing, these technical issues may not be as relevant for understanding the effects of man-made noise on animals in their natural habitats. Rather, the point to keep in mind is that it is the noise in the same frequency region as the signal that is critical in determining masking. Noise falling outside the frequency spectrum of the signal is a much less effective masker.

Critical ratios have now been measured in 14 species of birds and many other vertebrates including humans. Figure 2.2 summarizes these data. It shows the median critical ratio for 14 bird species compared with the critical ratios for humans. For birds and humans, these curves show how much a signal level must be above the spectrum level of the background noise in order to be heard. Human critical ratios are smaller than those measured in any other vertebrate, whereas the largest critical ratios measured in other terrestrial and marine mammals, fish, and amphibians are indicated in Fig. 2.2, *gray solid line* (summarized in Fay 1988).

There are several interesting things to note about these comparisons. First, the critical ratios for both birds and humans increase at about 3 dB/octave as frequency increases. The second is that critical ratios for birds are about 6 dB larger (worse) than those of humans over most of the frequency range. A smaller critical ratio indicates that the frequency region contributing to masking a given signal is narrower (lesser noise bandwidth). This means that humans hear better in noise than birds. In

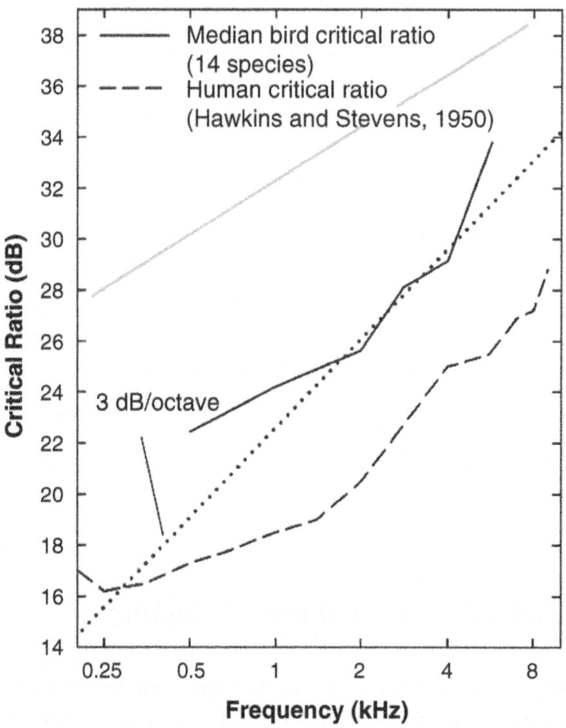

Fig. 2.2 Median critical ratios for 14 birds (*black solid line*) and human critical ratios (*black dashed line*) from Hawkins and Stevens, 1950. *Dotted line*, a slope of 3 dB/octave. The critical ratio at threshold is about 6 dB greater in the typical bird compared with humans over the frequency range of 1-5 kHz (Dooling et al. 2000). These median critical ratios for birds are data collected under rigorous experimental procedures using pure tones masked by broadband noise. *Gray solid line,* the largest critical ratios measured across animals (Fay 1988). Humans have the lowest critical ratios measured in any animal. In other words, all animals hear less well in noise than humans do. Modified from Dooling and Blumenrath (2013)

fact, humans hear as well or better, often much better, in noise than all other animals that have been tested (Fay 1988). Keeping the focus on birds, this has tremendous implications for judging how well birds hear in noisy environments. Imagine a human listener out in the field with a bird perched on its shoulder. The vocalizations of a distant bird that are just detectable by the human listener would be inaudible to the bird sitting on the human listener's shoulder. By the inverse square law, the sound level attenuates by 6 dB with every doubling of distance. This means that the human listener would have to cut the distance to the singing bird by half in order for the bird perched on the listener's shoulder to barely hear the singing bird. Otherwise said, masking noise in terms of communication distance is more detrimental to birds and other animals than to humans.

Even in relatively quiet natural environments, it is the low level of background noise that determines what can be heard. Figure 2.3 shows the relationship between

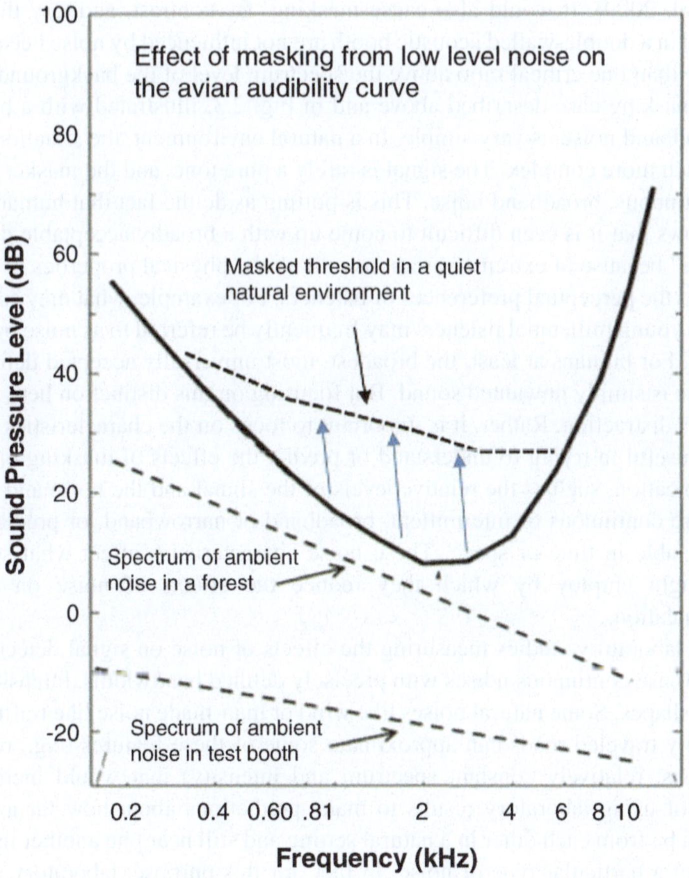

Fig. 2.3 Median bird hearing thresholds in a quiet test booth based on data from 49 bird species (*solid curved line*). Below the curve are typical spectrum levels of the background noise in a double-walled acoustic isolation testing chamber (*middle dashed line*) and a typical spectrum level of ambient noise that a bird might encounter in a quiet forest environment (*bottom dashed line*). Within the audiogram, the increased threshold due to masking from environmental noise in a quiet natural environment, e.g., a quiet forest, is shown (*top dashed line*). *Blue arrows* indicate the increase in threshold due to this low level of environmental noise (the amount of masking). Note that even an ambient-noise level less than 20 dB below the audiogram thresholds, which is the case in almost all natural environments, will raise the animal's thresholds (i.e., cause masking). An ambient-noise spectrum level at least 20 dB below the audiogram (i.e., in a test booth) will have no effect on hearing thresholds (i.e., no masking).

the bird audiogram, an idealized spectrum of ambient noise in a double-walled acoustic test booth, the idealized sound spectrum of ambient noise in a forest (schematically represented from Klump 1996), and the masked thresholds that result from this natural, low-level ambient noise (Dooling and Saunders 1975; Klump 1996). Natural ambient noise such as would occur in a quiet forest tends to have more energy at lower frequencies and less energy at high frequencies (Klump 1996;

Pohl et al. 2009). It would also cause masking. By contrast, auditory thresholds collected in a double-walled acoustic booth are not influenced by noise because they are more than one critical ratio above the spectrum level of the background noise.

The masking case described above and in Fig. 2.3, illustrated with a pure tone and broadband noise, is very simple. In a natural environment, the situation is typically much more complex. The signal is rarely a pure tone, and the masker is rarely flat, continuous, broadband noise. This is putting aside the fact that human experience shows that it is even difficult to come up with a broadly acceptable definition of "noise" because of extreme variations in both the physical properties of ambient noise and the perceptual preferences of listeners. For example, what may sound like music to young millennial listeners may frequently be referred to as noise by elderly listeners. For humans at least, the broadest, most universally accepted definition is that noise is simply unwanted sound. But focusing on this distinction here is somewhat of a distraction. Rather, it is important to focus on the characteristics of noise that are useful in trying to understand or predict the effects of masking on animal communication, such as the relative levels of the signal and the noise and whether noises are continuous or intermittent, broadband or narrowband, or predictable or unpredictable in time or space. These noise characteristics affect what strategies birds might employ by which they reduce the effects of noise on acoustic communication.

Most laboratory studies measuring the effects of noise on signal detection (see Sect. 2.3) use continuous noises with precisely defined bandwidths, intensities, and spectral shapes. Some natural noises like wind or man-made noise like traffic noises on heavily traveled roads can approximate some of these features (e.g., relatively continuous, relatively constant spectrum and intensity) that would increase the validity of using laboratory results to make predictions about how far away two birds can be from each other in a natural setting and still hear one another in a background of a particular type of noise. In fact, for this purpose, laboratory masking studies quantify the worst-case estimate of communication distance in the natural setting. This is because the animal being tested in the laboratory is in a fixed location with respect to the loudspeaker that is producing both the noise and the signal and, typically, head movement is restricted. Whenever these two conditions are not met, as would usually be the case in a natural setting, the amount of masking from traffic noise is likely to be less than that found in the laboratory for the same SNR and sometimes considerably less.

2.3.3 Masking of Bird Vocalizations by Noise

Lohr et al. (2003) examined the relationship between some of these key variables in considering the masking of communication signals in natural environments. The question was whether the SNR obtained from the masking of pure tones by noise in the laboratory (i.e., the critical ratio) can be used to predict the bird's ability to detect natural vocalizations in a background of noise. Related to this are two other questions. One is whether signal bandwidth has an effect on the amount of masking that occurs because some vocalizations are clearly more broadband than others. The other is how masker bandwidth and spectrum affect the masking of communication signals. Lohr and his colleagues created a masking noise in the laboratory modeled after typical highway noise, with a spectrum sloping downward from low to high frequencies, having more energy at frequencies below 1 kHz than above.

Lohr et al. (2003) then directly compared the detection and discrimination of contact calls by two different species of birds, the zebra finch (*Taeniopygia guttata*) and the budgerigar (*Melopsittacus undulates*), in flat noise and "traffic-shaped" noise. The bird vocalizations used as test stimuli, by contrast, contained more energy above 1 kHz than below even though some were broadband (zebra finch calls) and some were narrowband (budgerigar calls). For both zebra finches and budgerigars, discrimination thresholds were about 2-3 dB higher than detection thresholds, regardless of whether the birds were tested with conspecific or heterospecific calls. The masking noise having the spectral shape of traffic noise (with less energy in the frequency region of the bird calls) caused less masking of vocalizations than did a flat spectrum noise for both species even though both noises were presented at the same overall level. In other words, noise in the spectral region of the signal was more effective at masking the signal than noise in other frequency regions, which had little effect on thresholds. Importantly, the SNR between the level of the peak in the vocalization power spectrum and the spectrum level of the background noise was similar to the SNR of a just-masked pure tone compared with the spectrum level of noise surrounding the pure tone. Thus, a bird's critical ratio measured at the peak frequency in its vocalization can provide an estimate of whether the bird can detect a vocalization in noise.

With masking results such as these, it is possible to estimate the distance over which birds can detect and discriminate the vocalizations of other birds against a background of noise, the so-called active auditory space (Brenowitz 1982). The constraints on acoustic communication under natural environmental conditions are complex and involve a variety of factors, some of which are linear and some nonlinear with distance (Wiley and Richards 1982; Dooling and Blumenrath 2013). Two prominent sound propagation issues mentioned by Larsen and Radford (Chap. 5) concern level changes due to spherical spreading (i.e., the inverse square law) and excess attenuation (Marten and Marler 1977; Marten et al. 1977; Dooling 1982; Dooling and Blumenrath 2013). Under simple environmental conditions, spherical spreading amounts to a quartering of acoustic power or halving of pressure with each doubling of distance (i.e., a sound pressure change of −6 dB/doubling of distance). In homogeneous environments, there are also constant attenuation effects, which represent a deviation from the attenuation expected from the inverse square law. This excess attenuation is expected to be about 5 dB/100 meters for a sound source 10 meters above ground in an open field (e.g., Marten and Marler 1977).

2.3.4 *Masking, Communication Distance, and the Four Levels of Hearing*

There is enough laboratory data to provide a simple and comprehensive approach for predicting the effects of masking noise on the communication distance in birds that would include all four levels of hearing (i.e., detection, discrimination, recognition, and comfortable communication) described in Sect. 2.1.1. Knowing the spectrum and level of the masking noise (at the receiving bird's ear), the bird's hearing in quiet and in noise (i.e., critical ratio), and the peak spectrum level of a signaling bird's vocalizations reaching the ear of the receiver provides a quantitative and precise estimate of the animal's active acoustic space. Values for the signal and noise at the receiver's ear could be measured directly with a sound level meter or they could be estimated from source levels (e.g., a singing bird, noise at a highway) by applying sound propagation and attenuation algorithms to estimate both the signal and the noise arriving at the receiving bird (Dooling 1982; Nemeth and Brumm 2010).

Figure 2.4 illustrates the differential effects of traffic noise level on the four different auditory behaviors of birds. As the level of traffic noise is attenuated from 90 to 50 dB(A) , the communication distances increase, as indicated by the different colors of the curved lines. Figure 2.4, or nomogram, is constructed in such a way that by using a straightedge to connect known values on two lines, the unknown value can be read at the point of intersection with another line. Here, the *horizontal dashed line* at a traffic noise level of 60 dB indicates the distances intercepted by each of the four hearing levels. This example assumes a median bird critical ratio function and assumes that the singing bird is vocalizing at a maximum peak SPL approaching 100 dB (Brackenbury 1979) through an open area that would mean an excess attenuation of 5 dB/100 m, beyond the loss due to spherical spreading. The formula for calculating these effects is

$$\text{Drop} = 20\log\left(\frac{d_{\mathbf{mc}}}{d_0}\right) + \frac{EA * d_{\mathbf{mc}}}{100}$$

where Drop is the amount of signal attenuation from the source intensity to the signal level at which a signal can just be heard in noise. (i.e., critical ratio); d_{mc} is the maximum communication distance; d_0 is the distance at which the source intensity is measured; and EA is the amount of excess attenuation (linear attenuation, not due to spherical spreading).

As shown schematically in Fig. 2.5, in the acoustic scenario described here, a comfortable level of communication between two birds could occur up until the two birds are about 60 meters apart. Recognition and discrimination of a bird vocalization by the receiver are still possible at even greater interbird distances of up to about 220 meters and 270 meters, respectively. And finally, simple detection of another bird's vocalization can occur at even greater distances of up to 345 meters in this noise. To the extent that critical ratios differ across species, so will these

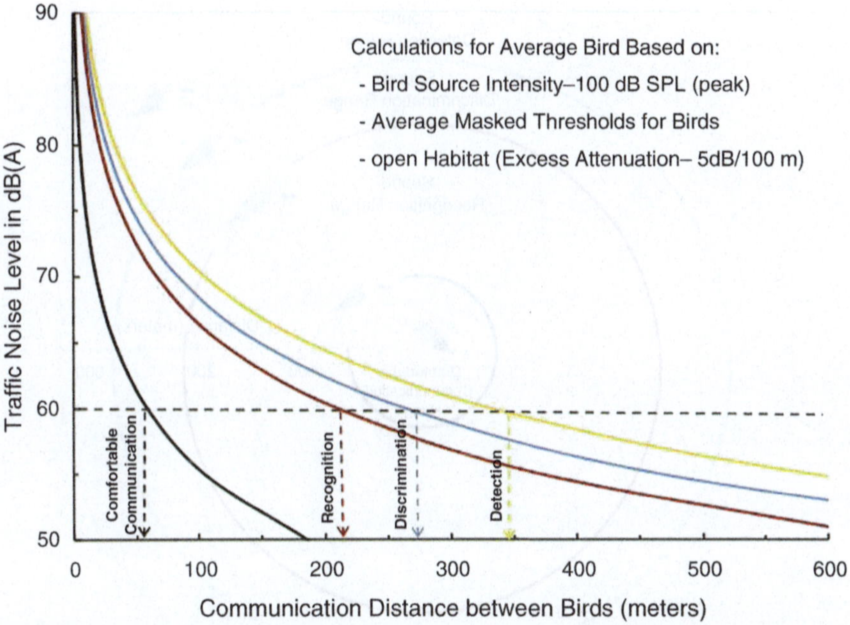

Fig. 2.4 The level of traffic noise affects acoustic communication between two birds in accordance with the distance between them. The effects on the four levels of hearing for different communication distances described in Sect. 2.1.1 are modeled. *Black line*, "comfortable" communication; *red line*, "recognition"; *blue line*, "discrimination"; *yellow line*, ability to "detect" a sound. *Horizontal dashed line*, a typical traffic noise spectrum at a level of 60 dB(A); *Down arrows* indicate the estimates of the limits of communication distance for each level of hearing: comfortable communication occurs up to 60 meters, recognition of a vocalization can occur up to about 210 meters, discrimination between two vocalizations occurs at about 270 meters, and detection occurs up to about 340 meters. Beyond this distance, a bird is not likely to detect the signal. This prediction is based on critical ratio data collected in the laboratory that doesn't allow any mitigating behaviors that might improve the signal-to-noise ratio at the bird's ear in a natural environment and therefore represents a worst-case scenario. Taken from Dooling and Blumenrath (2013)

maps. Birds with large critical ratios at all frequencies such as canaries (Okanoya and Dooling 1987) will show shorter communication distances (i.e., concentric circles); those with smaller critical ratios would show longer communication distances (Dooling and Blumenrath 2013).

The communication distance "limits" in Fig. 2.5 are unlikely, of course, to be perfectly concentric in a natural environment. In real-world situations, the acoustic dynamics of signal transmission are highly variable, both spatially and temporally, depending on the distribution and character of habitat types, prevailing meteorological conditions, and the behavior of both the caller and receiver in optimizing communication. Consequently, the shapes and sizes of the communication regions around the receiver will naturally vary in accordance with the physical conditions of the area, the species-specific hearing capabilities, the characteristics of the signal, and the mitigating strategies employed in communicating acoustically.

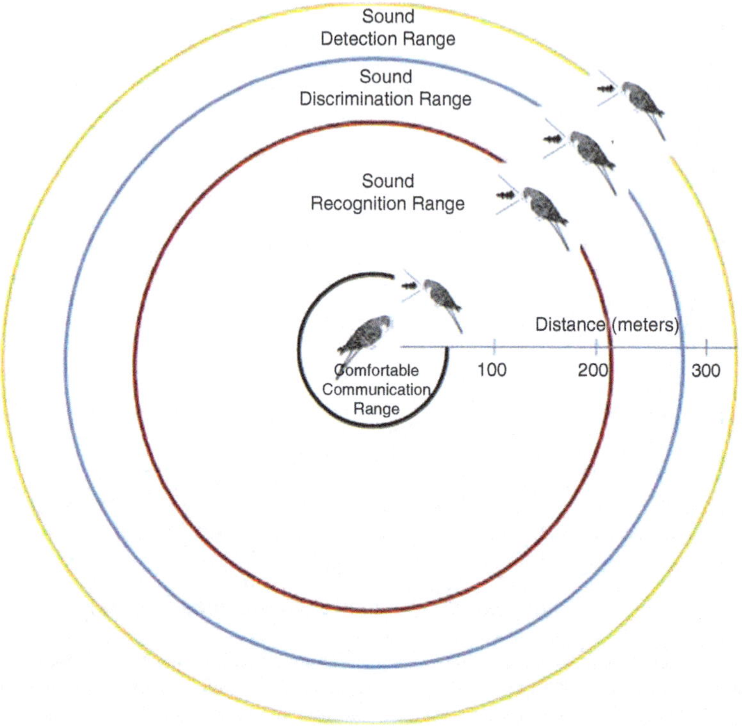

Fig. 2.5 A schematic representation of the data in Fig. 2.4, using the same colors, showing the level of hearing of a bird in noise located at different distances from a sound-emitting bird. A bird can just hear a vocalization (i.e., detect it; *yellow circle*) at a much greater distance than is required for comfortable communication (*black circle*). This represents the simple case based on critical ratio data from the laboratory and does not include short-term adaptation strategies described in the text, which would improve communication. Adapted from Dooling and Blumenrath (2013)

Some simple principles are apparent in this model. Communication distances for receiving birds closer to the noise source or with large critical ratios at 2 kHz would be represented by smaller concentric circles, whereas communication distances for birds further away from the noise source or with smaller critical ratios would be represented by larger concentric circles. Because the masking data on which this simple model is based are from the laboratory, they were collected under highly controlled conditions where the masking noise is a variant of continuous broadband white noise and the signals were either bird calls or pure tones. A reasonable question then is whether these results from the laboratory provide an accurate reflection of what occurs in the field? There are at least two parts to this question. One issue is whether critical ratio values reported for humans under laboratory conditions would also be obtained for humans listening to tones in natural environments with a measurable, quantifiable background noise. Preliminary results show that measuring pure-tone thresholds in a "quiet"

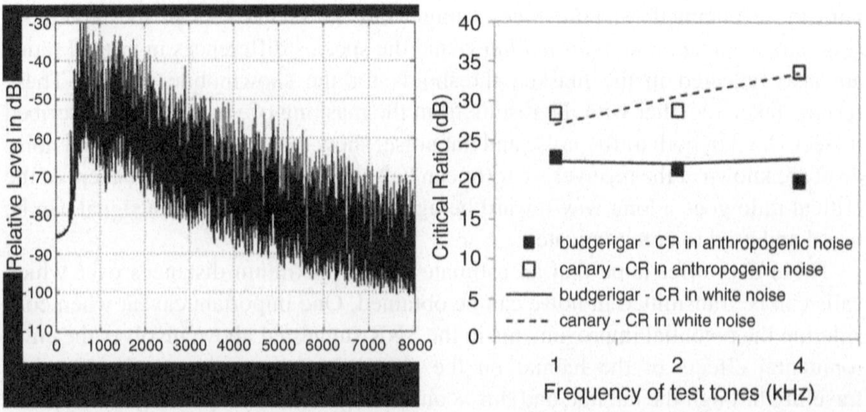

Fig. 2.6 *Left*: spectrum of a snowmobile recorded in Yellowstone National Park. *Right*: critical ratios (CR) measured at 1, 2, and 4 kHz in this noise. Masked thresholds for both species measured in snowmobile noise show a close approximation to those measured in white noise in the laboratory. From Dooling and Blumenrath (2016)

natural environment consisting of a low level of ambient noise gives critical ratio values in humans that are similar to those obtained in the laboratory under more rigorous conditions. In these testing conditions in the field, both the signal and background noise levels vary greatly moment to moment due to the uncontrolled acoustic environment. Nevertheless, over many trials at many different signal and noise levels, the SNR resulting in a performance of 50% detection closely tracks the published values for human critical ratios at several tone frequencies in ambient noise in the field (Dooling and Blumenrath 2016).

The second issue is whether noise that is generated from common man-made sources such as automobiles, motorcycles, aircraft, and snowmobiles will produce similar critical ratio values as those obtained with the continuous broadband noise used in the laboratory. If noise in the frequency region of the signal is the key feature that predicts the most effective masking of the signal, then complex man-made sounds that have energy in the spectral region of the signal should produce similar critical ratios as broadband white noise does. Obviously, it is impossible to test birds with psychophysical tests in a natural environment. But it is possible to bring man-made sounds as they occur in a natural environment into the laboratory and use them as maskers.

Critical ratios have been well established in canaries and budgerigars, with canaries showing critical ratios that are 3-4 dB larger than those of budgerigars (Okanoya and Dooling 1987). In a recent experiment, pure-tone thresholds in canaries and budgerigars were measured in a background of various man-made noises (Dooling and Blumenrath 2016). A long-term spectrum from one of these sound recordings from a snowmobile is shown in Fig. 2.6, *left*. The masking results from canaries and budgerigars tested with noise from a snowmobile are shown in Fig. 2.6, *right*, *open* and *solid squares*, respectively. Thresholds in both species agree well with those obtained for pure tones using white noise maskers in the laboratory (Fig. 2.6, *right*, *solid* and *dashed lines*) and the species differences in critical ratios are also reflected in the masked thresholds for the snowmobile masker. These results, taken together with the results from the masking of vocalizations described in Sect. 2.3.3 in both traffic noise and flat noise, show that when the signal and noise level are known at the receiver's ear (or can be calculated), the simple concept of the critical ratio goes a long way toward being able to predict whether a signal can be heard and used to communicate.

These data offer hope that an estimate of the maximum distances over which calls can be transmitted in noise can be obtained. One important caveat when considering the potential improvements in the SNR suggested above involves the environmental effects of the habitat on the signal. Vocalizations are altered when traveling through the habitat, and this acoustic degradation may change the characteristics important for determining detectability and discriminability (Wiley and Richards 1982; Blumenrath and Dabelsteen 2004). Signal degradation in the form of high-frequency attenuation, reverberation, and the "blurring" of amplitude and frequency patterns will alter a vocalization's spectral and temporal fine structure, affecting among other things the "peakiness" of the waveform and thus its informational content (Dabelsteen et al. 1993; Dabelsteen 2005; Mockford et al. 2011). Changes in signal characteristics may also result from changes in output levels such as what occurs from the well-known Lombard effect (Cynx et al. 1998; Brumm and Todt 2002; Brumm 2004; Brumm and Zollinger 2011).

Typically, in laboratory experiments examining the effect of noise on detection and discrimination thresholds, the vocalizations used as test stimuli are recorded under ideal conditions in a sound-attenuated environment that minimizes echoes. Using such stimuli in masking studies may result in overestimates of the actual effective communication distance compared with using degraded calls, i.e., calls rerecorded at biologically relevant distances in species-typical habitats and then brought into the laboratory and used as stimuli in psychophysical tests. This approach could the separate potentially synergistic effects of sound degradation and noise masking. Small-bird species tested with digitally reverberated stimuli suggest that discrimination of similar vocalizations from different individuals is significantly impaired when reverberation occurs with high noise levels, whereas neither reverberation nor noise alone had similarly detrimental effects (Blumenrath 2011). Also, the bird's ability to form auditory objects is affected along with the limitations on the bird's ability to segregate signals from multiple individuals (Blumenrath 2011). In other words, to predict whether and to what extent a given acoustic environment limits the communication range or interferes with the detection, discrimination, and recognition of biologically important sounds, it is important to consider the combined effects of sound-altering habitat characteristics and species differences in both auditory sensitivity and cognitive processing.

2.3.5 Laboratory Results Applied in an Environmental Context

Taken together, the results presented here provide a strategy and a model for determining whether a given noise in the environment has an effect on acoustic communication in birds and other animals. To return to the motivation for this chapter, without such a model, discussions among conservationists and regulators on whether noise affects a specific species often rely on broad, nonspecific comparative data or anecdotal evidence from uncontrolled observations in the field for how animals hear in noise, including whether a particular noise level may be detrimental or not. The goal of this chapter is to demonstrate that there are enough laboratory masking data on animals, especially birds, that reasonable, quantifiable estimates of the effects of masking noise on communication are now possible. Such information, properly applied, can guide the efforts of conservationists and regulators in making informed decisions for protecting wildlife.

There are three key variables to consider. One is the level of the signal (i.e., another bird's vocalization) at the receiver's ear. If the level at the ear cannot be measured directly, the level and spectrum of a bird's vocal output (i.e., at the source) are fairly easy to measure and these can be used to estimate the level at the receiving bird's ear. Commercially available sound mapping programs are now quite sophisticated in determining what happens to sound, both its level and spectrum, as it travels through the environment. This enables the estimation of the signal level at the receiver's ear. A second key variable is the level and spectrum of background noise at the receiver's ear. The third key variable is the animal's critical ratio. As far as is known, no animal has yet been shown to have critical ratios smaller than the critical ratios observed in humans and some are considerably larger (Fay 1988). The simple conclusion from this statement is that humans will hear better in noise in a natural environment than will animals. In other words, masking from noise is worse for animals than it is for humans.

The comparative approach above actually provides a very accurate sense of how noise affects animals in their natural habitat, Taking the median critical ratio function for birds as an example, critical ratios increase about 3 dB/octave from about 22 dB at 1 kHz to about 29 dB at 4 kHz. Humans show critical ratios that are 6 dB smaller than the median critical ratios for birds over this same range in similar laboratory conditions. This means that, without using any sound measuring or analysis equipment, human listeners can serve as a proxy for estimating the effects of noise on birds in a natural environment. To return to the illustration described in Sect. 2.3, a human listener in any natural environment will hear sounds at a distance, including bird vocalizations, much better than a bird would. The 6-dB difference in critical ratios between humans and birds means that humans could hear another bird at twice the distance than a bird could because a 6-dB level difference represents a doubling of distance. Communication distance functions for each level of hearing can then be interpreted on the basis of what is known about the species' biology, such as whether a recognition level of hearing in noise is required to defend a territory or which level of hearing in noise best matches with the distances to the song posts of neighboring birds. Presumably, for birds that spend time in large groups, the SNR required to function in such close proximity would require a higher SNR of 15 dB, similar to the SNR level at which humans would have comfortable communication.

2.4 Conclusions

Conservationists and regulators sometimes operate under the illusion that there is one number (i.e., SPL) that man-made noise should not exceed in order to protect wildlife. This is unrealistic. Instead, the comparative data reviewed in this chapter convey a simple message: when the SNR between call or song level and the spectrum level of the background noise is less than about 25 dB, communication may be affected. To hear better in such a level of noise, birds would need to engage in the various masking mitigation behaviors reviewed here that would improve a bird's hearing considerably just as they do for human listeners. These noise mitigation behaviors are widespread among animals, including humans. There is considerable variation between the critical ratios of mammals and other animals, but none, as far as we know, are smaller than those of humans (e.g., Fay 1988). Thus, human hearing provides an extraordinarily simple and efficient way of judging whether a given noise is problematical for birds and other animals. Most importantly, this can be accomplished by a human listener simply listening to biologically relevant sounds in the environment. Birds, because so many species have been tested and because they are so vocal, provide the strongest case for the simple, powerful value of the critical ratio approach for understanding communication in noise and related noise measurement and management strategies.

Acknowledgments This work was supported in part by National Institutes of Health grants to Robert J. Dooling and a Senior Research Career Scientist Award from the Department of Veterans Affairs Rehabilitation Research and Development Service to Marjorie R. Leek. The contents of this chapter do not represent the views of the Department of Veterans Affairs or the US Government.

Requirement Compliance with Ethics
Robert J. Dooling declares that he has no conflict of interest.
Marjorie R. Leek declares that she has no conflicts of interest.

References

American National Standards Institute (ANSI). (1999). *Maximum Permissible Ambient Noise Levels for Audiometric Test Rooms*. ANSI S3.1-1999 R2013, American National Standards Institute, Washington, DC.

American National Standards Institute (ANSI). (2013). *American National Standard Acoustical Terminology: Acoustic Terminology*. ANSI S1-1-2013, American National Standards Institute, Washington, DC.

Arbogast T. L., Mason, C. R., & Kidd, G., Jr. (2002). The effect of spatial separation on informational and energetic masking of speech. *The Journal of the Acoustical Society of America*, 112, 2086-2098.

Barber, J. R., Razak, K. A., & Fuzessery, Z. M. (2003). Can two streams of auditory information be processed simultaneously? Evidence from the gleaning bat *Antrozous pallidus*. *Journal of Comparative Physiology A: Neuroethology, Sensory, Neural, and Behavioral Physiology*, 189(11), 843-855.

Bee, M. A., & Klump, G. M. (2005). Auditory stream segregation in the songbird forebrain: Effects of time intervals on responses to interleaved tone sequences. *Brain, Behavior and Evolution*, 66(3), 197-214.

Bee, M. A., & Micheyl, C. (2008). The cocktail party problem: What is it? How can it be solved? And why should animal behaviorists study it? *Journal of Comparative Psychology*, 122(3), 235-251.

Blumenrath, S. H. (2011). *Communicating in Social Networks: Effects of Reverberation on Acoustic Information Transfer in Three Species of Birds*. Digital Repository at the University of Maryland, College Park, MD.

Blumenrath, S. H., & Dabelsteen, T. (2004). Sound degradation before and after foliation: Implications for acoustic communication in a deciduous forest. *Behaviour*, 141, 935-958.

Brackenbury, J. H. (1979). Power capabilities of the avian sound producing system. *Journal of Experimental Biology*, 78, 163-166.

Bregman, A. S. (1990). *Auditory Scene Analysis: The Perceptual Organization of Sound*. Cambridge, MA: MIT Press.

Bregman, A. S., & Campbell, J. (1971). Primary auditory stream segregation and perception of order in rapid sequences of tones. *Journal of Experimental Psychology*, 89, 244-249.

Brenowitz, E. A. (1982). The active space of red-winged blackbird song. *Journal of Comparative Physiology A: Neuroethology, Sensory, Neural, and Behavioral Physiology*, 147(4), 511-522.

Bronkhorst, A. W. (2000). The cocktail party phenomenon: A review of research on speech intelligibility in multiple-talker conditions. *Acta Acustica*, 86, 117-128.

Brumm, H. (2004). The impact of environmental noise on song amplitude in a territorial bird. *Journal of Animal Ecology*, 73(3), 434-440.

Brumm, H. (2013). *Animal Communication in Noise*. New York: Springer-Verlag.

Brumm, H., & Todt, D. (2002). Noise-dependent song amplitude regulation in a territorial songbird. *Animal Behaviour*, 63(5), 891-897.

Brumm, H., & Slabbekoorn, H. (2005). Acoustic communication in noise. *Advances in the Study of Behavior*, 35, 151-209.

Brumm, H., & Zollinger, S. A. (2011). The evolution of the Lombard effect: 100 years of psychoacoustic research. *Behaviour*, 148, 1173-1198.

Cynx, J., Lewis, R., Tavel, B., & Tse, H. (1998). Amplitude regulation of vocalizations in noise by a songbird, *Taeniopygia guttata*. *Animal Behaviour*, 56(1), 107-113.

Dabelsteen, T. (2005). Public, private or anonymous? Facilitating and countering eavesdropping. In P. K. McGregor (Ed.), *Animal Communication Networks* (pp. 38-62). Cambridge, UK: Cambridge University Press.

Dabelsteen, T., Larsen, O. N., & Pedersen, S. B. (1993). Habitat-induced degradation of sound signals: Quantifying the effects of communication sounds and bird location on blur ratio, excess attenuation, and signal-to-noise ratio in blackbird song. *The Journal of the Acoustical Society of America*, 93(4), 2206-2220.

Dent, M. L., Larsen, O. N., & Dooling, R. J. (1997). Free-field binaural unmasking in budgerigars (*Melopsittacus undulatus*). *Behavioral Neuroscience*, 111(3), 590-598.

Dooling, R. J. (1982). Auditory perception in birds. In D. E. Kroodsma & E. H. Miller (Eds.), *Acoustic Communication in Birds* (pp. 95-130). New York: Elsevier.

Dooling, R. J., & Saunders, J. C. (1975). Hearing in the parakeet: Absolute thresholds, critical ratios, frequency difference limens, and vocalizations. *Journal of Comparative and Physiological Psychology,* 88, 1-20.
Dooling, R. J., & Blumenrath, S. H. (2013). Avian sound perception in noise. In H. Brumm (Ed.), *Animal Communication in Noise* (pp. 229-250). New York: Springer-Verlag.
Dooling, R. J., & Blumenrath, S. H. (2016). Masking experiments in humans and birds using anthropogenic noises. In A. N. Popper & A. Hawkins (Eds.), *The Effects of Noise on Aquatic Life II* (pp. 239-243). New York: Springer-Verlag.
Dooling, R. J., & Popper, A. N. (2016). *Technical Guidance for Assessment and Mitigation of the Effects of Highway and Road Construction Noise on Birds.* Division of Environmental Analysis, California Department of Transportation, Sacramento.
Dooling, R. J., Lohr, B., & Dent, M. L. (2000). Hearing in birds and reptiles. In R. J. Dooling, R. R. Fay, & A. N. Popper (Eds.), *Comparative Hearing: Birds and Reptiles* (pp. 308-359). New York: Springer-Verlag.
Dooling, R. J., Leek, M. R., & West, E. (2009). Predicting the effects of masking noise on communication distance in birds. *The Journal of the Acoustical Society of America*, 125(4), 2517.
Fay, R. R. (1988). *Hearing in Vertebrates: A Psychophysical Databook.* Winnetka, IL: Hill-Fay Associates.
Fletcher, H. (1940). Auditory patterns. *Reviews of Modern Physics*, 12, 47-65.
Franklin, C. A., Thelin, J. W., Nabelek, A. K., & Burchfield, S. B. (2006). The effect of speech presentation level on acceptance of background noise in listeners with normal hearing. *Journal of the American Academy of Audiology,* 17, 141-146.
Freyaldenhoven, M. C., Smiley, D. F., Muenchen, R. A., & Konrad, T. N. (2006). Acceptable noise level: Reliability measures and comparison to preference for background sounds. *Journal of the American Academy of Audiology*, 17(9), 640-648.
Gil, D., & Brumm, H. (2014). *Avian Urban Ecology: Behavioural and Physiological Adaptations.* Oxford, UK: Oxford University Press.
Gross, K., Pasinelli, G., & Kunc, H. P. (2010). Behavioral plasticity allows short-term adjustment to a novel environment. *American Naturalist,* 176, 456-464.
Halfwerk, W., & Slabbekoorn, H. (2009). A behavioural mechanism explaining noise-dependent frequency use in urban birdsong. *Animal Behaviour*, 78, 1301-1307.
Hawkins, J. E., & Stevens, S. S. (1950). The masking of pure tones and of speech by white noise. *The Journal of the Acoustical Society of America*, 22, 6-13.
Hine, J. E., Martin, R. L., & Moore, D. R. (1994). Free-field binaural unmasking in ferrets. *Behavioral Neuroscience*, 108(1), 196-205.
Hulse, S. H. (2002). Auditory scene analysis in animal communication. *Advances in the Study of Behavior*, 31, 163-200.
Hulse, S. H., MacDougall-Shackleton, S. A., & Wisniewski, A. B. (1997). Auditory scene analysis by songbirds: Stream segregation of birdsong by European starlings (*Sturnus vulgaris*). *Journal of Comparative Psychology*, 111(1), 3-13.
Kidd, G., Jr., & Colburn, H. S. (2017). Informational masking in speech recognition. In J. Middlebrooks, J. Z. Simon, A. N. Popper, & R. R. Fay (Eds.), *The Auditory System at the Cocktail Party* (pp. 75-109). Cham, Switzerland: Springer International Publishing.
Klump, G. M. (1996). Bird communication in the noisy world. In D. E. Kroodsma & E. H. Miller (Eds.), *Ecology and Evolution of Acoustic Communication in Birds* (pp. 321-338). Ithaca, NY: Cornell University Press.
Klump, G. M., & Langemann, U. (1995). Comodulation masking release in a songbird. *Hearing Research*, 87(1-2), 157-164.
Lane, H. L., & Tranel, B. (1971). The Lombard sign and the role of hearing in speech. *Journal of Hearing and Speech Research*, (14), 677-709.
Lohr, B., Wright, T. F., & Dooling, R. J. (2003). Detection and discrimination of natural calls in masking noise by birds: Estimating the active space of a signal. *Animal Behaviour,* 65(4), 763-777.

MacDougall-Shackleton, S. A., Hulse, S. H., Gentner, T. Q., & White, W. (1998). Auditory scene analysis by European starlings (*Sturnus vulgaris*): Perceptual segregation of tone sequences. *The Journal of the Acoustical Society of America*, 103(6), 3581-3587.

Manabe, K., Sadr, E. I., & Dooling, R. J. (1998). Control of vocal intensity in budgerigars (*Melopsittacus undulatus*): Differential reinforcement of vocal intensity and the Lombard effect. *The Journal of the Acoustical Society of America*, 103(2), 1190-1198.

Marten, K., & Marler, P. (1977). Sound transmission and its significance for animal vocalization. I. Temperate habitats. *Behavioral Ecology and Sociobiology*, 2, 271-290.

Marten, K., Quine, D., & Marler, P. (1977). Sound transmission and its significance for animal vocalization. II. Tropical forest habitats. *Behavioral Ecology and Sociobiology* 2, 291-302.

Mathevon, N., Aubin, T., & Dabelsteen, T. (1996). Song degradation during propagation: Importance of song post for the wren *Troglodytes troglodytes*. *Ethology*, 102, 397-412.

Mathevon, N., Dabelsteen, T., & Blumenrath, S. H. (2005). Are high perches in the blackcap *Silvia atricapilla* song or listening posts? A transmission study. *The Journal of the Acoustical Society of America*, 117, 442-449.

Micheyl, C., Tian, B., Carlyon, R. P., & Rauschecker, J. P. (2005). Perceptual organization of tone sequences in the auditory cortex of awake macaques. *Neuron*, 48(1), 139-148.

Miller, G. A. (1947). The masking of speech. *Psychological Bulletin*, 44, 105-129.

Mockford, E. J., Marshall, R. C., & Dabelsteen, T. (2011). Degradation of rural and urban great tit song: Testing transmission efficiency. *PLoS ONE*, 6(12), e28242.

Moore, B. C. J. (2003). *An Introduction to the Psychology of Hearing*, 5th ed. San Diego, CA: Academic Press.

Nelson, D. A., & Marler, P. (1990). The perception of birdsong and an ecological concept of signal space. In W. C. Stebbins & M. A. Berkley (Eds.), *Comparative Perception, Vol. II: Complex Signals* (pp. 443-478). New York: John Wiley & Sons.

Nemeth, E., & Brumm, H. (2010). Birds and anthropogenic noise. Are urban songs adaptive? *The American Naturalist*, 176, 465-475.

Okanoya, K., & Dooling, R. J. (1987). Hearing in passerine and psittacine birds: A comparative study of absolute and masked auditory thresholds. *Journal of Comparative Psychology*, 101(1), 7-15.

Osmanski, M., & Dooling, R. J. (2009). The effect of altered auditory feedback on control of vocal production in budgerigars (*Melopsittacus undulatus*). *The Journal of the Acoustical Society of America*, 126, 911-919.

Pohl, N. U., Slabbekoorn, H., Klump, G., & Langemann, U. (2009) Effects of signal features and environmental noise on signal detection in the great tit, *Parus major*. *Animal Behaviour*, 78, 1293-1300.

Saberi, K., Dostal, L., Sadralodabai, T., Bull, V., & Perrott, D. R. (1991). Free-field release from masking. *The Journal of the Acoustical Society of America*, 90(3), 1355-1370.

Scharf, B. (1980). Critical bands. In J. V. Tobias (Ed.), *Foundations of Modern Auditory Theory* 1 (157-202). New York: Academic Press.

Slabbekoorn, H. (2013). Songs of the city: Noise-dependent spectral plasticity in the acoustic phenotype of urban birds. *Animal Behaviour*, 85, 1089-1099.

Slabbekoorn, H., & Peet, M. (2003). Ecology: Birds sing at a higher pitch in urban noise. *Nature*, 424(6946), 267.

Verzijden, M. N., Ripmeester, E. A. P., Ohms, V. R., Snelderwaard, P., & Slabbekoorn, H. (2010). Immediate spectral flexibility in singing chiffchaffs during experimental exposure to highway noise. *Journal of Experimental Biology*, 213(15), 2575-2581.

Vestergaard, M. D., Fyson, N. R., & Patterson, R. D. (2011). The mutual roles of temporal glimpsing and vocal characteristics in cocktail-party listening. *The Journal of the Acoustical Society of America*, 130, 429-439.

Vliegen, J., & Oxenham, A. J. (1999). Sequential stream segregation in the absence of spectral cues. *The Journal of the Acoustical Society of America*, 105(1), 339-346.

Wiley, R. H., & Richards, D. G. (1982). Adaptations for acoustic communication in birds: Sound transmission and signal detection. In D. E. Kroodsma & E. H. Miller (Eds.), *Acoustic Communication in Birds,* vol. 1 (pp. 131-181). New York: Academic Press.

Wisniewski, A. B., & Hulse, S. H. (1997). Auditory scene analysis in European starlings (*Sturnus vulgaris*): Discrimination of song segments, their segregation from multiple and reversed conspecific songs, and evidence for conspecific song categorization. *Journal of Comparative Psychology,* 111(4), 337-350.

Yost, W. A., & Shofner, W. P. (2009). Critical bands and critical ratios in animal psychoacoustics: An example using chinchilla data. *The Journal of the Acoustical Society of America,* 125(1), 315-323.

Chapter 3
Principles of Auditory Object Formation by Nonhuman Animals

Micheal L. Dent and Mark A. Bee

Abstract Early in the twentieth century, the Gestalt psychologists outlined principles governing the ability of the human visual system to construct integrated percepts of objects in visual scenes. By the close of the twentieth century, ample evidence suggested that the human auditory system follows similar principles of perceptual organization. Several Gestalt principles of grouping—proximity, similarity, common fate, good continuation, and familiarity, govern our ability to decompose complex mixtures of sounds into percepts of auditory objects in acoustic scenes. Auditory objects are perceptual groupings of sounds generated by the same source that are present at different times and in different parts of the frequency spectrum. The ability to form auditory objects likely plays an important role in allowing animals to navigate human-altered soundscapes. This chapter reviews studies of insects, fish, frogs, birds, and nonhuman mammals in which experimenters manipulated potential grouping cues and measured performance on behavioral tasks designed to reveal the animal's perception of auditory objects. These studies employed techniques ranging from measuring natural behaviors in response to communication signals to operant conditioning of responses to artificial sounds such as pure tones. The totality of the studies reviewed here unequivocally reveals that nonhuman animals not only form auditory objects but that they also follow the Gestalt principles of grouping. These principles and their underlying mechanisms allow animals to perceptually organize the often noisy and complex acoustic environments in which they live.

M. L. Dent (✉)
Department of Psychology, University at Buffalo, State University of New York (SUNY), Buffalo, NY, USA
e-mail: mdent@buffalo.edu

M. A. Bee
Department of Ecology, Evolution, and Behavior, University of Minnesota, St. Paul, MN, USA
e-mail: mbee@umn.edu

H. Slabbekoorn et al. (eds.), *Effects of Anthropogenic Noise on Animals*, Springer Handbook of Auditory Research 66,
https://doi.org/10.1007/978-1-4939-8574-6_3

Keywords Auditory illusions · Comparative psychoacoustics · Complex sound perception · Operant conditioning · Phonotaxis · Stream segregation · Temporal induction

3.1 Introduction

Imagine standing at the corner of a busy intersection in a large city like Berlin, Germany. There, one might witness a street vendor hawking Döner kebabs to passersby or a taxi zooming through the intersection trying to beat an ambulance approaching from the cross street. Construction workers might be using an excavator to dig up part of the street while a nearby dog behaves aggressively toward another dog on the opposite corner. Standing there, taking in this visual scene, perhaps one would also reflect on ideas advanced in the early part of the twentieth century by the Gestalt psychologists from the Berlin School of Experimental Psychology (Wagemans et al. 2012a,b). They were interested in identifying the principles governing the abilities to perceive whole, unitary objects in visual scenes, such as the people, animals, and vehicles in the imaginary cityscape. A major concern of the Gestalt psychologists was to identify the set of rules that the visual system follows to construct coherent percepts of physical entities from their constituent parts. That is, how are the separate visual elements composing an object grouped together to form the perception of a complete object instead of merely a collection of unrelated features?

Figure 3.1 uses street signs that one might also see in an imagined cityscape to illustrate some of the basic principles of visual grouping identified by the Gestalt psychologists (Palmer 2002; Goldstein 2010). According to the *Gestalt principle of*

Fig. 3.1 German street signs demonstrating the Gestalt principles of proximity (**a**), similarity (**b**), good continuation (**c**), common fate (**d**), and past experience (**a–d**)

proximity, spatial proximity between distinct elements in a visual scene promotes their perceptual grouping into a visual object. In Fig. 3.1a, for example, one readily recognizes the image of a person even though this person's head, upper body, and lower body are not connected. Nevertheless, the head is grouped with the upper body and the upper body is grouped with the lower body based on their close spatial proximity. One also tends to visually group elements of a visual scene according to the *Gestalt principle of similarity*. As illustrated in Fig. 3.1b, for example, a person in a crosswalk is easily identified. The lines forming the crosswalk are of similar size, shape, and orientation (and different from the shapes assigned to the person). The visual system also follows the *Gestalt principle of good continuation*. For example, in Fig. 3.1c, a portion of the inverted "U" appears to continue behind the diagonal slash, which itself appears to continue behind a different portion of the U. Each element is perceived as continuing through their points of intersection. According to the *Gestalt principle of common fate*, elements in the visual scene that appear to change in similar ways are grouped together. This principle is illustrated in Fig. 3.1d by the tracks left by a car that has just lost traction on a slippery road. Finally, the *Gestalt principle of past experience* (or familiarity) holds that elements in a visual scene that were frequently present together in past experience are more likely to be grouped together. According to this principle, past experiences with real people, real crosswalks, and real cars facilitate the perceptual grouping of related visual elements in street signs. In Sects. 3.3–3.7, these Gestalt grouping principles as they pertain to the perceptual organization of acoustic scenes are examined, which leads back to the imaginary cityscape.

Cities like Berlin can be noisy environments replete with biotic, abiotic, and especially human-generated sounds. For example, each of the visual objects in the imaginary cityscape described previously in this section is also a potential sound source: a shouting vendor, a barking dog, the racing engine of a taxi, the blaring siren of an ambulance, or the rumble of heavy machinery. Just like each physical entity is perceived as a distinct visual object, so too are the sounds they make perceived as perceptually distinct objects or *auditory objects* (Griffiths and Warren 2004; Bizley and Cohen 2013). In contrast to visual object perception, however, the formation of auditory objects requires overcoming a special challenge. Reflected light waves from one object can be occluded by a closer object. Not so for sound pressure waves, which instead are additive. In noisy acoustic scenes, multiple sound sources are active simultaneously. The sound pressure waves produced by these different sources, along with their echoes, add together to form a complex waveform representing the mixture of active sources. It is this complex sound pressure waveform that impinges on the ears. A primary function, some might say *the* primary function, of the auditory system is to correctly parse this sound mixture to perceive discrete auditory objects in the acoustic scene. This process of forming auditory objects is referred to as *auditory scene analysis* (Bregman 1990), and it is fundamental to hearing in noisy and acoustically cluttered environments. As the planet becomes noisier, primarily due to human-generated sounds, the perceptual organization of acoustic scenes becomes even more important for effective communication.

Auditory scene analysis by humans has become a productive area of research since the publication of Bregman's (1990) seminal book on the topic (e.g., Darwin and Carlyon 1995; Yost et al. 2008). Bregman drew attention to three distinctions that will be important throughout this chapter. First, auditory object formation involves processes of *integration* and *segregation*. That is, constituent sounds produced by the same source must be integrated or grouped together and segregated from the sounds produced by other sources. Whether one speaks in terms of integration or segregation often depends on the precise perceptual task at hand. Second, although all auditory objects necessarily include a temporal dimension because sound pressure always varies as a function of time, one can consider what Bregman (1990) referred to as *simultaneous integration* or *sequential integration*. Simultaneous integration refers to the binding together of sounds produced simultaneously by a given source (e.g., harmonics, formats) into a coherent auditory object (e.g., a particular vowel sound or a musical chord). In contrast, sequential integration refers to grouping together the sounds produce by a given source through time (e.g., syllables and words or musical notes) into a unitary sequence that is commonly called an *auditory stream* (Bregman 1990; Carlyon 2004). When two or more such sequential streams are concurrent, listeners can often segregate the mixture into separate streams in a process usually referred to as *auditory stream segregation*. Finally, Bregman (1990) drew an important distinction between *primitive* and *schema-based* auditory scene analysis. The former involves bottom-up, data-driven processes that contribute to grouping based on an analysis of the acoustic waveform. In contrast, the latter engages top-down processes related to attention, learning, and cognition.

One key contribution of Bregman's (1990) treatment was to highlight similarities in how the human auditory system follows some of the same Gestalt principles of grouping that characterize the perception of visual objects. The aim of this review is to highlight that adherence to Gestalt grouping principles in forming auditory objects is not a uniquely human ability. Although auditory scene analysis is commonly discussed in terms of speech and music perception, the ability to perceptually organize acoustic scenes no doubt evolved before either speech or music. In fact, current evidence from animal studies suggests that Gestalt principles of auditory object formation could have ancient evolutionary roots, perhaps dating back at least to our last common ancestor with fish (Fay and Popper 2000). Before diving into this literature, a brief review of early empirical studies demonstrating that nonhuman animals are also capable of auditory scene analysis is presented.

3.2 Auditory Scene Analysis in Nonhuman Animals

Studies of auditory scene analysis in animals have typically had one of three motivations, all of which are apparent in the studies reviewed in this chapter. First, some studies have sought to determine whether the fundamental principles of auditory scene analysis are shared across humans and nonhuman animals, for instance, as a result of our shared evolutionary ancestry. These studies typically employ simple stimuli, such as pure tones or trains of pulses, similar to those used in human psychoacoustic studies of auditory scene analysis (Fay 1998; MacDougall-Shackleton et al. 1998). Second, some studies have used animals as models to understand the neurophysiological basis of auditory scene analysis by humans. Such studies typically present animals (usually a mammal or bird) with stimuli known to probe specific aspects of auditory scene analysis in humans while simultaneously recording neural responses from the animal's brain (Fishman et al. 2001; Pressnitzer et al. 2008). Some studies have also sought to identify neural correlates of the animal's own perceptual organization abilities (Bee and Klump 2004; Middlebrooks and Bremen 2013). Finally, a number of studies have sought to understand the abilities of animals to perceptually organize acoustic scenes of ecological relevance to the animals themselves. Much of this work has been done in the context of animal acoustic communication (Farris and Taylor 2016; Klump 2016).

An implicit assumption in practically all studies of nonhuman acoustic communication prior to Bregman's (1990) book (and still in most current research on the topic) is that animals perform auditory scene analysis. The first studies to explicitly test this assumption were conducted with European starlings (*Sturnus vulgaris*). European starlings are excellent psychoacoustic subjects; much is known about their auditory systems from both behavioral and physiological studies, and they have proven to be excellent models for studies of auditory scene analysis (Hulse 2002; Klump 2016). They are also highly vocal songbirds, relying heavily on acoustic communication for survival and reproduction. Thus, it is very important to hear a conspecific among the dawn chorus and the myriad of other noise producers in their environment. The earliest European starling studies involved first training the birds using operant conditioning and positive reinforcement to discriminate between two or more easily distinguishable patterns and then testing their perception of untrained "probe" stimuli (Hulse et al. 1997; Wisniewski and Hulse 1997).

In 1997, Hulse and colleagues investigated auditory scene analysis by European starlings using their biologically relevant songs in two separate studies. Hulse et al. (1997) trained subjects to discriminate song combinations "containing a European starling song" from those "not containing a European starling song" (Fig. 3.2). The birds quickly learned this task to a high rate of performance (Fig. 3.3). In several transfer probe tests, performance remained high for classifying novel song pairs. Thus, the birds did not simply memorize the baseline song pairs presented to them in the initial training sessions. Instead, they segregated each song individually during training and discriminated the probes containing new song pairs accordingly. When the birds were later tested on never-before-heard single songs presented individually, performance remained high, further supporting the hypothesis that they segregated the training song mixtures into individual auditory objects. Finally, when the song pairs were embedded in a dawn chorus, subjects maintained high discrimination performance. The experiments in this study suggested that the temporal and spectral proximity of elements in the individual songs allowed the European starlings to segregate them into individual auditory objects following the Gestalt principle of grouping by proximity (see Sect. 3.3).

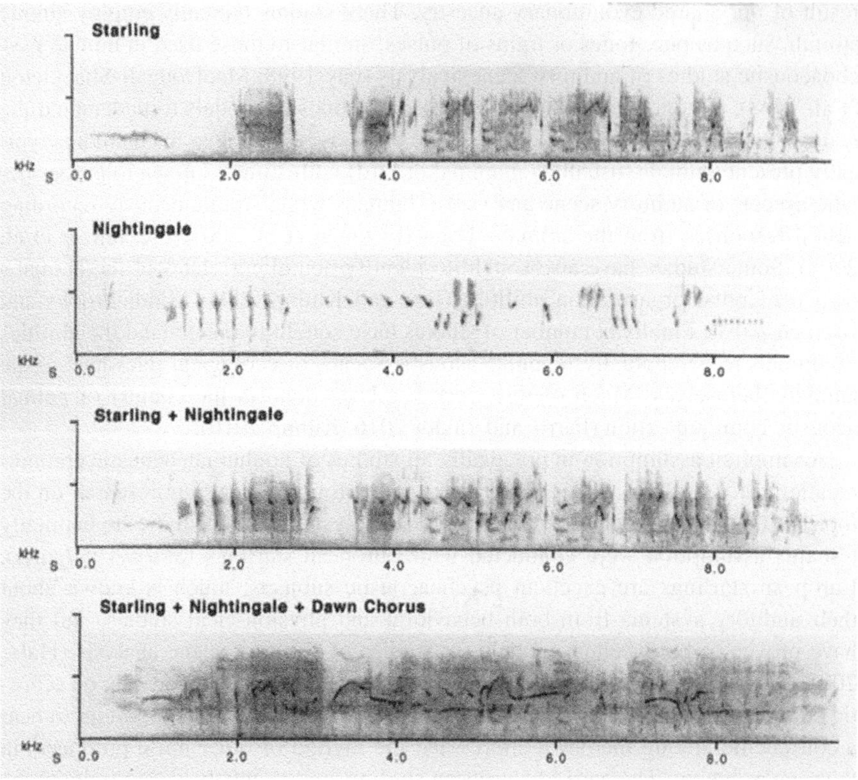

Fig. 3.2 Representative sonograms of the song of the European starling, nightingale, starling plus nightingale, and starling plus nightingale plus dawn chorus. Reprinted from Hulse et al. (1997), with permission

Wisniewski and Hulse (1997) expanded the investigation of auditory scene analysis in European starlings based on the species' ability to recognize other individuals by song. As in Hulse et al. (1997), performance remained high when the training songs were embedded with new songs, indicating that each song was perceived as a separate auditory object. These studies together suggested that, even in a very noisy environment, European starlings should be able to effectively communicate with one another in a manner similar to humans at a raucous cocktail party or out on a noisy street.

Benney and Braaten (2000) extended work on European starlings to two other songbirds, the zebra finch (*Taeniopygia guttata*) and the Bengalese finch (*Lonchura striata domestica*), to determine if the capacity for auditory scene analysis was specific to European starlings and their songs and to ascertain whether there were species-specific advantages to forming certain auditory objects over others. Individuals of each finch species were trained to detect either a conspecific or a heterospecific target song embedded in a mixture of songs from three species of birds. Birds trained on the conspecific song detection learned faster than birds trained on the heterospecific song detection, and both species were able to discriminate mix-

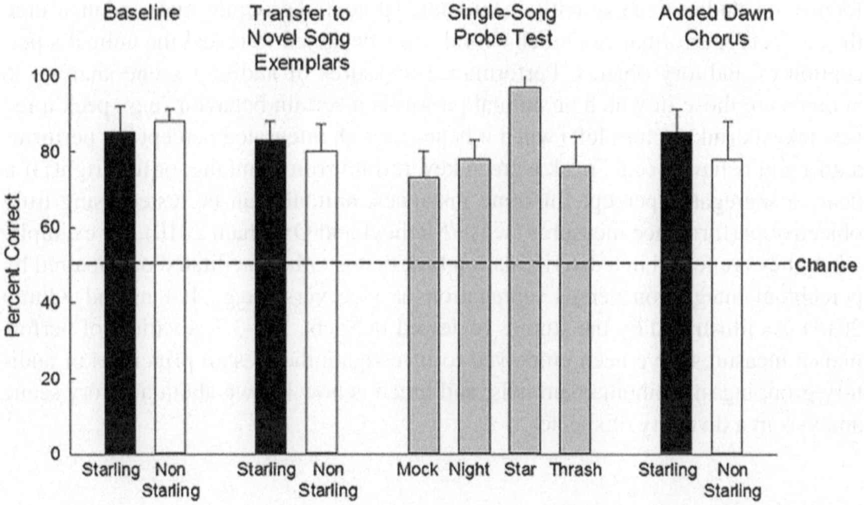

Fig. 3.3 Discrimination performance for European starlings trained on pairs of songs containing a starling song or not. Percent corrects are shown for the baseline test, transfer to novel song exemplars, and two probe experiments. In the single-song probe test, individual songs from mockingbirds, nightingales, starlings, and brown thrashers were presented. In the added dawn chorus probe experiment, pairs of songs with and without starling songs were embedded in a dawn chorus. Performance was high for all probe conditions (chance performance was 50%). Data from Hulse et al. (1997)

tures containing target songs. The added ability to discriminate novel probe targets confirmed that the birds were forming separate auditory objects of the songs rather than simply memorizing the stimuli. The zebra finches, but not the Bengalese finches, showed an enhanced ability to detect zebra finch song and had difficulty ignoring zebra finch song targets. Bengalese finches showed no such contrast, suggesting species differences in attention to conspecific song. These results indicated that auditory scene analysis was common among birds and laid the groundwork for future studies directly manipulating the features of signals important for auditory grouping.

In his review on auditory scene analysis and animal communication, Hulse (2002, p. 182) lamented that "because so little work has been done directly on auditory scene analysis in nonhuman animals, very little literature bears directly on the problem." The situation has improved over the last 15 years, however. Since the early work of Hulse and others, a number of studies have now more explicitly investigated auditory scene analysis and its role in communication and other behaviors in a diversity of animals, ranging from insects (Römer 2013) and frogs (Bee 2015; Farris and Taylor 2016) to birds (Dent et al. 2016; Klump 2016), bats (Moss and Surlykke 2001; Barber et al. 2003), and monkeys (Miller et al. 2001; Petkov et al. 2003). The remainder of this chapter reviews animal studies that have investigated the Gestalt principles of auditory object formation. Each major section covers a separate Gestalt principle, and within each section, studies are grouped by taxon, with the order of taxa in each section arranged according to chronology. The chapter focuses on studies that explicitly manipulated potential grouping cues and measured the subjects' performance on behavioral tasks designed to reveal the animal's perception of auditory objects. Performance measures of auditory scene analysis in animals are those in which an animal performs a certain behavior (e.g., peck a red key, take a drink, or turn left) when it hears, say, an integrated percept but performs a different behavior (e.g., peck a green key, refrain from drinking, or turn right) if it hears a segregated percept. In some instances, animals can be tested using truly objective performance measures (sensu Micheyl and Oxenham 2010), for example, when they are tested in a discrimination task that is either facilitated or impaired by perceptual integration versus segregation or vice versa (e.g., Itatani and Klump 2014). As illustrated by the studies reviewed in Sects. 3.3–3.7, a variety of performance measures have been employed to investigate the Gestalt principles of auditory grouping in nonhuman animals, and much is now known about auditory scene analysis in a diversity of species.

3.3 The Principle of Proximity

The Gestalt principle of proximity holds that elements in visual or acoustic scenes are more likely to be grouped when they are in close proximity. In the visual domain, it is *spatial* proximity that promotes grouping through stimulation of receptors on the retina that are also near to each other. In the auditory domain, however, it is proximity in *frequency* and *time* that promotes auditory grouping. Sound elements that stimulate nearby places along a tonotopically organized auditory papilla are close to each other in frequency but not necessarily in space. Studies on the effects of frequency on auditory stream segregation have shown that as frequency separation between stimuli increases, the perception of separate auditory objects also increases (reviewed in Bregman 1990). In humans, this form of streaming has been investigated by presenting sequences of two alternating tones, one with a low frequency (A) and one of a higher frequency (B), arranged into repeated triplets. At small frequency separations, listeners report hearing one auditory stream with a galloping rhythm composed of two tones alternating in frequency (ABA–ABA–ABA…). However, as frequency separation increases, the percept breaks into two streams with isochronous rhythms: one low-frequency stream (A–A–A–A–A–A–…) and one high-frequency stream (B–B–B–…). This effect is enhanced with faster repetition rates because tones of the same frequency (e.g., A–A–A–…) occur in closer temporal proximity. Frequency and time are arguably the most well-studied features of stimuli involved in auditory stream segregation in both humans and animals because the principle of grouping by proximity is so robust. The Gestalt psychologists felt that this type of perceptual organization was innate and automatic in the visual domain, and Bregman (1990) argued that the same was true in the auditory domain. The perceptual organization of auditory stimuli using the Gestalt principle of proximity has been well-known in humans for decades but has only been discovered fairly recently in animals.

3.3.1 *Birds*

MacDougall-Shackleton et al. (1998) used stimuli similar to those used in human-subject studies to determine some of the similarities between auditory streaming in humans and in nonhuman animals. European starlings were trained on two categories: galloping stimuli and isochronous stimuli (Fig. 3.4). Once the birds were

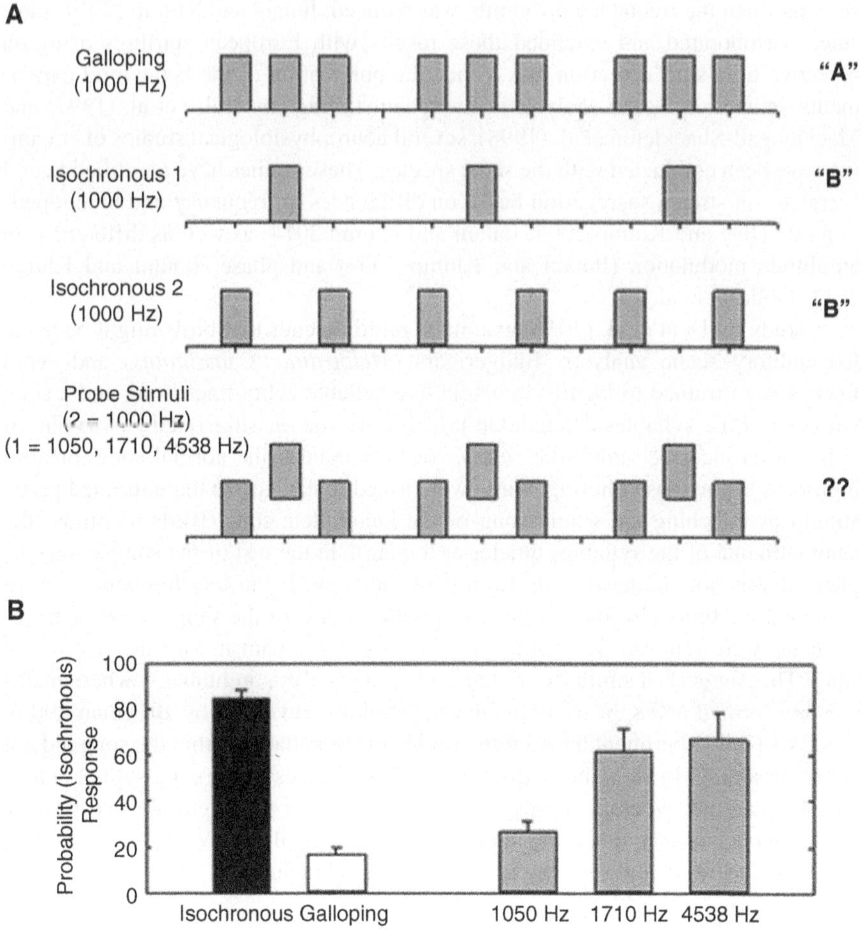

Fig. 3.4 (**A**): discrimination stimulus conditions. European starlings were trained to peck "A" for a galloping stimulus configuration and "B" for either isochronous stimulus. Once trained, the birds were presented with probe stimuli and their responses were recorded. The probe stimuli were presented in a galloping manner, with the center tone of the tri-tone complex presented at a higher frequency than the first and third tones. (**B**): mean isochronous responses to the isochronous training stimuli (*black*), galloping training stimuli (*white*), and three probe stimuli (*gray*). Error bars represent between-subject standard errors. European starlings were more likely to respond "galloping" when the probe frequencies were similar and more likely to respond "isochronous" when the probe frequencies differed. Data from MacDougall-Shackleton et al. (1998)

trained, probes were introduced on a small proportion of all trials. The probes were galloping stimuli in which the middle tones in the gallop were a higher frequency than the two flanking tones (ABA–ABA–ABA–…). Even though they could not verbalize their percept, the birds responded in a manner indicating a perceptual experience similar to that of humans. As the frequency separation increased, the probability that subjects responded as if the probes were isochronous also increased. Thus, when proximity in frequency was high, the tones were integrated into a single, galloping stream, but the tones were perceptually segregated into separate streams when the frequency proximity was reduced. Itatani and Klump (2014) have since corroborated and extended these results with European starlings using an objective time-shift detection task. Since the publication of the behavioral experiments on auditory scene analysis in European starlings by Hulse et al. (1997) and MacDougall-Shackleton et al. (1998), several neurophysiological studies of streaming have been conducted with the same species. These studies have identified neural correlates of stream segregation based on differences in frequency and tone repetition rate (Bee and Klump 2005; Itatani and Klump 2014) as well as differences in amplitude modulation (Itatani and Klump 2009) and phase (Itatani and Klump 2011; Dolležal et al. 2012).

A study by Dent et al. (2016) examined multiple cues that birds might be using for auditory scene analysis. Budgerigars (*Melopsittacus undulatus*) and zebra finches were trained to identify a whole five-syllable zebra finch song and a song with one of the syllables deleted. On probe trials, the missing syllable was played either at an incorrect amplitude, earlier or later than in the normal song, or after high-pass or low-pass filtering. Birds were asked to categorize the untrained probe stimuli as matching the whole song or the incomplete song. Birds identified the song with one of the syllables quieter or louder than the rest of the song as incomplete; it was not streamed with the rest of the song. If the low frequencies were removed, the birds also did not stream it with the rest of the song. If the high frequencies were removed from the syllable, they did stream it with the rest of the song. This suggests a similarity to musical scene analysis in humans, where melodies are formed and split by frequency separations (reviewed by Bregman 1990). Finally, in this experimental paradigm, timing of the syllable within the song did not affect streaming. In sum, the frequencies and amplitudes of the song syllable influenced streaming, whereas timing did not. This unique paradigm adds to what is known about auditory streaming in birds using ecologically relevant stimuli, finding more similarities to auditory stream segregation in humans.

3.3.2 Fish

Two studies by Fay (1998, 2000) investigated auditory stream segregation in the goldfish (*Carassius auratus*). Goldfish do not make sounds, yet they need to distinguish various sounds in their environments, making auditory stream segregation critical for survival. To test streaming, goldfish were classically conditioned to withhold respiration to isochronous or alternating pulse trains with two spectral profiles and repetition rates (Fay 1998). During a subsequent test on the single-pulse trains presented alone, the goldfish generalized more to the single-pulse trains presented at the original paired rates. This suggests that the goldfish identified the single pulses in training even though they never heard them in isolation during training. This is similar to the results of MacDougall-Shackleton et al. (1998) in their study of auditory stream segregation in European starlings (see Sect. 3.3.1). In a follow-up study, Fay (2000) extended his results to measuring generalization to various frequency separations. As in humans and European starlings, the goldfish showed greater stream segregation to pulse trains with a greater frequency separation and a faster presentation rate. Fay's comparative demonstrations of auditory stream segregation in goldfish, matching those from MacDougall-Shackleton et al. (1998) in several ways, support the idea that auditory scene analysis is a common feature of auditory processing across vertebrates.

3.3.3 Mammals

A series of studies on macaques have examined auditory streaming, correlating behavioral findings with physiological results. The auditory system of nonhuman primates is similar to that of humans, resulting in similar psychophysical temporal and spectral thresholds across species (e.g., Stebbins 1973). This makes macaques important animal models for identifying cortical and subcortical processes involved in auditory stream segregation. Behaviorally, stream segregation has been measured using a number of methods. Izumi (2002) required Japanese macaques (*Macaca fuscata*) to discriminate a rising sequence of tones from a nonrising sequence. Those sequences were later presented with no auditory distractors, with nonoverlapping (in frequency) distractors, and with overlapping (in frequency) distractors. The discrimination performance of the targets in quiet and with nonoverlapping distractors was good, whereas discrimination of the sequences within the overlapping distractors was poor. These results suggest the importance of frequency proximity in determining whether or not overlapping sounds are integrated together or segregated into separate auditory streams in yet another animal model.

Other behavioral paradigms have included reaction time measurements in crab-eating macaques (*Macaca fascicularis*) presented with ambiguous and unambiguous multimodal stimuli (Selezneva et al. 2012) and the reporting of one or two streams by rhesus macaques (*Macaca mulatta*; Christison-Lagay and Cohen 2014). These fairly direct measures allow for greater comparisons of results between humans and primates and have demonstrated the importance of frequency separation, repetition rate, and signal duration on auditory stream segregation. These behavioral results have been correlated with neurophysiological findings from the auditory cortex in awake macaques (Fishman et al. 2001, 2004).

Ferrets (*Mustela putorius*) and guinea pigs (*Cavia porcellus*) have also shown quantitative and qualitative similarities to humans and other animals tested on the effects of frequency and time on auditory stream segregation. Ferrets were trained to detect shifts in the frequencies of tones in a varying multitone background in one experiment and to detect a stream of regularly repeating targets embedded within varying multitone backgrounds in another (Ma et al. 2010). The ferrets had qualitatively similar results to humans; as frequency differences between distractors and targets increased, detectability increased, suggesting that targets and distractors were segregated into different streams. Ferrets needed a much larger frequency separation than humans to detect the target changes, suggesting differences in informational masking or selective attention across species. Scholes et al. (2015) also found qualitative similarities to behavioral stream segregation in the anesthetized auditory cortex of guinea pigs.

3.3.4 Insects

Possibly the greatest evidence that auditory stream segregation is a universal feature of acoustic communication in animals comes from a study of a katydid, the round-tipped conehead *Neoconocephalus retusus*, by Schul and Sheridan (2006). Although not a behavioral study, it is included here to emphasize the taxonomic breadth of mechanisms for auditory stream segregation. Male round-tipped coneheads produce loud sexual advertisement songs consisting of 15-kHz pulses produced at a rate of 140/s. Females approach singing males on the wing but can avoid bat echolocation calls produced at the same time but at higher frequencies (e.g., 40 kHz) and at slower rates (e.g., 7/s). A broadly tuned auditory neuron, the TN-1, can be driven by both the low frequencies in the male song and the high frequencies of bat echolocation pulses. However, it quickly adapts to the fast pulse rates of a male's low-frequency song but will continue to respond to high-frequency bat echolocation pulses interleaved occasionally with pulses of the male's song. Interestingly, similar responses were obtained when the acoustic frequencies of the two sounds were switched such that the neuron responded to low-frequency (15-kHz), slow-rate (7/s) sounds but not to high-frequency (40-kHz), fast-rate (140/s) sounds. Thus, at the single-neuron level, round-tipped coneheads are able to segregate simultaneous male song and predator cues into separate auditory streams based on differences in the temporal proximity of repeated sound elements from each source. These findings demonstrate a very primitive form of auditory scene analysis, but one that is vital for survival of these katydids.

3.3.5 Frogs

Female frogs commonly rely on their auditory system to analyze a male's advertisement calls to determine whether he is a male of her own species, whether he would be a high-quality mate, and where he is located (Gerhardt and Huber

2002). What is more, females often must make these determinations in social environments ("breeding choruses") that are noisy and cluttered with the calls of numerous individuals of multiple species (Bee 2012, 2015). When females hear a conspecific male or a speaker broadcasting real or synthetic conspecific calls, they approach and investigate at a much higher rate than they would approach a heterospecific male or any other signal. The performance of phonotaxis by female frogs has been used extensively to study auditory scene analysis (see Sects. 3.4, 3.5, and 3.7).

Inspired by studies of auditory stream segregation in humans (Bregman 1990) and birds (see Sect. 3.3.1) that used sequences of interleaved sounds as stimuli (e.g., ABA–ABA–ABA–…), Nityananda and Bee (2011) presented females of Cope's gray treefrog (*Hyla chrysoscelis*) with mixtures of synthetic pulse sequences designed to simulate mixed-species breeding choruses. A short, pulsatile target that simulated a conspecific call was temporally interleaved with a continuous, pulsatile distractor designed to simulate the pulsatile calls of another species. Whenever the target sound was presented, the composite pulse rate of the target and distractor together was designed to be unattractive unless subjects could perceptually segregate them into two different auditory streams. Indeed, when the frequency separation between the target and distractor was small, the females did not respond. However, when the frequency separation between the two sounds was greater than three semitones, the females approached the target sound, suggesting that they perceptually segregated the target from the distractor. These results indicated that the naturally occurring frequency separations found by female frogs in mixed-species choruses aid in auditory stream segregation of calls in noisy social environments.

3.3.6 Summary

Grouping based on proximity is a dominant principle guiding the formation of both visual and auditory objects. Overall, studies of many species of nonhuman animals published by many different laboratories using many different methodologies all support the notion that auditory grouping based on proximity in frequency and time is a robust and universal auditory phenomenon. Importantly, these studies have uncovered both behavioral and neural evidence for exact correlates to auditory scene analysis in humans.

3.4 The Principle of Similarity

According to the Gestalt principle of similarity, the elements of visual and acoustic scenes that are more similar to each other tend to be grouped together. Two sources of similarity that have been studied in the auditory domain are timbre and spatial location. Timbre refers to the "multidimensional attribute of auditory sensation which enables a listener to judge that two non-identical sounds, similarly presented and having the same loudness, pitch, spatial location, and duration, are dissimilar" (American National Standards Institute 2013). Differences in timbre are part of the reason that we perceive different instruments, such as a violin and a saxophone, playing the same note to nevertheless sound different. Sounds from the same source are typically similar in timbre, and similarity in timbre facilitates auditory grouping (e.g., Bregman 1990).

Because sounds generated by a single source typically originate from the same location, space might be expected to play an important role in auditory grouping, and indeed it can (Darwin 2008). Although the use of common spatial location by the visual system is an example of Gestalt grouping by proximity, auditory systems must compute spatial location using binaural comparisons of one or both of two cues: interaural time differences (ITDs) and interaural level differences (ILDs). For this reason, grouping based on similar spatial cues (i.e., ITDs or ILDs) is best considered as an example of the Gestalt principle of similarity. Although it may be intuitive that spatial location should be a particularly potent cue for grouping sounds produced by the same source, psychophysical data from humans often suggest that spatially mediated grouping is weaker than grouping based on other perceptual cues. Thus far, only one study of a nonhuman animal has investigated grouping by similarity in timbre, whereas several studies have investigated grouping by similarity in spatial cues.

3.4.1 Birds

In what is perhaps the earliest study designed explicitly to investigate a Gestalt principle of auditory grouping in a nonhuman animal, Braaten and Hulse (1993) measured performance for the perception of series of tone pips of two different timbres, such as XXXXXOOO or OXOXOOXX (with X representing one timbre and O representing a second timbre). In probe trials, deleting half of the tone pips of a particular timbre (e.g., the Xs) in the patterns did not disrupt performance; the European starlings correctly identified the two incomplete patterns, suggesting that they perceived the Xs as one stream with a specific rhythm and the Os as a second stream with a different rhythm. This was the earliest known evidence for stream segregation in a nonhuman animal, and it laid the groundwork for the series of experiments by Hulse and colleagues on auditory scene analysis in European starlings (see Sects. 3.2 and 3.3.1).

The role of space in auditory scene analysis has also been investigated in two species of birds not known to be sound localization specialists, the budgerigar and the zebra finch. In contrast to owls, which hunt in the dark and have excellent sound localization acuity, the birds tested here have very high sound localization thresholds. Thus, it was thought that spatial location might not be a robust cue for the creation of auditory streams. Dent et al. (2016) trained birds to identify a category

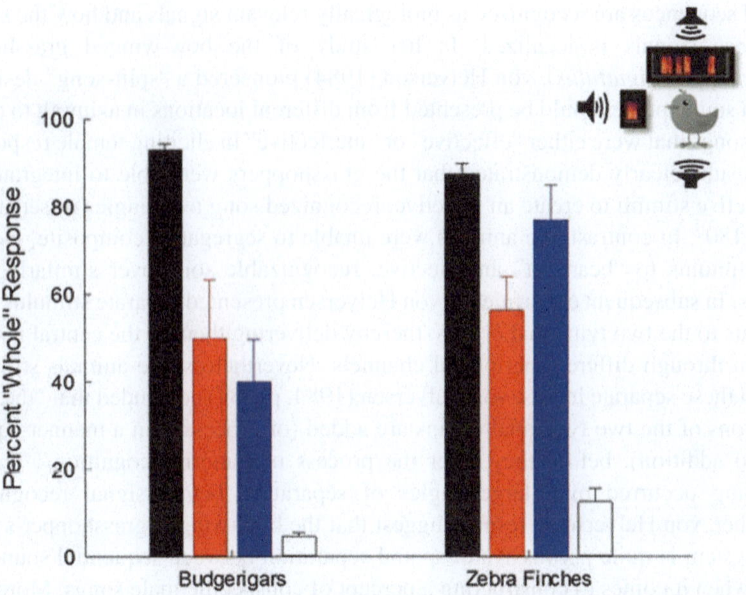

Fig. 3.5 Stimulus conditions (*inset*) and results for the influence of location on streaming in budgerigars and zebra finches. Error bars represent between-subject standard errors. Birds reliably identified the whole zebra finch songs as whole (*black bars*), rarely identified the incomplete (broken) zebra finch songs as whole (*white bars*), and responded to the probe stimulus in an intermediate manner. When the missing syllable was presented from 90° (*red bars*) or 180° (*blue bars*) from the rest of the songs, the birds were less likely to stream that syllable with it. Data from Dent et al. (2016)

containing a complete zebra finch song from an incomplete zebra finch song where one syllable was deleted. On probe test trials, the missing syllable was presented at the proper time but from another location in space, either 90° or 180° away from the rest of the song. The birds correctly categorized the whole and incomplete songs with high (over 90%) accuracy and responded intermediately to the probes (Fig. 3.5), suggesting a breakdown in streaming when spatial cues were unreliable. Thus, in this task, location was a dependable cue for streaming. Interestingly, there were no differences between the two species, even though there is a threefold difference in localization thresholds between budgerigars and zebra finches (Park and Dooling 1991). Smaller location separations between the probe syllables and the incomplete song may have teased out some species differences.

3.4.2 Insects

Two studies are relevant to consideration of spatial cues in auditory grouping by insects (von Helversen 1984; Weber and Thorson 1988). The primary goal motivating both studies was to elucidate the relationship between how temporally patterned sound sequences are recognized as biologically relevant signals and how the source of those signals is localized. In her study of the bow-winged grasshopper (*Chorthippus biguttulus*), von Helversen (1984) pioneered a "split-song" design in which sound pulses could be presented from different locations in azimuth to create male songs that were either "effective" or "ineffective" in eliciting female responses. Her results clearly demonstrated that the grasshoppers were able to integrate two ineffective stimuli to create an effective, recognized song over angles of separation up to 180°. In contrast, the animals were unable to segregate a composite, ineffective stimulus to "hear out" an effective, recognizable song over similarly large angles. In subsequent experiments, von Helversen presented separate stimulus components to the two tympanal organs, thereby delivering them to the central nervous system through different peripheral channels. Nevertheless, the animals still integrated these separate inputs. von Helversen (1984, p. 837) concluded that "the input functions of the two tympanal organs are added (or processed in a manner equivalent to addition), before they enter the process of pattern-recognition." That is, grouping occurred over large angles of separation *before* signal recognition. Together, von Helversen's results suggest that the bow-winged grasshopper's auditory system is quite permissive of spatial separation between sequential sounds, at least when it comes to constructing a percept of conspecific male songs. Moreover, these results indicate that peripheral channeling for stream segregation (cf. Hartmann and Johnson 1991) may not apply to some invertebrates.

Weber and Thorson (1988) used the split-song method to investigate the role of spatial cues in song recognition and localization by female field crickets (*Gryllus bimaculatus*). They measured phonotaxis by females as they walked on a spherical treadmill designed to compensate for the subjects' movements, thereby keeping subjects in the same position relative to two speakers spatially separated in azimuth. In some conditions, two ineffective stimuli, each with half the pulse rate of the male's normal calling song, were temporally interleaved and delivered from the two speakers such that the composite stimulus had the effective, recognized pulse rate that equaled that of the male's calling song. Even when the two speakers were separated by 135°, females exhibited phonotaxis, indicating recognition, but their movements were directed toward a phantom source midway between the two speakers. Thus, the crickets readily grouped two interleaved, ineffective sounds into an effective, recognized song but incorrectly estimated its location. This willingness of female crickets to group over large angles of separation to create a recognizable signal is similar to that of von Helversen's (1984) grasshoppers. But another test revealed an interesting difference between these two species. When two effective stimuli were interleaved to create a composite stimulus with an ineffective pulse rate that was twice that of a male's normal song, females again exhibited phonotaxis. In this condition, however, their responses were directed toward one or the other speaker separated by 135° and not toward a phantom source midway between them. This was an important result because it demonstrated that the crickets could segregate the two effective signals into separate auditory streams based on their spatial locations. The limits of spatially mediated stream segregation were not investigated, but efforts to do so could be informative.

3.4.3 Frogs

Like many orthopteran insects, frogs commonly produce communication sounds consisting of sequences of pulses produced at species typical rates (Gerhardt and Huber 2002). This is true, for example, for the closely related eastern gray treefrog (*Hyla versicolor*) and Cope's gray treefrog (*H. chrysoscelis*), which have been the focus of two studies using a split-song paradigm to investigate the role of spatial cues in auditory grouping (Schwartz and Gerhardt 1995; Bee and Riemersma 2008). The results of these studies suggested that frogs are also like insects in their willingness to group sequential sounds across large angles of spatial separation.

The pulse rate of *H. chrysoscelis* calls is about twice (2×) that of *H. versicolor* (1×). Females of both species are highly selective for conspecific pulse rates and discriminate strongly against pulse rates typical of the other species (Bush et al. 2002). Discrimination based on this twofold difference in pulse rate was key to experiments on spatially mediated grouping in these two species, which used interleaved stimuli with 1× pulse rates to create composite stimuli having a 2× pulse rate. For one of the two species (*H. versicolor*, 1×) perceptually *segregating* the two interleaved stimuli would promote call recognition, whereas for the other (*H. chrysoscelis*, 2×), *integrating* them would promote call recognition. Hence, the prediction was that spatial separation between the two interleaved sequences would promote phonotaxis behavior in one species (*H. versicolor*; Schwartz and Gerhardt 1995) and disrupt it in the other (*H. chrysoscelis*; Bee and Riemersma 2008). However, in both species, the effects of increasing spatial separation had relatively weak effects on shifting phonotaxis behavior in the predicted directions (Fig. 3.6). Thus, gray treefrogs appear to be similar to bow-winged grasshoppers and field crickets when it comes to their limited use of spatial cues in sequential auditory grouping.

Subsequent work by Bee (2010) investigated the importance of spatial coherence in simultaneous grouping in *H. chrysoscelis*. Each pulse of the species' call has a bimodal frequency spectrum characterized by two prominent spectral peaks, each of which is primarily transduced by a different sensory papilla in the anuran inner ear. The experimental design exploited the general preference that females of several treefrog species have for calls with bimodal spectra over those lacking one of the two peaks (Gerhardt 2005; Gerhardt et al. 2007). Using two-alternative choice tests, Bee (2010) examined preferences for calls with spectral peaks that were either spatially coherent or spatially separated. The prediction was that spatial separation between the two peaks should render calls relatively less attractive given female preferences for bimodal over unimodal calls. Somewhat surprisingly, in light of the permissiveness exhibited by females in the sequential grouping task described in the previous paragraph, even small degrees of spatial separation (e.g., between 7.5° and 60°) between the two spectral peaks were sufficient to shift preferences toward the spatially coherent call. Thus, in contrast to their willingness to group sequential sound elements across space, gray treefrogs appear to be far less tolerant of spatial separation when it comes to grouping simultaneous components of the spectrum.

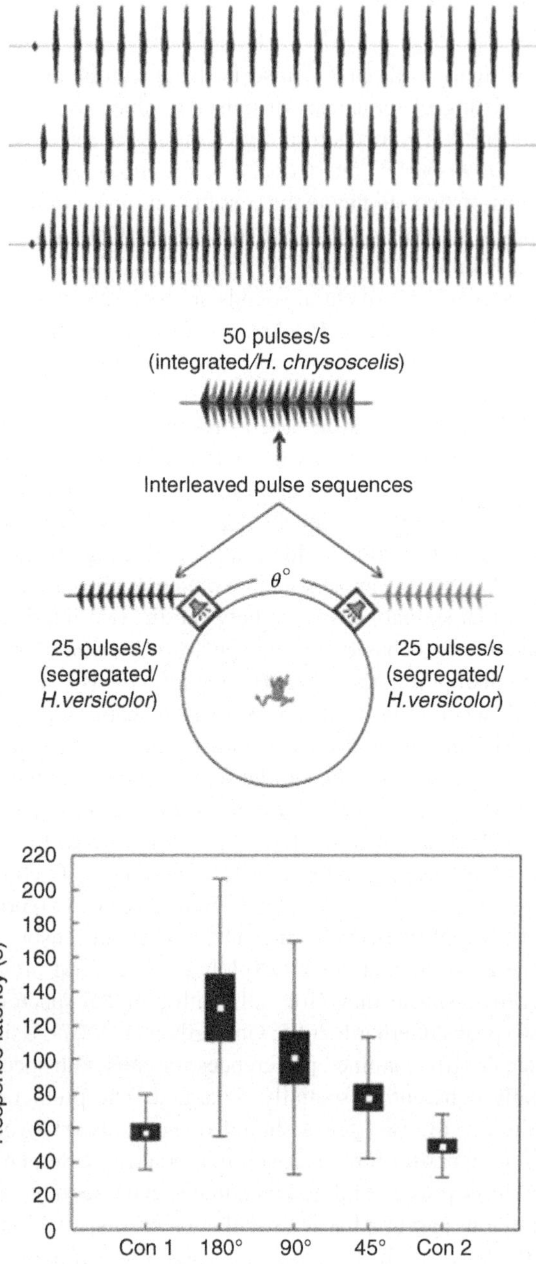

Fig. 3.6 Stimulus configurations (*top*) and results (*bottom*) from tests of spatially mediated stream segregation in Cope's gray treefrog. Females were presented with two temporally interleaved 25 pulse/s sequences from various locations and their responses were analyzed. Response latencies are shown for the two control conditions (Con 1 and Con 2) as well as for the spatially incoherent conditions where the speakers were separated by 180°, 90°, and 45°. Points depict mean values, boxes depict ±1 SE, and whiskers depict ±1 SD. Reprinted from Bee and Riemersma (2008), with permission from Elsevier

Spatially mediated grouping of sound sequences has also been investigated in the túngara frog (*Physalaemus pustulosus*; Farris and Taylor 2016). Túngara frogs produce a two-note call consisting of a downward frequency-modulated (FM) sweep (the "whine") followed by a brief harmonic stack (the "chuck"). Although the whine is necessary and sufficient for call recognition, females prefer whines followed by one or more chucks, although chucks by themselves do not elicit phonotaxis. Farris et al. (2002, 2005) found that females were willing to group a whine and a chuck that were separated by up to 135°. Based on this result, túngara frogs would appear to be as willing as gray treefrogs, bow-winged grasshoppers, and field crickets to group temporally separated call components across large spatial separations. In subsequent work, however, Farris and Ryan (2011) demonstrated a somewhat better performance when the animals were forced to make decisions about which of two spatially separated chucks to group with a preceding whine. For example, females preferentially grouped chucks that were 45° away from the whine over a chuck that was either 135° or 180° away but not one that was 90° away. Given that túngara frogs often communicate in noisy choruses of signaling males, in which calls often overlap in time, relative comparisons may better reflect the animals' performance in real-world listening situations.

3.4.4 Mammals

The use of spatial cues for auditory object formation has also been measured in two mammals, cats (*Felis catus*) and rats (*Rattus norvegicus*). Cats have excellent sound localization ability (e.g., Populin and Yin 1998), and in a behavioral task, they also show high spatial acuity for object recognition for stimuli separated by more than 8° on both the horizontal and vertical planes (Middlebrooks and Onsan 2012). These results showed a much higher spatial acuity for streaming objects in space relative to results from insects, frogs, and birds, and the authors suggested task differences as a possible reason. Cats were measured using a "voluntary" stream segregation task where listeners can exploit interaural cues of stimuli for streaming rather than perceived differences in the locations of stimuli, as in the typically used "obligatory" stream segregation task. Middlebrooks and Bremen (2013) extended the behavioral auditory stream segregation results in cats by correlating them with electrophysiological recordings from the auditory cortex, and Yao et al. (2015) later recorded from multiple neurons along the ascending auditory pathway of rats using a spatial stream segregation paradigm.

3.4.5 Summary

The studies reviewed in this section suggest that the Gestalt principle of similarity also guides auditory object formation in many nonhuman animals. Specifically, these studies demonstrate that similarity in both timbre and spatial location can be used as cues for auditory grouping. However, the limits of these two cues are not yet widely known across animals, and a comparative perspective indicates a high likelihood of evolutionary diversity regarding the utility of similarity in some sound features, such as common location, for auditory grouping. Exploring these limits and this diversity in future studies would be worthwhile.

3.5 The Principle of Common Fate

According to the principle of common fate, visual object formation is promoted when elements in the visual scene change in a coherent fashion, for example, when they simultaneously begin moving in the same direction and at the same speed. In acoustic scenes, sound elements that start and stop at the same time and that undergo similar patterns of amplitude modulation through time probably arose from same source and should therefore be grouped together according to the principle of common fate (Bregman 1990). Temporal coherence proves to be a particularly powerful auditory grouping cue that can override other grouping cues such as frequency proximity. For example, van Noorden (1975) and Elhilali et al. (2009) found that frequency separation has no impact on streaming in humans when the stimuli presented overlap in time either partially or completely. Streams of tones with wide frequency separations but presented simultaneously create the percept of one auditory object. A few studies have measured this type of scene analysis, also referred to as temporal coherence, in animals.

3.5.1 Mammals

Thus far, only one behavioral study has measured the perception of overlapping sounds in mammals (Christison-Lagay and Cohen 2014). Rhesus macaques, who require about a 5-semitone separation to report 2 tone streams instead of 1 when the high and low tones alternate in time, report hearing only one stream with as much as a 10-semitone separation when the high and low tones are presented simultaneously. This suggests that, as in humans, frequency separation of auditory objects plays a vital role in sequential stream segregation but not in simultaneous stream segregation (see also results from ferrets by Elhilali et al. 2009).

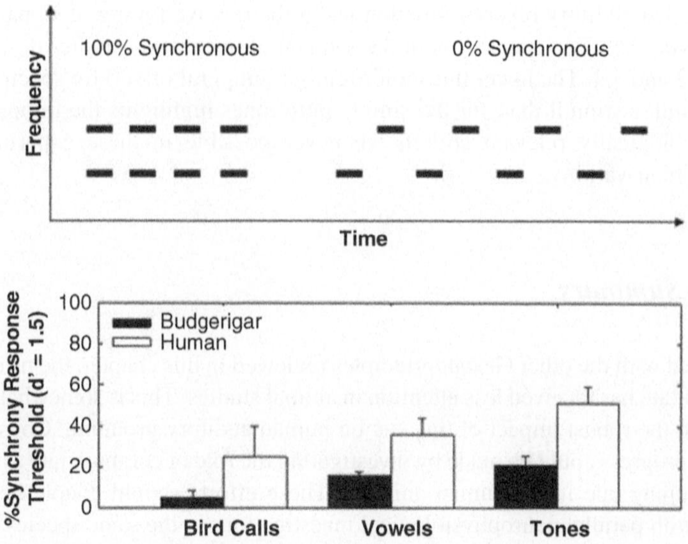

Fig. 3.7 Stimulus conditions (*top*) and results (*bottom*) for temporal coherence in budgerigars and humans. Subjects were trained to identify 100% synchronous stimuli as one category and 0% synchronous as the other category. Once trained, they were presented with intermediate probes ranging from 10 to 90%. d′ thresholds were determined for birds and humans for budgerigar contact calls, human vowels, and pure tones. Thresholds represent the change in percept from overlapping to alternating. Thresholds were lower (earlier) for spectrotemporally complex stimuli and for birds compared to humans. Data from Neilans and Dent (2015b)

3.5.2 Birds

The perception of temporally overlapping stimuli was characterized in a series of behavioral experiments in budgerigars (Neilans and Dent 2015a,b). When presented with a series of high- and low-frequency tones, the birds were trained to peck one key when the high and low signals overlapped (100% synchrony) and another key when they alternated in time (0% synchrony; Fig. 3.7). Once trained on those stimuli, intermediate probe signals with differing magnitudes (10–90% synchrony) of temporal overlap were presented and the birds were required to categorize them. Similar to humans and macaques, signals with very large frequency separations were still reported by budgerigars as synchronous when they overlapped in time. As the temporal asynchrony between signals increased, the responses of the birds changed independent of frequency separation of the signals. When the signals were pure tones, responses of humans tested on the same apparatus with the same procedures did not differ from those of the birds. Segregation thresholds for complex signals were much lower than those for pure tones in both birds and humans (Fig. 3.7). Furthermore, the thresholds for birds were lower than those for humans, meaning that the birds required significantly less temporal separation between sounds to segregate them into separate auditory objects. This result suggests that auditory object formation using the relative timing of signals differs qualitatively between species, as in the sequential integration studies described in Sects. 3.3 and 3.4. The lower thresholds (earlier temporal offset) for spectrotemporally complex stimuli than for the simple pure tones highlights the importance of using ecologically relevant stimuli whenever possible in these experiments to increase their validity.

3.5.3 *Summary*

Compared with the other Gestalt principles reviewed in this chapter, the principle of common fate has received less attention in animal studies. This is somewhat surprising given the robust impact of this cue on human auditory grouping. Considerable forward progress could be made by investigating the role of common fate as an auditory grouping cue in nonhuman animals. These efforts should couple behavioral studies with parallel neurophysiological investigations in the same species.

3.6 The Principle of Good Continuation

In the visual domain, when one object partially occludes another, the occluded object is still recognized as continuing behind the foreground object. This ability demonstrates the Gestalt principle of good continuation. The analogous situation in the auditory domain is when a sound is obscured by a noise masker and no noticeable interruption of that signal is perceived. This was first studied by Warren (1970, 1984), who discovered several types of gapped auditory signals that become completed when a noise is placed in the gap. This phenomenon, known as auditory (or temporal) induction, has been demonstrated in humans with numerous signal types, including pure tones, FM sweeps, speech, and music. As a result of induction, an illusion of auditory continuity is created such that listeners report hearing sounds that simply are not present in the sound waveform impinging on the ears. As discussed in this section, some, but not all, nonhuman animals are capable of perceptually restoring missing sound and thus also appear to follow the Gestalt principle of good continuation (see also Petkov and Sutter 2011).

3.6.1 *Mammals*

A number of studies have now demonstrated that a variety of nonhuman mammals, including rodents, cats, and nonhuman primates, experience auditory continuity illusions using a diversity of stimulus paradigms and behavioral responses. The first such demonstration was by Sugita (1997), who found that domestic cats experienced auditory induction using a phonotaxis assay. In a two-alternative choice test, cats approached a speaker broadcasting a continuous FM sweep instead of one broadcasting a discontinuous FM sweep with a silent gap in the middle. On test trials, noise was added to the middle of both stimuli such that the temporal gap in the discontinuous sweep was filled with noise. As the sound level of the noise increased, the cats' discrimination performance declined, suggesting that they could no longer discriminate between the continuous and discontinuous sweeps. Parallel results in performance decline were obtained with a small set of human listeners tested on the same stimuli. The interpretation of this pattern of results was that both the cats and humans experienced the continuity illusion as a result of auditory induction. Sugita (1997) reported neural correlates of auditory induction in the firing rates of neurons in primary auditory cortex, but these results have been questioned (Micheyl et al. 2003; Petkov and Sutter 2011).

Miller et al. (2001) exploited the natural behavior of cotton-top tamarins (*Saguinus oedipus*) to provide the first demonstration of auditory induction in a nonhuman primate. Their study was also the first to demonstrate the phenomenon using an animal's natural behavioral repertoire instead of an operant task. Native to Colombia, cotton-top tamarins live in dense, tropical forests and produce a species-specific contact call when separated from other members of their social group. Other nearby tamarins, on hearing a contact call, respond by producing an "antiphonal" contact call. Miller et al. (2001) showed that if the middle of the "whistle" portion of a contact call was replaced with silence, the animals no longer produced a typical number of antiphonal calls in response. When the silence was replaced with white noise, antiphonal responses were restored, suggesting that the tamarins experienced a continuity illusion and perceived the sound as their species-typical contact call. If the silent portion replaced either the beginning or ending of the call instead of the middle portion, the animals failed to produce antiphonal calls in response. Thus, the illusion was not created simply by the presence of continuous acoustic energy but instead relied on contextual information provided by the beginning and ending of a normal whistle portion of the call.

Auditory induction has also been demonstrated in rhesus macaques by Petkov et al. (2003, 2007). Subjects were trained to hold a lever and continue holding the lever for two identical sounds but to release that lever if two different sounds were presented. Those sounds were either 2-kHz tones or macaque "coo" calls. On probe trials, the second stimulus was identical to the first except that it contained a silent gap. On "interrupting-noise" trials, the gaps were filled with a broadband noise of various amplitudes. On "surrounding-noise" trials, the noise encompassed the duration of the entire stimulus. In humans, the interrupting noise leads to auditory induction, whereas the surrounding noise leads to masking. Frequency characteristics of the noise were also varied and could be broadband, broadband but with a notch at the 2-kHz tone frequency, or broadband with a notch at 8 kHz. Induction in humans only occurs for noise with frequencies encompassing those of the gap. In monkeys, as in humans, surrounding noise led to masking and interrupting noise led to induction. As the amplitude of the noise increased, induction was more pronounced for both the coo calls and the pure tones. Interestingly, similar to humans tested on

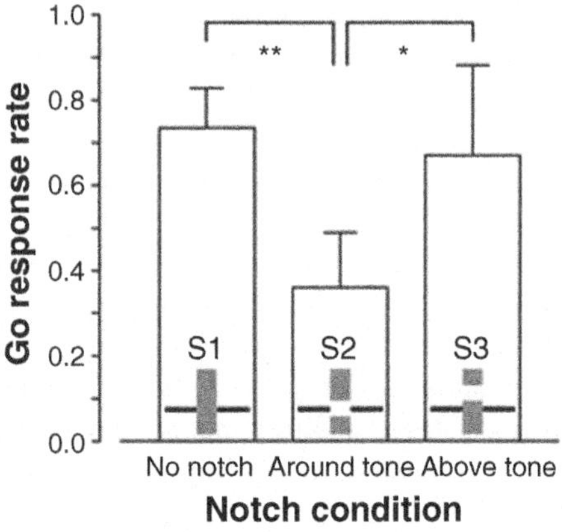

Fig. 3.8 Mean Go response rate (+SD) in gerbils to discontinuous tones filled with interrupting noise without a notch (S1), with a bandstopped notch around the tone's frequency (S2), and with a bandstopped notch above the tone's frequency (S3). Gerbils responded at high rates for the no-notch and above-tone conditions but not the around-tone condition. Reprinted with permission from Kobayasi et al. (2012)© 2012, Acoustical Society of America

speech sounds, induction was more pronounced for the coo calls than it was for the pure tones. Finally, the control notch-noise conditions led to no induction when the notch frequencies did not contain the same frequency as the tone and led to induction when the notch frequencies spanned that of the tone. These behavioral results laid the foundation for studies looking for neural correlates of induction in the auditory cortex of macaques (Petkov et al. 2007).

Studies on auditory induction in rodents indicated the pervasiveness of auditory induction across mammals. Kobayasi et al. (2012) trained Mongolian gerbils (*Meriones unguiculatus*) to discriminate a continuous from a gapped tone. As in the macaque study (Petkov et al. 2003), probe trials were later presented to the gerbils where the gap was filled with a broadband noise, a notched noise, or a surrounding noise, and intensities of the noise were varied. The "Go" response rate, indicating that the rats had perceived a complete tone, was high in the broadband noise condition and the condition with the notch in the noise at frequencies above the tone, not those encompassing the tone (Fig. 3.8), similar to results from macaques (Petkov et al. 2003). Noto et al. (2016) found correlates of auditory induction in several regions of the auditory cortex of rats, as in macaques, using optical imaging techniques.

3.6.2 *Birds*

Two near-simultaneous studies of auditory induction reported that European starlings experience a continuity illusion using pure tones (Klump et al. 1999; cited in Klump 2016) and songs (Braaten and Leary 1999) as stimuli. In an operant task,

Klump et al. (1999) trained European starlings to discriminate between a long tone and a second stimulus comprising two short tones separated by a silent gap. On probe trials, a broadband noise was presented in the center of each stimulus, that is, in the middle of the long tone or in the silent gap between the two short tones. At low noise levels, the birds could still discriminate between the two stimuli, but as the level of the noise was increased, they failed to do so. Thus, when the two tones separated by a noise burst of sufficiently high amplitude were presented, the birds responded as if they had heard a single long tone continuing through the noise. This result was interpreted as evidence of a continuity illusion because the birds behaved as if they had perceptually restored a missing element between the two short tones.

Braaten and Leary (1999) demonstrated auditory induction in European starlings using complex stimuli. Their subjects were trained in an operant task to press one key when presented with a European starling song and another key when presented with budgerigar warble. Once trained to criterion, song and warble stimuli were presented on probe trials in which the middle portion of the song or the warble was removed and replaced with either silence or noise. Subjects reported hearing European starling song when noise filled the gap in the European starling song but not when noise filled the gap in the budgerigar warble. This result was consistent with the idea that the birds were able to perceptually restore missing portions of conspecific song. There was no indication of perceptual restoration when a silent gap was introduced in either stimulus. Seeba and Klump (2009) corroborated and extended the findings of Braaten and Leary (1999) by demonstrating that auditory induction of song was only possible if the birds were already familiar with the sounds (see Sect. 3.7.1).

3.6.3 Frogs

In contrast to studies with birds and mammals, no evidence has thus far shown that frogs adhere to the principle of good continuation in the perceptual restoration of communication sounds. The phenomenon has been investigated in the three species discussed in Sects. 3.3 and 3.4, Cope's gray treefrog, the eastern gray treefrog, and the túngara frog. All three studies have used broadly similar methodology based on inserting silent gaps in conspecific calls to disrupt call recognition and then inserting noise in the gaps to ask whether call recognition is perceptually restored.

In their study with Cope's gray treefrog, Seeba et al. (2010) exploited the fact that females prefer calls that mimic the natural sequence of pulses in a male's calls over alternatives in which brief gaps are introduced by removing pulses. Their main experimental question was whether inserting bursts of noise in the silent gaps would restore the performance of phonotaxis based on females perceiving illusory pulses continuing through the noise. Results from some initial experiments looked promising. Females strongly preferred continuous stimuli over stimuli of equivalent overall duration with silent gaps inserted (Fig. 3.9). In addition, they also strongly preferred stimuli having gaps filled with noise over stimuli of equivalent duration having silent gaps. However, although naive human listeners reported hearing illusory pulses during the bursts of noise, the question remained as to whether the frogs themselves experienced illusory pulses. To answer this question, Seeba et al. (2010) took advantage of the preference females have for relatively longer calls with more pulses. For example, females strongly prefer a 35-pulse call over a 20-pulse call when both are continuous. But when given a choice between two stimuli containing 20 pulses, one of which was continuous and one of which was made to have the longer duration of a 35-pulse stimulus by introducing several silent gaps, females preferred the shorter, continuous call (Fig. 3.9). Therefore, if introducing noise into the silent gaps of the longer stimulus resulted in the perception of illusory pulses, then females were predicted to prefer the longer call because they would perceive it as having more pulses (20 real + 15 illusory) over the shorter, continuous call, even though both stimuli had the same number of real pulses. The data were not consistent with this prediction. Based on these data, Seeba et al. (2010) concluded that it was unlikely that females of Cope's gray treefrogs were experiencing auditory induction. Schwartz et al. (2010) reached a similar conclusion in their study of auditory induction in the eastern gray treefrog.

Seeba et al. (2010) suggested that gray treefrogs may not experience a continuity illusion because the sound of interest was not itself continuous but instead was a sequence of temporally discrete pulses separated by brief intervals of silence. Before accepting the conclusion that frogs (in general) do not experience auditory induction, they suggested that other frog species with continuous signals should be tested. Baugh et al. (2016) addressed this issue in their study of auditory induction in túngara frogs. Recall that túngara frogs produce a FM whine that is necessary and sufficient for call recognition. Given a choice between continuous whines and whines interrupted by a brief silent gap, female túngara frogs preferred continuous whines. When the silent gaps in whines were filled with noise, however, there was no evidence to suggest that females actually heard a complete whine continuing through the noise. (Interestingly, human listeners experience the continuity illusion with these stimuli.) A number of different gap durations, noise bandwidths, and noise levels were tested, but none appeared able to induce the illusory percept of a complete whine.

3.6.4 Summary

As illustrated by the studies reviewed in this section, birds and nonhuman mammals adhere to the Gestalt principle of good continuation in forming auditory objects. Moreover, this principle can be demonstrated in both groups, as in humans, using simple, artificial sounds as well as spectrotemporally complex communication sounds. In contrast, the available evidence suggests that frogs differ from mammals and birds in not following the principle of good continuation. The underlying neurophysiological basis for this apparent evolutionary difference remains unknown.

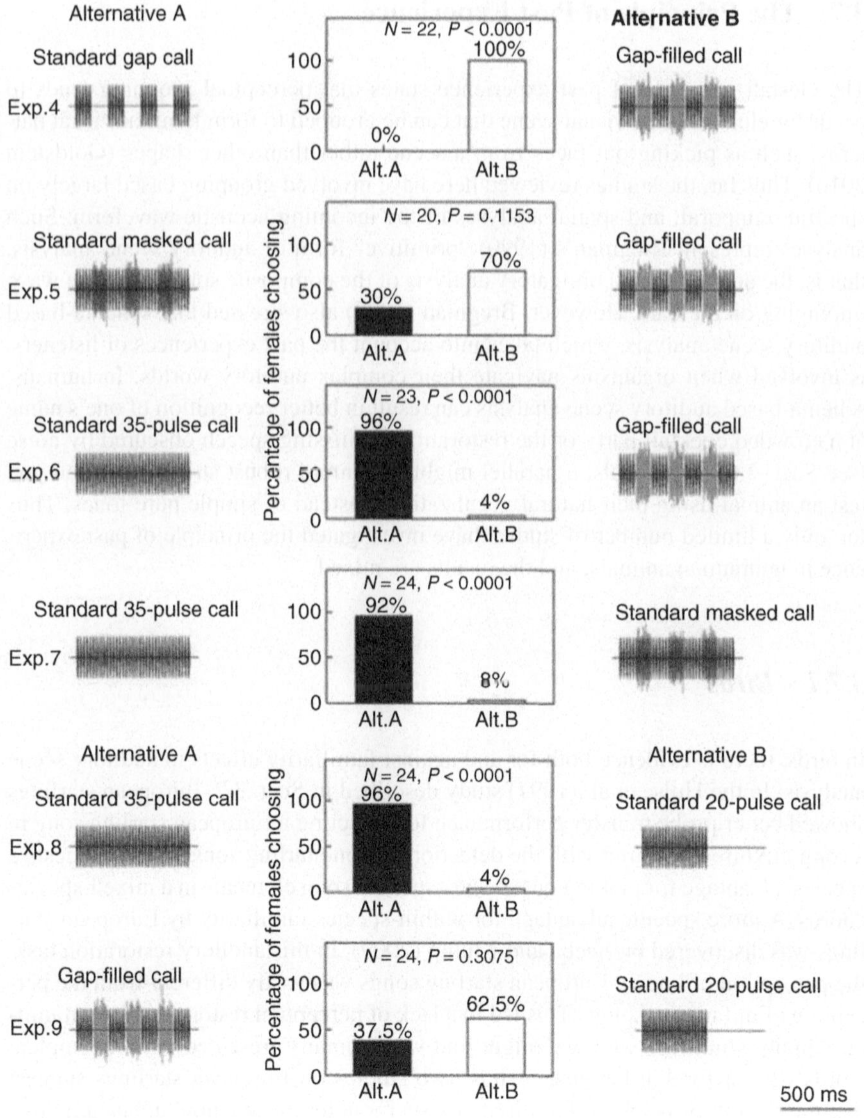

Fig. 3.9 Results from experiments 4–9 of Seeba et al. (2010). Waveforms of the two paired alternatives A and B (Alt. A and Alt. B, respectively) in each two-choice experiment are shown along with the percentage of females choosing each alternative. Reprinted from Seeba et al. (2010), with permission from Elsevier

3.7 The Principle of Past Experience

The Gestalt principle of past experience states that perceptual grouping tends to occur for elements in a visual scene that can be grouped to form familiar visual patterns, such as picking out faces from a scene rather than other shapes (Goldstein 2010). Thus far, the studies reviewed here have involved grouping based largely on spectral, temporal, and spatial analyses of the incoming acoustic waveform. Such analyses represent Bregman's (1990) "primitive" form of auditory scene analysis, that is, the automatic and obligatory analysis of the composite sound pressure wave impinging on the ears. However, Bregman (1990) also stressed that schema-based auditory scene analysis, which takes into account the past experiences of listeners, is involved when organisms navigate their complex auditory worlds. In humans, schema-based auditory scene analysis can result in better recognition of one's name at a crowded cocktail party or the restoration of missing speech obscured by noise (see Sect. 3.6). In animals, a parallel might be a more robust streaming when you test an animal using their natural vocalizations instead of simple pure tones. Thus far, only a limited number of studies have investigated the principle of past experience in nonhuman animals, and the results are mixed.

3.7.1 Birds

In birds, there is evidence both for and against familiarity effects in auditory scene analysis. In the Hulse et al. (1997) study described in Sect. 3.2, European starlings showed better probe transfer performance for detecting a European starling song in a song mixture compared with the detection of non-starling songs. This suggests a species advantage for picking out species-specific song elements in a mixed-species chorus. A more specific advantage for within-species familiarity by European starlings was discovered by Seeba and Klump (2009). In this auditory restoration task, the perception of known European starling songs was vastly different from the perception of unfamiliar songs. This led to a lack of perceptual restoration for unfamiliar stimuli, similar to what we might find with humans presented with incomplete words from a foreign language. Thus, two studies on European starlings suggest that the Gestalt principle of familiarity exists for at least some nonhuman animals.

Two studies of budgerigars, on the other hand, showed that the spectrotemporal make up of sounds led to differences in auditory streaming but that familiarity did not play a role. In the temporal coherence study described in Sect. 3.5, Neilans and Dent (2015b) found that thresholds were not better for familiar sounds (budgerigars tested on budgerigar calls, humans tested on speech phonemes) but that thresholds simply improved as spectrotemporal complexity increased from pure tones to speech and speech-like sounds and bird calls. Similarly, Dent et al. (2016) found that streaming of an incorrect syllable within a zebra finch song did not differ between zebra finches and budgerigars. It also did not matter whether the replaced syllable was from a zebra finch or from another species of bird. The probe syllable was simply more likely to be streamed with the rest of the song when it was more spectrotemporally similar to the missing syllable, not when it was produced by a zebra finch.

3.7.2 Frogs

In contrast to previous work on humans and birds, work on frogs related to the principle of past experience has taken a much different view of what it means for something to be "familiar." In a colloquial sense, an object, such as a person's face in a visual scene or the sound of their voice in an acoustic scene, would typically be considered to be "familiar" if it had been encountered before, that is, if information about it had been encoded in memory because something about it had been learned through past experience. There is now clear evidence that some frogs can also learn to recognize other individuals by voice (Bee 2016), but how this experienced-based form of familiarity relates to auditory grouping has not been investigated. Instead, research on one species, the túngara frog, has considered grouping based on "stored information" in the form of call recognition templates created over the course of evolutionary time (Farris and Ryan 2011). That is, "past experience" is construed as a mechanism of grouping based on the fate of ancestors in the evolutionary past. Recall that túngara frogs produce a two-part call consisting of a whine that can be followed by one or more chucks. The first chuck is morphologically constrained to occur within a fairly narrow time window following a preceding whine. Farris and Ryan (2011) have proposed that grouping based on an unlearned template that evolved to recognize the natural call sequence (i.e., whine then chuck) represents an example of schema-based grouping in action. Their argument is that the template is a form of stored information, albeit not a learned memory, that represents a schema that can be exploited for grouping. The data in support of this view are somewhat mixed.

As discussed in Sect. 3.4.3, Farris et al. (2002) demonstrated that female túngara frogs will readily group a single chuck with a preceding whine across large angles of separation up to 135° when the two sounds occur in their normal temporal sequence. In a follow-up study, Farris et al. (2005) discovered that females would continue to do so at angles as large as 90° even when the chuck occurred simultaneously with the onset of the whine, which is a signal that males are morphologically unable to produce. This result clearly indicates that females can be quite permissive of violations of natural signal timing when the acoustic scene consists of a single whine and a single chuck, similar to how they are permissive of violations of spatial coherence under similar conditions (see Sect. 3.4.3). In subsequent work, however, Farris and Ryan (2011) showed that when females had to decide which of two chucks to group with a whine, they preferentially grouped the chuck that occurred at a time when the first chuck would normally occur in a natural call. Thus, as with spatial grouping, a permissiveness for violations of temporal sequence was generally reduced when listeners were forced to make a decision about grouping based on a relative comparison. Curiously, however, tolerance for temporal sequence violations could still be found in some cases when a relative comparison would have predicted a different outcome. For example, in one test, a whine was presented with two different chucks, one of which violated the normal sequence by preceding the whine by 80 ms and the second of which occurred at the appropriate time in the natural sequence. According to a schema-based analysis, the subjects should have preferentially grouped the whine with the subsequent chuck, but they did not. At present, there is no compelling explanation for why female túngara frogs would apply their evolutionarily encoded schema to solve grouping problems in some conditions but not others.

3.7.3 *Summary*

The small number of studies reviewed in this section highlights the need for additional investigations into the role of past experience in auditory object formation in nonhuman animals. The incompatible results from studies of birds suggest that more needs to be done to tease apart the role of spectrotemporal complexity and familiarity in auditory scene analysis in this group. Research on frogs has not yet investigated the impact on auditory grouping of learning from past experience, but the potential to do so exists in species that learn to recognize other individuals' calls. The importance of past experience in auditory object formation in nonhuman mammals, fish, and insects also remains largely unexplored. It also remains to be determined whether neural templates created not as a result of an individual's past experience but by its species' evolutionary past should generally be regarded as representing schema-based grouping.

3.8 Conclusions

In an earlier review on auditory scene analysis, Fay (2008, p. 321) suggested that "all animals that hear are likely capable of some sort of source segregation and scene analysis in perception." The most important conclusion from the present review is that many of the Gestalt principles of auditory object formation identified in human studies—the principles of proximity, similarity, common fate, good continuation, and past experience—are in no way specific to human auditory scene analysis. Instead, research conducted after the pioneering studies of auditory scene analysis in European starlings (Hulse et al. 1997) and also in goldfish (Fay 1998) reveals unequivocally that these principles also guide the perceptual organization of acoustic scenes in a wide diversity of nonhuman animals, including both vertebrates and invertebrates. Although all of the studies reviewed in this chapter were conducted under controlled laboratory settings, these principles are no doubt involved in allowing animals to adaptively navigate the often noisy and acoustically complex environments they encounter in the real world, for example, in attempting to find a mate, identify a rival, or detect a predator. Moreover, the diversity of species in which the Gestalt principles of auditory object formation appear to operate suggests some of their underlying mechanisms could have ancient evolutionary roots, as already suggested by others (Fay and Popper 2000).

Although there has been significant progress in revealing the principles of auditory object formation in nonhuman animals, significant questions remain to be answered. Perhaps the most important of these concerns is the role of attention (Shamma et al. 2011) and other schema-based processes in auditory object formation. Studies show, for example, that there can be mismatches between neurophysiological measures of auditory object formation and the perceptual decisions of listeners, possibly reflecting attention and other top-down processes (Itatani and Klump 2014). Related to attention is the well-known bistability of some perceptual organizations, which can be influenced in humans by directing the attention of a listener to one or the other percept. Bistability generally has not been considered in studies of perception in nonhuman animals. Other features of auditory organization identified in humans, such as the buildup of auditory stream segregation and the principle of exclusive allocation (Bregman 1990), have also received little attention in animal studies. Although neural correlates of the buildup of streaming have been identified in macaques (Micheyl et al. 2005) and European starlings (Bee et al. 2010), behavioral correlates have yet to be measured in any nonhuman species. Virtually nothing is known about the developmental trajectory of auditory scene analysis in nonhuman animals and how much of a role the environment plays in the development of normal auditory object formation. Furthermore, aside from the studies reviewed here, research on nonhuman animal communication systems has been slow to adopt a view of acoustic signals as potential auditory objects in a receiver's acoustic scene (Miller and Bee 2012; Bee and Miller 2016). Neuroethological studies of auditory scene analysis in animal communication are needed to understand the diversity of underlying neural mechanisms. Forward progress will be made on all of these questions by integrating behavioral experiments with neurophysiological studies at a cellular level, not only in the same nonhuman species (e.g., Middlebrooks and Onsan 2012; Middlebrooks and Bremen 2013) but also in the same individuals (e.g., Itatani and Klump 2014).

Finally, there is much work to be done in determining how these Gestalt principles work when the auditory objects are man-made instead of natural. Humans wandering the noisy streets of Berlin do not have trouble segregating the passing ambulance from the barking dog. Virtually nothing is known about the grouping of anthropogenic sounds versus animal sounds. Anthropogenic noise is known to impact animal communication and survival in many ways, but how animals are able to overcome noise using Gestalt grouping is still a mystery. The studies summarized in this chapter suggest that animals have the tools to effectively communicate under most of the same conditions as humans because so many of the studies summarized here demonstrate strikingly similar parallels.

Compliance with Ethics Requirements

Micheal L. Dent declares that she has no conflict of interest.

Mark A. Bee declares that he has no conflict of interest.

References

American National Standards Institute (ANSI). (2013). *American National Standard Acoustical Terminology*. ANSI S1.1, American National Standards Institute for the Acoustical Society of America, Washington, DC.

Barber, J., Razak, K., & Fuzessery, Z. (2003). Can two streams of auditory information be processed simultaneously? Evidence from the gleaning bat *Antrozous pallidus*. *Journal of Comparative Physiology A: Neuroethology, Sensory, Neural, and Behavioral Physiology,* 189(11), 843–855.

Baugh, A. T., Ryan, M. J., Bernal, X. E., Rand, A. S., & Bee, M. A. (2016). Female túngara frogs do not experience the continuity illusion. *Behavioral Neuroscience*, 130(1), 62–74.

Bee, M. A. (2010). Spectral preferences and the role of spatial coherence in simultaneous integration in gray treefrogs (*Hyla chrysoscelis*). *Journal of Comparative Psychology*, 124(4), 412–424.

Bee, M. A. (2012). Sound source perception in anuran amphibians. *Current Opinion in Neurobiology*, 22(2), 301–310.

Bee, M. A. (2015). Treefrogs as animal models for research on auditory scene analysis and the cocktail party problem. *International Journal of Psychophysiology*, 95(2), 216–237.

Bee, M. A. (2016). Social recognition in anurans. In M. A. Bee & C. T. Miller (Eds.), *Psychological Mechanisms in Animal Communication* (pp. 169–221). New York: Springer International Publishing.

Bee, M. A., & Klump, G. M. (2004). Primitive auditory stream segregation: A neurophysiological study in the songbird forebrain. *Journal of Neurophysiology*, 92(2), 1088–1104.

Bee, M. A., & Klump, G. M. (2005). Auditory stream segregation in the songbird forebrain: Effects of time intervals on responses to interleaved tone sequences. *Brain, Behavior and Evolution*, 66(3), 197–214.

Bee, M. A., & Riemersma, K. K. (2008). Does common spatial origin promote the auditory grouping of temporally separated signal elements in grey treefrogs? *Animal Behaviour*, 76(3), 831–843.

Bee, M. A., & Miller, C. T. (2016). Signaler and receiver psychology. In M. A. Bee & C. T. Miller (Eds.), *Psychological Mechanisms in Animal Communication* (pp. 1–16). New York: Springer International Publishing.

Bee, M. A., Micheyl, C., Oxenham, A. J., & Klump, G. M. (2010). Neural adaptation to tone sequences in the songbird forebrain: Patterns, determinants, and relation to the build-up of auditory streaming. *Journal of Comparative Physiology A: Neuroethology, Sensory, Neural, and Behavioral Physiology,* 196(8), 543–557.

Benney, K. S., & Braaten, R. F. (2000). Auditory scene analysis in Estrildid finches (*Taeniopygia guttata* and *Lonchura striata domestica*): A species advantage for detection of conspecific song. *Journal of Comparative Psychology*, 114(2), 174–182.

Bizley, J. K., & Cohen, Y. E. (2013). The what, where and how of auditory-object perception. *Nature Reviews Neuroscience*, 14(10), 693–707.

Braaten, R. F., & Hulse, S. H. (1993). Perceptual organization of auditory temporal patterns in European starlings (*Sturnus vulgaris*). *Perception & Psychophysics*, 54(5), 567–578.

Braaten, R. F., & Leary, J. C. (1999). Temporal induction of missing birdsong segments in European starlings. *Psychological Science*, 10(2), 162–166.

Bregman, A. (1990). *Auditory Scene Analysis: The Perceptual Organization of Sound*. Cambridge, MA: MIT Press.

Bush, S. L., Gerhardt, H. C., & Schul, J. (2002). Pattern recognition and call preferences in treefrogs (Anura: Hylidae): A quantitative analysis using a no-choice paradigm. *Animal Behaviour*, 63(1), 7–14.

Carlyon, R. P. (2004). How the brain separates sounds. *Trends in Cognitive Sciences*, 8(10), 465–471.

Christison-Lagay, K. L., & Cohen, Y. E. (2014). Behavioral correlates of auditory streaming in rhesus macaques. *Hearing Research*, 309, 17–25.

Darwin, C. J. (2008). Spatial hearing and perceiving sources. In W. A. Yost, A. N. Popper, & R. R. Fay (Eds.), *Auditory Perception of Sound Sources* (pp. 215–232). New York: Springer US.

Darwin, C. J., & Carlyon, R. P. (1995). Auditory grouping. In B. C. J. Moore (Ed.), *The Handbook of Perception and Cognition, Vol. 6: Hearing* (pp. 387–424). New York: Academic Press.

Dent, M. L., Martin, A. K., Flaherty, M. M., & Neilans, E. G. (2016). Cues for auditory stream segregation of birdsong in budgerigars and zebra finches: Effects of location, timing, amplitude, and frequency. *The Journal of the Acoustical Society of America*, 139(2), 674–683.

Dolležal, L.-V., Itatani, N., Gunther, S., & Klump, G. M. (2012). Auditory streaming by phase relations between components of harmonic complexes: A comparative study of human subjects and bird forebrain neurons. *Behavioral Neuroscience*, 126(6), 797–808.

Elhilali, M., Ma, L., Micheyl, C., Oxenham, A. J., & Shamma, S. A. (2009). Temporal coherence in the perceptual organization and cortical representation of auditory scenes. *Neuron*, 61(2), 317–329.

Farris, H. E., & Ryan, M. J. (2011). Relative comparisons of call parameters enable auditory grouping in frogs. *Nature Communications*, 2, 410.

Farris, H. E., & Taylor, R. C. (2016). Mate searching animals as model systems for understanding perceptual grouping. In M. A. Bee & C. T. Miller (Eds.), *Psychological Mechanisms in Animal Communication* (pp. 89–118). New York: Springer International Publishing.

Farris, H. E., Rand, A. S., & Ryan, M. J. (2002). The effects of spatially separated call components on phonotaxis in túngara frogs: Evidence for auditory grouping. *Brain, Behavior and Evolution*, 60(3), 181–188.

Farris, H., Rand, A. S., & Ryan, M. J. (2005). The effects of time, space and spectrum on auditory grouping in túngara frogs. *Journal of Comparative Physiology A: Neuroethology, Sensory, Neural, and Behavioral Physiology*, 191(12), 1173–1183.

Fay, R. R. (1998). Auditory stream segregation in goldfish (*Carassius auratus*). *Hearing Research*, 120(1), 69–76.

Fay, R. R. (2000). Spectral contrasts underlying auditory stream segregation in goldfish (*Carassius auratus*). *Journal of the Association for Research in Otolaryngology*, 1(2), 120–128.

Fay, R. R. (2008). Sound source perception and stream segregation in nonhuman vertebrate animals. In W. A. Yost, A. N. Popper, & R. R. Fay (Eds.), *Auditory Perception of Sound Sources* (pp. 307–323), New York: Springer US.

Fay, R. R., & Popper, A. N. (2000). Evolution of hearing in vertebrates: The inner ears and processing. *Hearing Research*, 149(1–2), 1–10.

Fishman, Y. I., Reser, D. H., Arezzo, J. C., & Steinschneider, M. (2001). Neural correlates of auditory stream segregation in primary auditory cortex of the awake monkey. *Hearing Research*, 151(1), 167–187.

Fishman, Y. I., Arezzo, J. C., & Steinschneider, M. (2004). Auditory stream segregation in monkey auditory cortex: Effects of frequency separation, presentation rate, and tone duration. *The Journal of the Acoustical Society of America*, 116(3), 1656–1670.

Gerhardt, H. C. (2005). Acoustic spectral preferences in two cryptic species of grey treefrogs: Implications for mate choice and sensory mechanisms. *Animal Behaviour*, 70(1), 39–48.

Gerhardt, H. C., & Huber, F. (2002). *Acoustic Communication in Insects and Anurans: Common Problems and Diverse Solutions*. Chicago, IL: The University of Chicago Press.

Gerhardt, H. C., Martínez-Rivera, C. C., Schwartz, J. J., Marshall, V. T., & Murphy, C. G. (2007). Preferences based on spectral differences in acoustic signals in four species of treefrogs (Anura: Hylidae). *Journal of Experimental Biology*, 210(17), 2990–2998.

Goldstein, E. B. (2010). Introduction to perception. In *Sensation and Perception* (pp. 3–20). Belmont, CA: Wadsworth Cengage Learning.

Griffiths, T. D., & Warren, J. D. (2004). What is an auditory object? *Nature Reviews Neuroscience*, 5(11), 887–892.

Hartmann, W. M., & Johnson, D. (1991). Stream segregation and peripheral channeling. *Music Perception: An Interdisciplinary Journal*, 9(2), 155–183.

Hulse, S. H. (2002). Auditory scene analysis in animal communication. *Advances in the Study of Behavior*, 31, 163–200.
Hulse, S. H., MacDougall-Shackleton, S. A., & Wisniewski, A. B. (1997). Auditory scene analysis by songbirds: Stream segregation of birdsong by European starlings (*Sturnus vulgaris*). *Journal of Comparative Psychology*, 111(1), 3–13.
Itatani, N., & Klump, G. M. (2009). Auditory streaming of amplitude-modulated sounds in the songbird forebrain. *Journal of Neurophysiology*, 101(6), 3212–3225.
Itatani, N., & Klump, G. M. (2011). Neural correlates of auditory streaming of harmonic complex sounds with different phase relations in the songbird forebrain. *Journal of Neurophysiology*, 105(1), 188–199.
Itatani, N., & Klump, G. M. (2014). Neural correlates of auditory streaming in an objective behavioral task. *Proceedings of the National Academy of Sciences of the United States of America*, 111(29), 10738–10743.
Izumi, A. (2002). Auditory stream segregation in Japanese monkeys. *Cognition*, 82(3), B113-B122.
Klump, G. M. (2016). Perceptual and neural mechanisms of auditory scene analysis in the European starling. In M. A. Bee & C. T. Miller (Eds.), *Psychological Mechanisms in Animal Communication* (pp. 57–88). New York: Springer International Publishing.
Klump, G. M., Fichtel, C., Hamann, I., & Langemann, U. (1999). Filling in the gap: Evidence for apparent continuity in the songbird auditory system. *ARO Midwinter Research Meeting*, Abstract 108.
Kobayasi, K. I., Usami, A., & Riquimaroux, H. (2012). Behavioral evidence for auditory induction in a species of rodent: Mongolian gerbil (*Meriones unguiculatus*). *The Journal of the Acoustical Society of America*, 132(6), 4063–4068.
Ma, L., Micheyl, C., Yin, P., Oxenham, A. J., & Shamma, S. A. (2010). Behavioral measures of auditory streaming in ferrets (*Mustela putorius*). *Journal of Comparative Psychology*, 124(3), 317–330.
MacDougall-Shackleton, S. A., Hulse, S. H., Gentner, T. Q., & White, W. (1998). Auditory scene analysis by European starlings (*Sturnus vulgaris*): Perceptual segregation of tone sequences. *The Journal of the Acoustical Society of America*, 103(6), 3581–3587.
Micheyl, C., & Oxenham, A. J. (2010). Objective and subjective psychophysical measures of auditory stream integration and segregation. *Journal of the Association for Research in Otolaryngology* 11(4), 709–724.
Micheyl, C., Carlyon, R. P., Shtyrov, Y., Hauk, O., Dodson, T., & Pullvermüller, F. (2003). The neurophysiological basis of the auditory continuity illusion: A mismatch negativity study. *Journal of Cognitive Neuroscience*, 15(5), 747–758.
Micheyl, C., Tian, B., Carlyon, R. P., & Rauschecker, J. P. (2005). Perceptual organization of tone sequences in the auditory cortex of awake macaques. *Neuron*, 48(1), 139–148.
Middlebrooks, J. C., & Onsan, Z. A. (2012). Stream segregation with high spatial acuity. *The Journal of the Acoustical Society of America*, 132(6), 3896–3911.
Middlebrooks, J. C., & Bremen, P. (2013). Spatial stream segregation by auditory cortical neurons. *The Journal of Neuroscience*, 33(27), 10986–11001.
Miller, C. T., & Bee, M. A. (2012). Receiver psychology turns 20: Is it time for a broader approach? *Animal Behaviour*, 83(2), 331–343.
Miller, C. T., Dibble, E., & Hauser, M. D. (2001). Amodal completion of acoustic signals by a nonhuman primate. *Nature Neuroscience*, 4(8), 783–784.
Moss, C. F., & Surlykke, A. (2001). Auditory scene analysis by echolocation in bats. *The Journal of the Acoustical Society of America*, 110(4), 2207–2226.
Neilans, E. G., & Dent, M. L. (2015a). Temporal coherence for pure tones in budgerigars (*Melopsittacus undulatus*) and humans (*Homo sapiens*). *Journal of Comparative Psychology*, 129(1), 52–61.
Neilans, E. G., & Dent, M. L. (2015b). Temporal coherence for complex signals in budgerigars (*Melopsittacus undulatus*) and humans (*Homo sapiens*). *Journal of Comparative Psychology*, 129(2), 174–180.

Nityananda, V., & Bee, M. A. (2011). Finding your mate at a cocktail party: Frequency separation promotes auditory stream segregation of concurrent voices in multi-species frog choruses. *PLoS ONE*, 6(6), e21191.
Noto, M., Nishikawa, J., & Tateno, T. (2016). An analysis of nonlinear dynamics underlying neural activity related to auditory induction in the rat auditory cortex. *Neuroscience*, 318, 58–83.
Palmer, S. E. (2002). Perceptual organization in vision. In H. Pashler & S. Yantis (Eds.), *Stevens' Handbook of Experimental Psychology, Vol. 1: Sensation and Perception,* 3rd ed. (pp. 177–234). New York: John Wiley & Sons.
Park, T. J., & Dooling, R. J. (1991). Sound localization in small birds: Absolute localization in azimuth. *Journal of Comparative Psychology*, 105(2), 125–133.
Petkov, C. I., & Sutter, M. L. (2011). Evolutionary conservation and neuronal mechanisms of auditory perceptual restoration. *Hearing Research*, 271(1), 54–65.
Petkov, C. I., O'Connor, K. N., & Sutter, M. L. (2003). Illusory sound perception in macaque monkeys. *The Journal of Neuroscience*, 23(27), 9155–9161.
Petkov, C. I., O'Connor, K. N., & Sutter, M. L. (2007). Encoding of illusory continuity in primary auditory cortex. *Neuron*, 54(1), 153–165.
Populin, L. C., & Yin, T. C. (1998). Behavioral studies of sound localization in the cat. *The Journal of Neuroscience*, 18(6), 2147–2160.
Pressnitzer, D., Sayles, M., Micheyl, C., & Winter, I. M. (2008). Perceptual organization of sound begins in the auditory periphery. *Current Biology*, 18(15), 1124–1128.
Römer, H. (2013). Masking by noise in acoustic insects: Problems and solutions. In H. Brumm (Ed.), *Animal Communication and Noise* (pp. 33–63). Berlin Heidelberg: Springer-Verlag.
Scholes, C., Palmer, A. R., & Sumner, C. J. (2015). Stream segregation in the anesthetized auditory cortex. *Hearing Research*, 328, 48–58.
Schul, J., & Sheridan, R. (2006). Auditory stream segregation in an insect. *Neuroscience*, 138(1), 1–4.
Schwartz, J. J., & Gerhardt, H. C. (1995). Directionality of the auditory system and call pattern recognition during acoustic interference in the gray treefrog, *Hyla versicolor. Auditory Neuroscience*, 1, 195–206.
Schwartz, J. J., Huth, K., Jones, S. H., Brown, R., & Marks, J. (2010). Tests for call restoration in the gray treefrog, *Hyla versicolor*. *Bioacoustics*, 20, 59–86.
Seeba, F., & Klump, G. M. (2009). Stimulus familiarity affects perceptual restoration in the European starling (*Sturnus vulgaris*). *PLoS ONE*, 4(6), e5974.
Seeba, F., Schwartz, J. J., & Bee, M. A. (2010). Testing an auditory illusion in frogs: Perceptual restoration or sensory bias? *Animal Behaviour*, 79(6), 1317–1328.
Selezneva, E., Gorkin, A., Mylius, J., Noesselt, T., Scheich, H., & Brosch, M. (2012). Reaction times reflect subjective auditory perception of tone sequences in macaque monkeys. *Hearing Research*, 294(1), 133–142.
Shamma, S. A., Elhilali, M., & Micheyl, C. (2011). Temporal coherence and attention in auditory scene analysis. *Trends in Neurosciences*, 34(3), 114–123.
Stebbins, W. C. (1973) Hearing of old world monkeys (Cercopithecinae). *American Journal of Physical Anthropology*, 38(2), 357–364.
Sugita, Y. (1997). Neuronal correlates of auditory induction in the cat cortex. *Neuroreport*, 8(5), 1155–1159.
van Noorden, L. P. A. S. (1975). *Temporal Coherence in the Perception of Tone Sequences.* Unpublished doctoral dissertation, Technische Hogeschool Eindhoven, Eindhoven, The Netherlands.
von Helversen, D. (1984). Parallel processing in auditory pattern recognition and directional analysis by the grasshopper *Chorthippus biguttulus* L.(Acrididae). *Journal of Comparative Physiology A: Neuroethology, Sensory, Neural, and Behavioral Physiology,* 154(6), 837–846.
Wagemans, J., Elder, J. H., Kubovy, M., Palmer, S. E., Peterson, M. A., Singh, M., & von der Heydt, R. (2012a). A century of Gestalt psychology in visual perception: I. Perceptual grouping and figure-ground organization. *Psychological Bulletin,* 138(6), 1172–1217.

Wagemans, J., Feldman, J., Gepshtein, S., Kimchi, R., Pomerantz, J. R., van der Helm, P. A., & van Leeuwen, C. (2012b). A century of Gestalt psychology in visual perception: II. Conceptual and theoretical foundations. *Psychological Bulletin*, 138(6), 1218–1252.

Warren, R. M. (1970). Perceptual restoration of missing speech sounds. *Science*, 167(3917), 392–393.

Warren, R. M. (1984). Perceptual restoration of obliterated sounds. *Psychological Bulletin*, 96(2), 371–383.

Weber, T., & Thorson, J. (1988). Auditory behavior of the cricket. *Journal of Comparative Physiology A: Neuroethology, Sensory, Neural, and Behavioral Physiology,* 163(1), 13–22.

Wisniewski, A. B., & Hulse, S. H. (1997). Auditory scene analysis in European starlings (*Sturnus vulgaris*): Discrimination of song segments, their segregation from multiple and reversed conspecific songs, and evidence for conspecific song categorization. *Journal of Comparative Psychology*, 111(4), 337–350.

Yao, J. D., Bremen, P., & Middlebrooks, J. C. (2015). Emergence of spatial stream segregation in the ascending auditory pathway. *The Journal of Neuroscience*, 35(49), 16199–16212.

Yost, W. A., Popper, A. N., & Fay, R. R. (Eds.). (2008) *Auditory Perception of Sound Sources.* New York: Springer US.

Chapter 4
Characteristics of Temporary and Permanent Threshold Shifts in Vertebrates

James C. Saunders and Robert J. Dooling

Abstract Studies of noise-induced threshold shift to acoustic over exposure, conducted in the laboratory, employ a simple and rigorous paradigm. First, hearing is measured usually as a series of thresholds for pure tones distributed throughout the range of hearing. Then the subject is exposed to a loud sound, after which estimates of threshold are repeated. A difference between the pre- and postthreshold values represents the consequences of the sound exposure. From this approach, much is known about the effects of intense sound exposure on hearing. This chapter reviews the data on temporary and permanent threshold shifts. Examples of noise-induced threshold shift (NITS) are drawn from the extensive animal literature obtained from a mammal (chinchilla) and several avian (budgerigar and chicken chick) species. The conclusions from these examples is that when the parameters of exposure are well characterized and thresholds carefully determined, a set of reliable and valid observations on the magnitude and duration of NITS emerge. Most importantly, certain aspects of hearing loss appear to be consistent across species. Given that a set of repeatable patterns of NITS are known, it is suggested that meaningful predictions can be made on the susceptibility of hearing loss in wild animals exposed to noise in natural environments. This suggestion assumes that the parameters of the exposure such as the sound pressure level and duration are reasonably well-known. The validity of these estimates has limitations and these are also considered.

Keywords Adaptation · Asymptotic threshold shift · Continuous exposure · Decibel · Fatigue · Hearing loss · Impulse/impact exposure · Intermittent exposure · Noise-induced threshold shift · Permanent threshold shift recovery · Sound field · Sound pressure level · Temporary threshold shift · Threshold shift

J. C. Saunders
Department of Otorhinolaryngology, Head and Neck Surgery, The Perelman School of Medicine, University of Pennsylvania, Philadelphia, PA, USA
e-mail: saunderj@mail.med.upenn.edu

R. J. Dooling (✉)
Department of Psychology, University of Maryland, College Park, MD, USA
e-mail: rdooling@umd.edu

H. Slabbekoorn et al. (eds.), *Effects of Anthropogenic Noise on Animals*, Springer Handbook of Auditory Research 66,
https://doi.org/10.1007/978-1-4939-8574-6_4

4.1 Introduction

With increases in the human population and attending urbanization, the world is becoming a noisier place. Understanding the detrimental effects of increased noise on both humans and animals is important. From decades of work on humans (Rabinowitz 2012), it is possible to describe at least four general overlapping effects of exposure to loud sounds: (1) a permanent hearing loss from which there is never any recovery, (2) a temporary hearing loss that recovers over a period of time from the end of a noise exposure, (3) masking, and (4) other physiological and/or behavioral responses (Salvi et al. 1986). The effects of loud sound depend primarily on the level of sound arriving at the ear and its duration.

This chapter focuses on the two major consequences of acoustic overexposure, temporary threshold shift (TTS) and permanent threshold shift (PTS), with the goal of characterizing the various parameters associated with noise-induced threshold shift (NITS). Data are drawn from highly controlled animal studies in the laboratory where the acoustic parameters of the exposure (at the ear) were well-defined. The results reveal that NITS can be organized in an orderly manner that follows a set of principles (see Clark 1991). These principles appear to be consistent across different modes of exposure and across various laboratory species and likely apply across all vertebrate species. They provide a basis for understanding the extent to which NITS is a concern for wild animals living in natural environments.

Exposing the ear to loud sounds for a sufficient duration will cause a temporary elevation in threshold for subsequent sound detection. The difference between the pre- and postexposure threshold is called a threshold shift (TS), which is usually measured in decibels sound pressure level (SPL). After an intense sound exposure, postexposure thresholds are typically measured repeatedly over time and if the decibel value of the TS decreases over time (i.e., if hearing systematically improves) and eventually returns to the preexposure threshold level, the hearing loss is referred to as a TTS.

The array of sound exposure conditions capable of inducing a TTS are extensive and accompanied by many underlying biological mechanisms that contribute to the temporary hearing loss. Conventionally, the fact that the hearing loss is temporary implies that auditory system function was altered but then underwent repair, returning it to its original state. Recently, however, the idea that recovery from NITS represents a completely healed auditory system has come into question with the finding of residual long-term structural changes in the inner ear (Kujawa and Liberman 2009). The biological basis for the complete recovery of hearing with lingering anatomical damage remains to be understood.

Increasing the exposure SPL and/or the duration inevitably results in greater levels of TTS. As the level of TTS increases, there is corresponding increase in the recovery time required to reach stable postexposure thresholds. At some level of exposure, a sufficiently severe shift in threshold occurs, resulting in an incomplete hearing recovery. In other words, the preexposure threshold is never reached. This incomplete recovery constitutes a PTS, a hearing loss that for mammals, at least, will be carried for the rest of their life. As might be expected, a positive relationship exists between the severity of the exposure and the magnitude of TTS and PTS (Carder and Miller 1972).

4.2 Stimulus-Induced Auditory Fatigue: Adaptation and Threshold Shifts

All sensory systems are subject to sensory fatigue, a condition where exposure to stimulation itself results in a diminished capacity to respond to subsequent stimuli (Hood 1956). The consequences of sensory fatigue lie along a continuum, and at one end, sustained mild sound levels result in a transitory shift in threshold referred to as adaptation. Adaptation occurs when a constantly repeating or continuous sound that carries no behavioral relevancy, like that of a ticking clock or the steady rumble of an air conditioner, simply disappears from a listener's awareness. The adapted sound reappears immediately with a change in the sound such as an interruption or sudden intensity shift. Synaptic fatigue (see Spassova et al. 2004) may be the physiological basis for adaptation.

At the other end of this continuum are loud and sustained sound exposures that can temporarily or permanently disrupt cellular metabolic and physiological processes or physically damage/destroy inner ear tissues and are the focus of this chapter. The interested reader will find a well-documented older literature describing the mechanical, metabolic, physiological, and neural changes to the inner ear after intense, long-duration sound exposure (Saunders et al. 1985a, 1991).

4.3 Variables Affecting Temporary Threshold Shift and Permanent Threshold Shift

There are various parameters that define the characteristics of an intense sound exposure and many methods are available for defining thresholds. Among these are sound level, sound field homogeneity, spectral properties, mode, and the duration or other temporal properties of the exposure. The variables employed to measure TS, include physiological and behavioral estimation methods. Observations of individual differences in the susceptibility to NITS is another important consideration.

4.3.1 Sound Pressure Level

The degree of TTS is most dependent on the intensity of the exposure. It is important to understand the logarithmic nature of sound level measurement to appreciate the link between intensity and the potential for detrimental impact. The sound level is expressed as the decibel SPL (Davis 2006), where SPL is referenced to a standard pressure of 20 μPa (dB re 20 μPa). A 20-dB difference between two sounds represents a 10-fold difference in pressure. This means that an exposure at 120 dB SPL exerts a pressure on the tympanic membrane (20 Pa), which is 100 times greater than an 80-dB exposure (0.2 Pa). This may explain, in part, why a relatively small numerical increase in decibel SPL, at high SPLs, could shift a TTS to a PTS exhibiting substantial inner ear damage.

Typically, in experiments investigating NITS, the exposure SPL is measured in an open sound field, where an average decibel level is defined for a specific spatial area in the vicinity of an animal's head. It might also be measured within a closed field, as, for example, in the ear canal directly in front of the tympanic membrane. The difference between the exposure SPL for animals restrained within the sound field, compared with those that are free to move about, can be large and represents a serious source of variability because an accurate estimate of the exposure SPL is essential in understanding the degree of TS.

4.3.2 Exposure Spectral Properties

The spectrum of the sound exposure is also important. The spectrum may be a single frequency (a pure tone) or a complex sound containing an array of frequencies occurring simultaneously, often referred to as a noise. Noises may be narrow- or wideband and presented as either continuous or intermittent sounds. The bandwidth of a noise is typically referenced as being centered on a specified frequency. In contrast to the situation in the laboratory, where sound characteristics can be controlled, the specification of sounds in the natural environment is challenging for two reasons. First, natural and anthropogenic sounds are rarely as simple as pure tones. Second, sounds encountered in the natural environment are rarely constant in intensity level, spectrum, or duration and may be further influenced by ambient sound levels capable of contributing to overall noise levels (Dooling et al. 2009). An excellent tutorial for measuring sound properties can be found in Davis (2006).

4.3.3 Modes of Exposure

In addition to steady-state sounds such as pure tones and bands of noise, there are other classes of sounds with transient properties, referred to as impulse sounds. Generally, impulse sounds derive from a rapid gas expansion, taking the form of a blast wave, or via the occurrence of one object striking another, creating an impact sound. Each of these impulsive sounds has unique waveforms and spectral properties. They are characterized by short durations, on the order of a dozen to hundreds of milliseconds, and can achieve extremely high pressures levels. They might occur as a single event or as a train of impulses occurring either randomly or periodically. The various acoustic variables associated with impulsive sounds have been described elsewhere (Akay 1978; Henderson and Hamernik 1986).

4.3.4 Exposure Temporal Schedule

The schedule of the exposure in a NITS experiment may be continuous, having durations for as long as hours, days, weeks, or months. Some of these conditions were approximated in the earlier literature that evaluated NITS and hearing loss in factory workers assigned to fixed work stations in acoustic environments that were unchanged for decades (Taylor et al. 1965). Other exposures may be intermittent, having a schedule of time on and time off, cycling for varying durations (see Saunders et al. 1977). Continuous and intermittent schedules can be applied to steady-state or impulsive modes of noises.

Aside from the duration of the exposure, another critical issue related to sound duration is the constancy of the SPL throughout the exposure. These variability problems are likely to be common for animals in natural environments, particularly those prone to move about freely within a sound field. The problem is exacerbated if the SPL has poor spatial homogeneity for fixed distances from the source (e.g., greater than ±5 dB). Another factor contributing to variability in TTS measurements is the time window used for determining recovery. A definition of recovery based on a fixed postexposure interval, as opposed to recovery based on a stable threshold level, may give quite different results.

4.3.5 The Threshold Estimation Method

The predictive value of laboratory NITS data depends on the validity of the techniques used to measure hearing and TS. Threshold estimates can be obtained by either physiological or behavioral methods. Physiologically, techniques frequently employed might be threshold estimates of the distortion product otoacoustic emission (DPOAE), a field-evoked response from a peripheral or central auditory brain center, or the sound-driven patterns of individual neuron discharges. Similarly, a pinna twitch, an eye blink, a change in heart rate or respiration, or some other reflexive (autonomic) response elicited by sound could also serve to estimate the threshold. The threshold estimate would be defined by the stimulus level that elicited a criterion level of DPOAE response, a just-detectable evoked response, or a just-detectable autonomic response. An advantage of these physiological threshold estimates is that the response is derived from a restricted part of the auditory pathway. For example, DPOAE activity emerges from the cochlea, evoked activity arises from a particular auditory brain region, and a reflex response, such as a pinna twitch, may be controlled by brainstem mechanisms. A disadvantage is that the response criteria used to define threshold are defined by the investigator as opposed to a behavioral method that involves the whole organism and actually defines what is meant by hearing.

Behaviorally measured thresholds defined as an animal's conscious response to sound involve the full cognitive/perceptual power of the organism. The response itself is usually a learned behavior (e.g., a shuttle behavior, a bar press, a pecking behavior) that is instrumental in producing a reward or avoiding a punishment (Carder and Miller 1972; Dooling and Saunders 1975). The initiation of the instrumental response (sometimes referred to as an operant response) is related to SPL through one of several psychometric procedures (e.g., method of constant stimuli, method of limits, tracking) that involves the systematic presentation of stimuli at different sound levels and, depending on the pattern of responses, results in a statistically defined threshold, which is typically that the SPL is correctly responded to 50% of the time.

Decades of animal research show a high correlation among results obtained from various threshold-estimating measures and among different laboratories. Thus, with careful control over experimental variables, either a physiological, a reflex, or an instrumental behavioral response can all yield a valid estimate of threshold and TS. Generally, thresholds obtained by physiological or reflexive measures in either awake or anesthetized animals are typically higher (i.e., less sensitive) than those obtained by instrumental behavioral procedures using rigorous psychophysical techniques.

4.3.6 Individual Differences in Susceptibility to Overstimulation

Individual differences in susceptibility to loud sound are a vexing problem and occur even if all experimental variables are controlled to the fullest extent (Henderson et al. 1993). Why individuals within a species, including humans, show variations in hearing loss to identical exposures remain unknown, although genetic factors are suspected (Davis et al. 2001). For this reason, most NITS studies use as large a sample of subjects as practical to lessen the consequences of an outlier. Such an approach is difficult when trying to estimate NITS in animals in their natural environment where it may be a problem to sample numerous individuals (see Finneran 2015).

4.4 Characteristics of Temporary Threshold Shift and Permanent Threshold Shift with Steady-State Noise Exposure

The general principles of NITS to steady-state sound exposures are well illustrated with three extensively studied species: the chinchilla (*Chinchilla lanigera*), the budgerigar (*Melopsittacus undulatus*), and the chicken chick (*Gallus domesticus*).

Chinchillas have been employed in studies of NITS because the hearing range and sensitivity as well as the anatomy of the auditory receptor show many similarities with the human auditory system. Moreover, chinchillas can readily be conditioned to respond to sound in a instrumental behavioral testing paradigm.

Birds also make excellent subjects for both behavioral and physiological testing because they are easily trained, their auditory peripheral apparatus is readily accessible for study, and they can regenerate hair cells (after auditory trauma) in which the physiology operates on the same principle as mammalian hair cells (Köppl 2011).

In the studies cited here with the chinchilla and budgerigar, a learned instrumental response was used in combination with a psychophysical procedure (the method of limits) to obtain pure-tone thresholds (Saunders and Dooling 1974; Saunders and Rosowski 1979). The physiological threshold estimates in the chick, however, were determined from just-detectable tone-burst evoked activity recorded from electrodes placed in brainstem auditory nuclei. These thresholds methods were used to describe the growth and recovery of NITS after exposure to an intense noise or pure tone.

4.4.1 *Growth and Recovery of Threshold Shift in the Chinchilla*

Figure 4.1 illustrates threshold shifts averaged over four chinchillas sampled at various intervals after exposure onset (Mills 1973). Chinchillas were exposed to an octave-band noise centered at 4.0 kHz. The frequency of maximum hearing loss in mammals was a half octave above the center frequency of the noise band; thus thresholds were measured at 5.7 kHz. The offset between the center frequency of the exposure and the frequency of maximum TS has been attributed to nonlinear mechanical behavior of the organ of Corti and hair cell stereocilia during intense stimulation (Ramamoorthy and Nuttall 2012).

The growth of threshold shift curves in Fig. 4.1 was obtained by turning off the noise at selected exposure intervals and then obtaining a threshold estimate. There is no evidence that removal from the noise for short periods of a few minutes influences the systematic growth of TS with exposure time.

The parameters in Fig. 4.1 are four different sound exposure levels: 80, 86, 92, and 98 dB SPL. The earliest indication of hearing loss occurred after about 15 min. With continued exposure, the level of TS grew. This growth then slowed and remained relatively constant over the next eight days. The flattening of the growth curve, followed by steady levels of TS, has been referred to as asymptotic TS (ATS).

The observation of ATS was remarkable for several reasons. First, it remained constant. Indeed, other studies have demonstrated that ATS remains largely unchanged with exposure durations lasting months (Mills 1973, 1976). Second, the level of ATS appeared to be dependent on exposure intensity, not on duration. As Fig. 4.1 illustrates, the ATS level increased as the exposure SPL increased.

The biological basis of ATS remains unknown, but it may represent a situation where the damaging effects of overstimulation come into balance with inherent

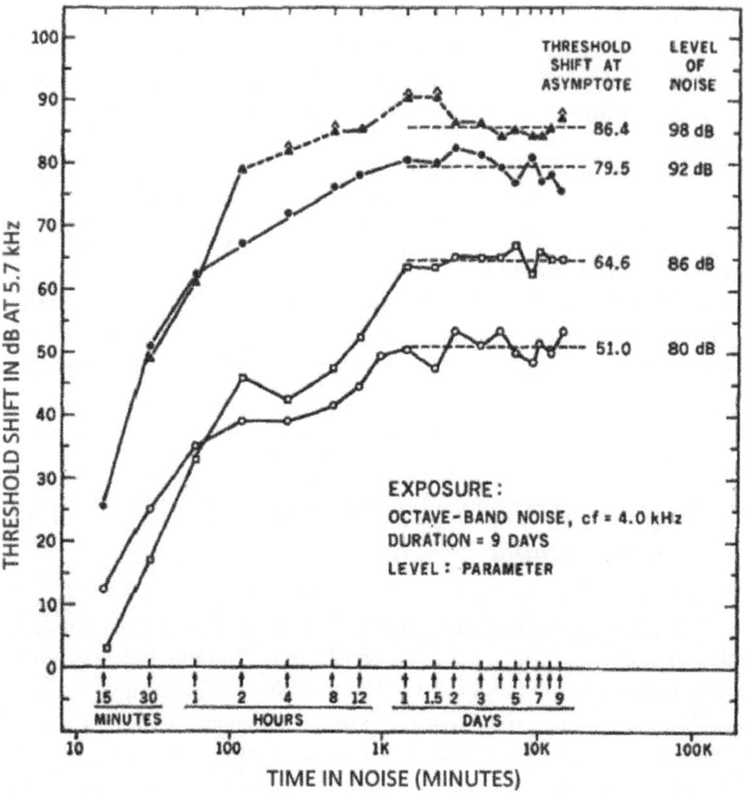

Fig. 4.1 The growth of noise-induced threshold shift (NITS) at 5.7 kHz in the chinchilla is shown for four levels of an octave-band noise centered at 4.0 kHz. Exposure duration was continuous and each data point is the average of four animals. Data are from Mills (1973)

processes of repair. This state of homeostasis is universal and has been observed over a wide range of continuous sound exposure parameters in chinchilla and other species, including human listeners, and for both steady-state and impulse exposures (Clark 1991; Salvi et al. 2011).

The postexposure recovery of TS is illustrated in Fig. 4.2 and is characterized by relatively little change during the first 40 minutes of quiet. Beyond 40 minutes, the TS diminished at a generally linear rate with the logarithmic passage of postexposure time. Preexposure thresholds were achieved within 3-4 days of recovery in the group exposed at 80 dB SPL. However, animals experiencing more intense exposures (and greater threshold shifts) did not show a recovery plateau until 15 days postexposure. Most importantly, the recovery for exposures between 86 and 98 dB all showed a PTS between 11 and 37 dB.

The NITS in chinchilla caused by an intermittent sound exposure is shown in Fig. 4.3. The exposure schedule cycled the noise for 6 hours on and then for 18 hours off. The sequence was repeated over 9 days, and this schedule mimicked a pattern of noise exposure that might be found in a workplace over the course of 24 hours. It

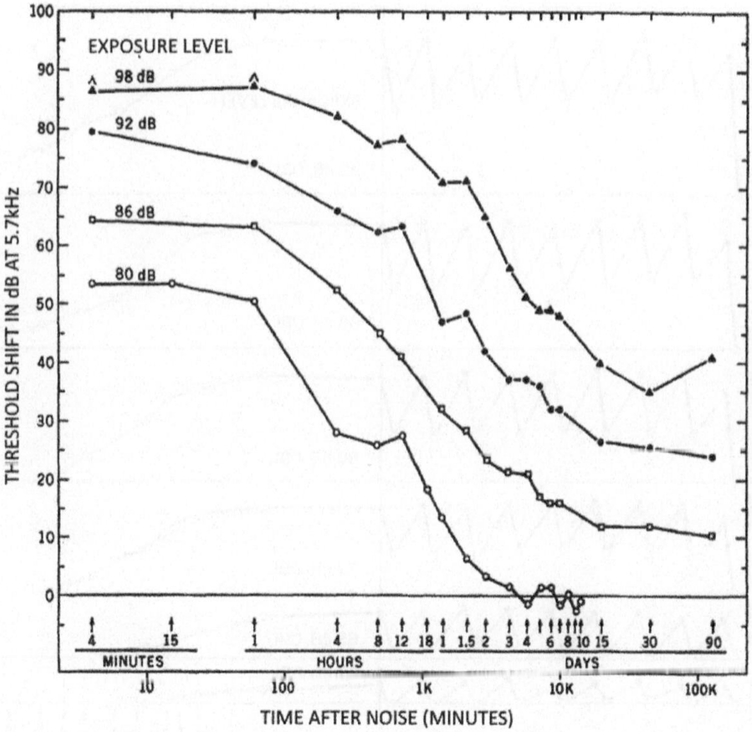

Fig. 4.2 The recovery of NITS for the same exposure and test conditions as shown in Fig. 4.1. After 1 hour of recovery, the reduction in NITS progresses in an approximate linear manner. A permanent threshold shift (PTS) occurs for exposures between 86 and 94 dB sound pressure level (SPL). Each data point is the average of four animals. Data are from Mills (1973)

is also a pattern that could be experienced by animals living near a source of anthropogenic noise, such as found at a construction site (Dooling et al. 2009). The noise band and threshold test frequency were the same as in Fig. 4.1, and the parameter evaluated in Fig. 4.3 is the exposure SPL, which occurred in 6 steps between 57 dB and 92 dB (Saunders et al. 1977).

Figure 4.3, *left,* revealed a series of "switchbacks" because the 6-hour exposure produced a hearing loss, with the subsequent 18 hours in quiet resulting in a degree of recovery. The sawtooth appearance cycled over nine successive days. A level of ATS was reached within the first two cycles and was proportional to the exposure SPL.

Figure 4.3, *right,* shows the recovery of TS. The shifts in threshold remained relatively constant over the first hour or so, after which they declined linearly in the logarithm of time. Preexposure thresholds were reached within 13 hours to 5 days for exposures between 57 and 80 dB SPL. Recovery time lengthened to 10 days or more at exposure levels of 86 and 92 dB SPL. At these two highest exposures, PTSs of 8 and 12 dB, respectively, occurred.

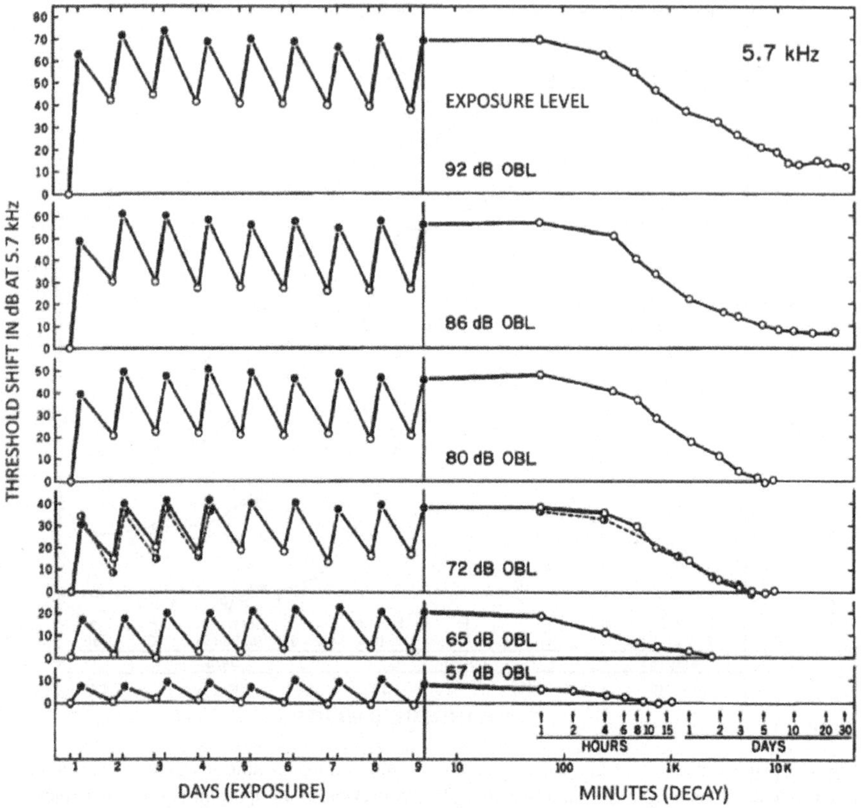

Fig. 4.3 The growth and recovery of NITS are shown for six noise levels using an intermittent exposure (6 hours on and 18 hours off) cycling over 9 days. Testing occurred when the noise was turned off and just before it was turned on for each cycle. Each data point is the average of four chinchillas. Data are from Saunders et al. (1977)

Figure 4.4 shows ATS level plotted as a function of exposure SPL. Two sets of data are depicted, and each shows a linear increase in the relationship between exposure SPL and the level of ATS in decibels. Figure 4.4, *open circles*, represents the continuous exposure results from Fig. 4.1, with additional ATS data points reported by Mills and Talo (1972).

Figure 4.4, *solid circles*, represents the ATS levels for the intermittent exposure data plotted in Fig. 4.3. When the two functions are extrapolated to the 0-dB TS axis, it reveals that continuous and intermittent exposures are separated by 5 dB. This would be predicted by an intermittency schedule with a one-to-three ratio of sound on to sound off (Clark 1991). There are two ways of interpreting Fig. 4.4. First, continuous noise exposures always cause more ATS than intermittent exposures at the same exposure level. Second, an intermittent exposure will always require a higher SPL to achieve the same level of ATS as a continuous noise. The linear fit to

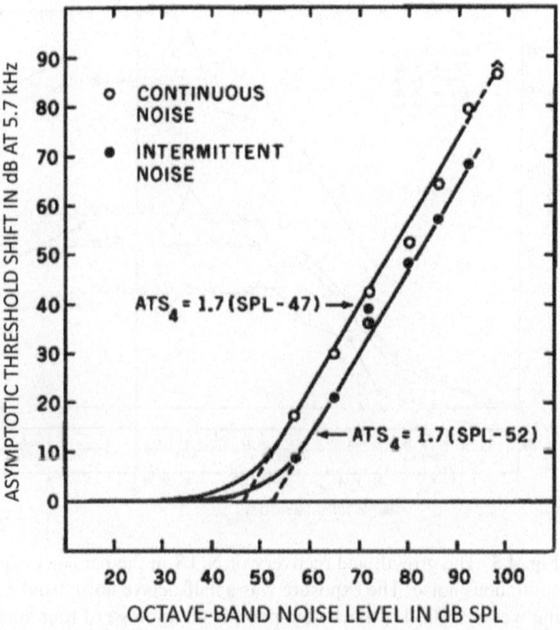

Fig. 4.4 The functions depict the growth of NITS with increasing levels of continuous or intermittent noise exposures. Each dataset was fit by a regression line and both show identical slopes. It requires a 5-dB more intense noise for the intermittent exposure to achieve the same level of NITS as seen with continuous exposures. ATS_4, average asymptotic threshold shift of four animals. Figure modified from Saunders et al. (1977)

both those datasets (Fig. 4.4) revealed a slope where a 1.7-dB increase in ATS occurred with every decibel increase in exposure level.

4.4.2 Growth and Recovery of Threshold Shift in Birds

The effects of intense sound exposure have been studied in the budgerigar, quail (*Coturnix coturnix*), canary (*Serinus canaria*), chaffinch (*Fringilla coelebs*), and chicken chicks (Ryals et al. 1999; Saunders 2010). The characteristics of threshold shift in these three avian species show important similarities and differences with those reported for mammals.

Figure 4.5 illustrates data in the budgerigar after exposure to a one-third octave band of noise centered at 2.0 kHz for 72 hours. The exposure SPLs were 76, 86, 96, and 106 dB (Saunders and Dooling 1974; Saunders and Dooling, unpublished observations). Operant behavior coupled to a method of limits was used to estimate thresholds, and all thresholds were measured at 2.0 kHz. This was the same center frequency of the exposure stimulus. The half octave upward shift between exposure frequency and maximum threshold shift frequency, as seen in mammals (see Sect. 4.4.1) does not occur in birds. The explanation for this is not completely clear but is likely related to the fact that nonlinear processes along the sensory epithelium of the avian basilar papilla are different from those found along the basilar membrane of mammals (Köppl 2011).

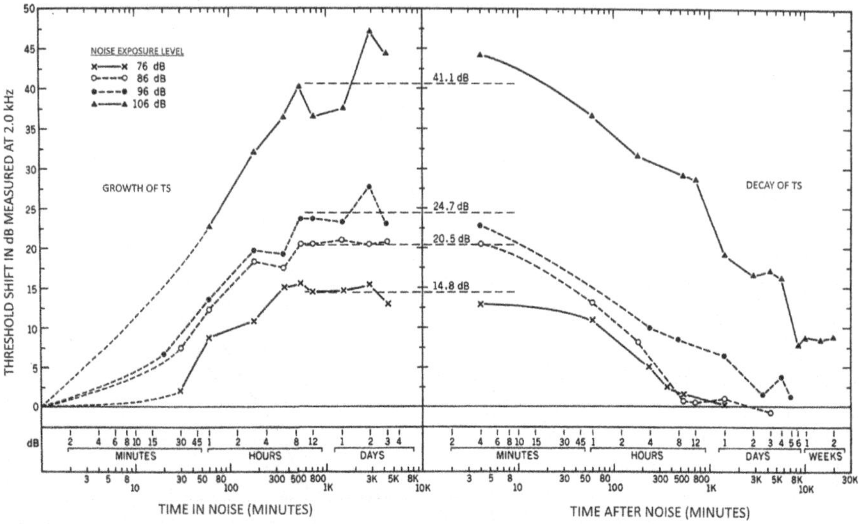

Fig. 4.5 The growth and recovery of NITS in the parakeet exposed to different levels of a 4-day continuous noise. The exposure was a half-octave noise band centered at 2.0 kHz. Threshold testing was at 2.86 kHz. Each data point is the average of four animals. Modified from Saunders and Dooling (1974)

The ATS levels were reached after only after 8-12 hours of overstimulation. The recovery functions are shown in Fig. 4.5, *right*. Complete recovery was realized about 8 hours postexposure for the 76, 86, and 96 dB SPL noises. A PTS of 17 dB occurred after the 106 dB SPL exposure. The relationship between ATS and exposure SPL was not as orderly as that seen in the chinchilla (see Fig. 4.4), perhaps because of repeated testing at successively higher sound levels in the same animals.

The role of hair cell regeneration in contributing to budgerigar (parakeet) NITS was not addressed in Saunders and Dooling (1974) because it was unknown at the time. There is ample evidence now that destroyed hair cells are fully replaced on the avian papilla within 4 weeks of recovery (Saunders 2010). Thus, it is possible, had Saunders and Dooling (1974) followed recovery for a longer period, there might have been less PTS than that shown in Fig. 4.5 for the 106-dB exposure.

Another example of overstimulation in a bird used a just-detectable level of cochlear nucleus evoked responses to estimate TSs in chicken chicks after an intense pure-tone exposure (Fig. 4.6). The evoked responses were recorded from indwelling brainstem electrodes in anesthetized animals (McFadden and Saunders 1989). Each data point in Fig. 4.6 represents a different group of 8 subjects in which the TSs were averaged over three test frequencies (0.7, 0.9, and 1.2 kHz). Figure 4.6 shows the results for exposure durations of either 48 or 200 hours of a continuous 0.9-kHz tone at 120 dB SPL (Saunders et al. 1993; Saunders 2010). The results from this

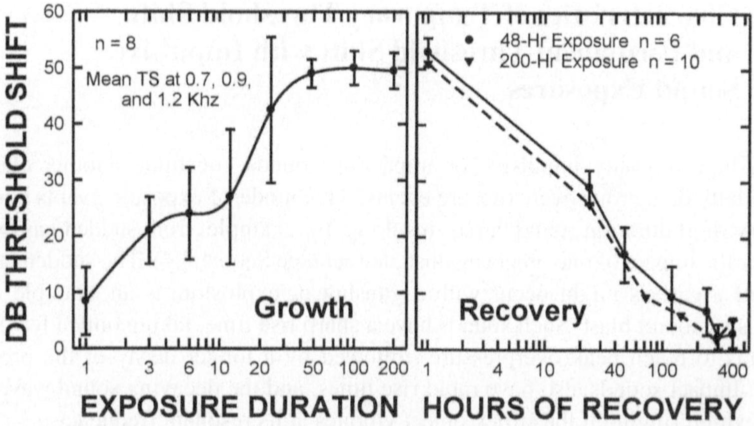

Fig. 4.6 The growth and recovery of NITS are shown for 4- to 7-day-old chicks exposed to a 120 dB SPL pure tone at 0.9 kHz for either 48 or 200 hours. Each data point is the average of eight different animals, and values are means ± SD. Threshold shifts were obtained from tone-burst evoked response activity in the avian brainstem for test frequencies at 0.7, 0.9, and 1.2 kHz. Data are from Saunders et al. (1993) and Saunders (2010)

labor-intensive study can be compared with the mammalian growth and recovery in TSs shown in Figs. 4.1 and 4.2.

An 8-dB TS occurred after an hour of exposure. The growth curve also had two inflections, the first occurring after 5 hours and the second after 40 hours. TSs after exposures of 48 or 200 hours grew to an ATS of 50 dB. The variance between subjects, revealed by the vertical bars (±1 SD), is large in the period preceding the asymptotic ceiling, but once ATS was achieved, variability among subjects decreased. Recovery from ATS in these two groups proceeded linearly with the logarithm of time, returning to preexposure levels nearly 300 hours postexposure. Despite a 152-hour difference in exposure time, the rates of recovery were essentially the same for both durations (Pugliano et al. 1993; Saunders et al. 1993). A family of curves evaluating growth and recovery for different exposure levels was not available for the chick given the substantial number of subjects required just to generate the data in Fig. 4.6.

The comparison between avian and mammalian TSs reveals both similarities and differences. Some of the differences might be attributed to species-specific features of the peripheral receptor organ. For example, the middle ears of birds are interconnected by an air passage, producing complex acoustic interactions between the ears. Mammalian middle ears, in contrast, are separate and isolated from each other. The avian middle ear has a single ossicle and the basilar papilla of the inner ear lacks the distinct rows of inner and outer hair cells found in mammals. As mentioned in Sect. 4.4.2, all birds have the capacity to regenerate hair cells lost to trauma, whereas mammals do not. Additional unique aspects of the avian auditory system have been discussed in detail elsewhere (Köppl 2011).

4.5 Characteristics of Temporary Threshold Shift and Permanent Threshold Shift with Impulsive Sound Exposures

Brief, high-intensity impulsive or impacting sounds constitute another class of potentially dangerous overexposure events. This mode of exposure events consists of very short duration sound bursts, resulting, for example, from sudden gas expansion or the impact of one object against another (see Sect. 4.3.4). The sudden expansion of a gas, as might occur with a gunshot or explosion, is an example of an impulsive sound blast. Such sounds have a sharp rise time, taking only a few milliseconds to reach peak overpressure, followed by a longer decay of the pressure wave. Impact sounds also have rapid rise times, and the decaying sound waveform may exhibit ringing if the struck object vibrates at its resonant frequency.

The characteristics of hearing loss from impulse or impact sounds show both similarities and differences from those described for continuous steady-state sound exposures. As a general point, they tend to produce more mechanical disruptions of inner ear structures than anything else (Henderson et al. 1991). Figure 4.7 shows the growth in TS at 0.5 and 8.0 kHz after exposure to a repetitive hammer impact on a steel plate. The strike produced a peak sound level of 120 dB SPL and was presented at the rate of 1 per second for 10 days. Instrumental behavioral thresholds were used to assess TS at 0.5 and 8.0 kHz.

Two features of the results in Fig. 4.7 are representative of impulse sound exposures. As with continuous sound, TSs at both test frequencies demonstrated ATS. In this example, the ATS was 56 dB at 0.5 kHz and 65 dB at 8.0 kHz. The second feature in Fig. 4.7 is the short exposure duration needed to reach ATS, on the order of 2 hours or less. This was quite different from the 24 hours needed to achieve ATS

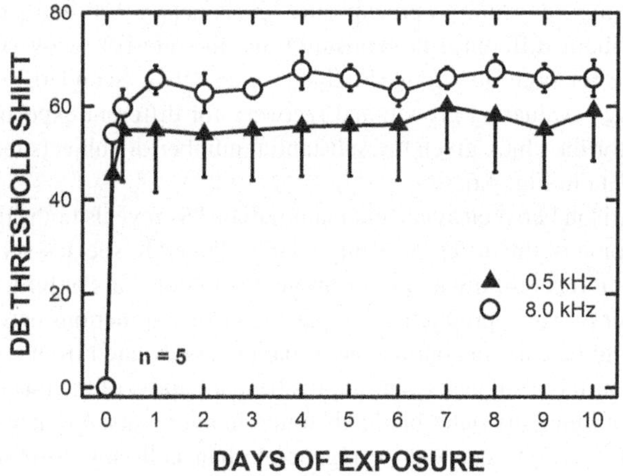

Fig. 4.7 The growth of NITS is shown for two frequencies during a reverberant impact noise exposure: one 120 dB SPL impact per second over 10 days. Note that ATS occurs after several hours of exposure. Values are means ± SD and are plotted in only one direction for the 0.5-kHz function to avoid confusion. Data are redrawn from Henderson and Hamernik (1986)

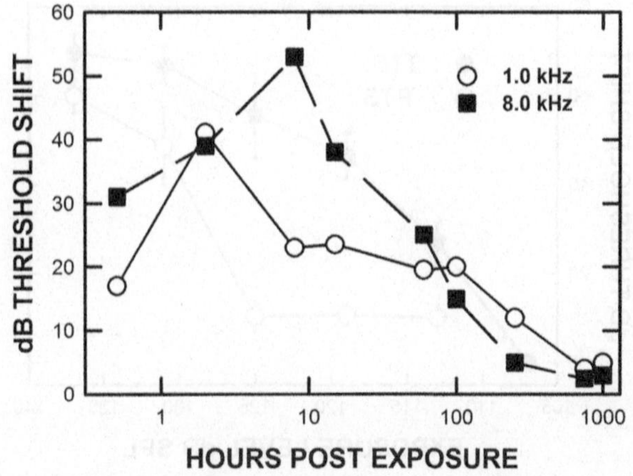

Fig. 4.8 The recovery from temporary threshold shift (TTS) produced by 50 1-ms impulse blasts at 155 dB SPL is shown for two frequencies. Note that the magnitude of TTS grew between 1 and 10 hours of recovery. Thereafter, the NITS followed a linear return to preexposure thresholds in the logarithm of time. Data are redrawn from Henderson and Hamernik (1986)

with the steady-state noise depicted in Fig. 4.1 and has been replicated in other studies with various impulse sound parameters. It is not entirely clear what accounts for this rapid growth in TS, but it may be related to the sharp rise time of the stimulus and the subsequent damage caused by high velocity displacements of the cochlear partition (Henderson and Hamernik 1986).

Figure 4.8 shows the recovery from impulse sounds at two frequencies after exposure conditions that are much the same as in Fig. 4.7 (Henderson and Hamernik 1986). Complete recovery occurred around 1,000 hours postexposure (about 42 days). The recovery time course, however, was nonmonotonic, and this was quite different from the monotonic recovery after the steady-state exposures in Fig. 4.2. The nonmontonicity in Fig. 4.8 appears in the increased postexposure TS between 1 and 10 hours of recovery. After 10 hours of recovery, the TS then declined linearly in the logarithm of time, much like that seen in Fig. 4.2.

An example of the relationship between impact-sound SPL and the degree of TS is shown in Fig. 4.9 (Henderson et al. 1991). The exposure was a simulated hammer striking a steel plate, with a rise time of 12 ms, a duration of 200 ms, and a repetition rate of 1 strike every 4 seconds. The exposure level increased in 6 steps between 107 dB and 137 dB SPL, and exposure duration varied at each level. By way of example, it was 80 days at 107 dB and 1.8 hours at 137 dB SPL. These duration differences were designed to produce exposures with equal energy at each SPL. If equal exposure energy were a critical aspect in determining the degree of TS, then the amount of TS would be the same at each level (Roberto et al. 1985).

TSs in the Henderson et al. (1991) study were obtained by estimating the SPL that produced a just-detectable tone-evoked response from an electrode placed in the chinchilla inferior colliculus. Two plots are shown in Fig. 4.9. The *solid circles* represent NITS values averaged between one-quarter hour and 240 hours (10 days) postexposure and were referred to as the TTS data by Henderson et al. (1991). These

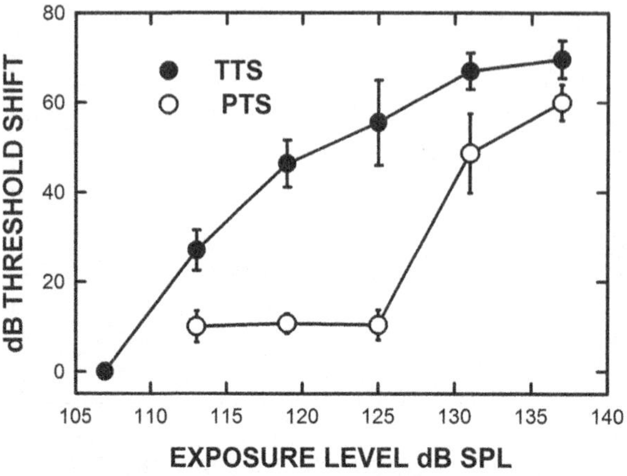

Fig. 4.9 The relationship between impact sound intensity and NITS is shown. Data have been averaged for shifts measured at 0.5, 2.0, and 8.0 kHz. The exposure was 5 days, and shifts levels were averaged across impact rates of 0.25, 1, and 4 per second. The two parameters are the level of TTS defined by NITS estimates over a period of time shortly after the impulse exposure was turned off. The PTSs were obtained from threshold measures ~30 days after the end of the exposure. Data are redrawn from Henderson et al. (1991)

results indicated that 80 days of exposure at 107 dB SPL produced no change in threshold. However, an exposure level of 137 dB for only 1.8 hours resulted in a TTS of 72 dB. The TTS from impact noise between 113 and 131 dB SPL grew by 38 dB and represented a 2.1-dB growth in NITS for every decibel increase in exposure level. This was close to the 1.7-dB slope reported in Fig. 4.4 for continuous or interrupted noise exposures.

The PTS results were determined after approximately 30 days of postexposure recovery. Figure 4.9 shows a 10-dB PTS for exposures between 113 dB and 125 dB SPL. Above 125 dB, the PTS rapidly increased, reaching a maximum of 60 dB for the 137 dB SPL exposure. Although the PTS values for exposures between 113 and 125 dB SPL were much the same, the remaining data in Fig. 4.9 show little support of an equal energy conclusion. The interested reader can explore the equal energy hypothesis in relation to NITS elsewhere (Roberto et al. 1985; Henderson and Hamernik 1986).

Figure 4.9 also reveals an abrupt increase in PTS when the exposure equaled or exceeded 125 dB SPL. This exposure level has led to the notion of a critical intensity where the mammalian organ of Corti suddenly becomes irreversibly damaged from acute mechanical injury to the inner ear (Henderson and Hamernik 1986).

4.6 Species Differences in Susceptibility to Noise-Induced Hearing Loss

Currently, there are laboratory data describing the consequences of loud sound exposure in primates (human and monkeys [*Macaca*]; Moody et al. 1976; Melnick 1990) as well as in other mammals such as dogs (Harvey and ter Haar 2016), cats

(Miller et al. 1963), rabbits (*Oryctolagus cuniculus*; Borg et al. 1995), chinchilla (Mills 1973), pigmented and nonpigmented guinea pigs (Conlee et al. 1986; Syka and Popelár 1980), various rat strains (Chen et al. 2014), gerbils (Ryan and Bone 1978), hamsters (Heffner and Koay 2005), and numerous mice strains (Ou et al. 2000). A recent review also considered the extensive literature on NITS in marine mammalian species (Finneran 2015). In birds, sound-induced hearing loss data are available for quail, budgerigar, canary, finch, and young chickens (Ryals et al. 1999; Saunders 2010).

This diversity of animals raises an interesting thought experiment. What would be the result if all these species were exposed to the same sound level as measured by a common criteria (e.g., SPL at the surface of the tympanic membrane) for an equal duration, where the spectrum of the exposure was set to the frequency region of best hearing sensitivity for each respective species? Would each show the same level of ATS (using ATS as a common criterion), or would there be substantial differences in this metric across species? The evidence already presented suggests that similar qualitative characteristics will emerge. Every species will demonstrate a TS growth function, a stable level of ATS after sufficient exposure duration, and a degree of postexposure TS recovery. However, the specific quantitative values associated with growth rate, ATS level, and recovery rate would most likely differ among species, indicating that not all species are equally susceptible to hearing loss from loud sound exposure.

4.7 Estimating Threshold Shift from Loud Sound Exposure for Animals in Their Natural Habitat

The data considered thus far were all collected from animals commonly used in laboratory studies. Moreover, the anatomy and physiology of the auditory systems of these species were well-known, and this provided a firm foundation for understanding the biological processes underlying normal hearing and hearing loss after intense sound exposure.

Federal or state wildlife agencies often identify loud sound sources in the natural environment and have concerns as to whether it has an effect on the hearing of animals in the immediate vicinity of the sound. This is an important question whose answer has an impact on wildlife management strategies and policies as well as environmental impact statements. At first glance, it might seem this is not really an issue because wouldn't it be logical to assume that "free-ranging" animals would simply remove themselves from any sound perceived as uncomfortably loud? That logic is challenged by evidence that wild animals may remain in intense sound fields for various reasons, including approaching loud sounds out of sheer curiosity (Busnel 1978). Moreover, humans are free-ranging creatures that repeatedly subject themselves to intense continuous or impulsive environmental noises sufficient to cause TSs. These sounds come from sources such as motorcycles, heavy equipment, rock bands, movies, music players, guns, explosives, and noisy sports venues, to name a few. Indeed, in the modern industrialized world, loud and potentially dangerous workplace and recreational loud sounds abound, and humans rarely flee these when they are perceived as uncomfortably loud.

It is difficult, if not impossible, to obtain direct measures of NITS from wild animals in their natural habitat of similar quality to those described in Sects. 4.4 and 4.5. Absent such direct TS data, it is hard to reach a definitive conclusion that animals living in the wild are at risk for hearing loss from loud sound exposure. Nevertheless, an excellent example of addressing these challenges comes from studies on marine mammals (Finneran 2015). This review highlights the unique difficulties of studying this class of animals and demonstrates how stimuli, test conditions, and threshold estimates can be ingeniously configured to study the consequences of overstimulation. Although the complexities of testing marine mammals can be overcome, the issue of sufficient sample size remains a problem. Despite small sample sizes, it is remarkable that many of the parameters of TS for the chinchilla have been replicated in marine mammals. This observation is important because it implies that the basic consequences of loud sound exposure are consistent across mammalian species, and it is this observation that offers the possibility of understanding the risk to hearing from acoustic overexposure for animals living in their natural environment. What follows is an exploration of how it might be possible to model the susceptibility to hearing loss from loud sounds in native species.

4.7.1 Sound Field Acoustics

Acoustic properties such as SPL and spectrum can be measured accurately, but the distribution of SPL within a sound field may not be uniform because of interactions between the propagating sound wave and physical objects within the field (see Larsen and Radford, Chapter 5). Depending on frequency and object size, sound diffracts or reflects in complex patterns and, along with sound shadows, can create localized regions of higher or lower SPL. These localized perturbations occur in addition to the overall reduction in sound SPL as it propagates away from its source.

It is worthwhile to consider how distance alone from a sound source affects SPL. For example, assume a sound is 120 dB SPL at 0.5 meters from its source. If it propagates outward as a circular plane wave with no barriers in its path, then the inverse distance law can be used to calculate the SPL at any point from the source. This law is independent of frequency and whether the sound is continuous or impulsive (see Tontechnik-Recher-sengplelaudio 2016). Using that law, the distance from the source where the SPL will be 86 dB occurs at 25 meters. An exposure level of 86 dB was chosen from the observations in Fig. 4.2. Assuming our wild animal had the hearing capacity and an auditory apparatus similar to those of a chinchilla, if it remained at the 25-meter distance for 24 hours or more, it should suffer a 65-dB ATS (see Fig. 4.1). After recovery, it would exhibit a small PTS of 10 dB (see Fig. 4.2). Of course, if it moved closer to the source than 25 meters (for 24 hours), the hearing loss would be greater, and conversely if it were further than 25 meters, there would be less hearing loss.

4.7.2 *Constancy of Exposure Sound Pressure Level and Animal Movements*

So, how realistic is the prediction in Sect. 4.7.1? The answer depends on the accuracy in specifying the SPL at the animal's ears and how long that exposure level was maintained. Sound fields are rarely if ever "simple." Complex sound fields, where the propagating waves encounter objects (bushes, trees, rocky outcroppings, buildings, or any other imaginable object) will result in a nonhomogeneous sound field. Furthermore, the interaction between exposure frequencies and sound field objects further complicate the patterns of sound reflections, refractions, and shadows. Consequently, specification of SPL at the ear becomes difficult, devolving into a statistical estimate of SPL measured at different locations. All these aspects need to be considered in creating a comprehensible distribution of sound pressure at different distances from the source.

Remember, in contrast to humans or animals studied in the laboratory, it is unlikely that any free-roaming animal would remain in a loud, potentially painful, acoustic field for hours on end. Still, there are anecdotal observations and scattered reports over the years of wild animals actually seeking out loud sounds and noisy environments, seemingly undisturbed by acoustic overexposure. Examples would include porpoises swimming in the bow wave of high-speed motor boats and birds or grazing animals adjacent to airport runways, completely ignoring aircraft noises and the proximity of large, fast-moving, aircraft. As mentioned in Sect. 4.7, some animals, perhaps through curiosity, even seek out sources of intense sound (Busnel 1978).

The accuracy of specifying exposure SPL at the ear is critical in establishing a relationship between exposure conditions and threshold shifts. Thus, the ability of an animal to move within or in and out of a sound field adds substantially to the problem of determining hearing risk in the natural habitat.

4.7.3 *The Inherent Susceptibility to Noise-Induced Threshold Shift by Animals in Their Natural Habitat*

In Sect. 4.7.1, there was a description of how exposure level changed with respect to its proximity to the sound source, and an example used an 86 dB SPL exposure at the animal's ear. The extent to which hearing loss from an 86 dB SPL exposure in the chinchilla can be generalized to other species is a question that depends on species susceptibility to intense sound. There are very few studies comparing the susceptibility of different species to NITS; however, those few that do exist (see, e.g., Ryals et al. 1999; Duan et al. 2008) concluded that near-identical exposure conditions resulted in different levels of TTS and PTS. It would seem that species are not equally susceptible to the effects of loud sound. This should not be surprising given the structural and functional aspects of the of the auditory receptor organ that has unique aspects for every species within and among animal classes.

Nevertheless, what follows is an attempt to compare studies where the exposure and testing conditions were sufficiently similar (e.g., exposure was a band of noise of sufficient duration to achieve ATS, behavioral TSs were measured, recovery was traced to determine PTS levels, and where the extent of inner or outer hair cell loss was determined after long recovery durations) to see if various species could be ranked on an ordinal scale according to their susceptibility to loud sound. Although, admittedly, this is a crude attempt and is based on rather subjective considerations of behavioral threshold shift at ATS, the magnitude of PTS, and/or the degree of hair cell damage and loss. The ranking of susceptibility to noise exposure from most to least might look something like this: chinchilla > guinea pig > rabbit > cat > primate > hamster > rat > mouse (see Sect. 4.6 for references to TS data in different species). Also interesting in this regard, albino guinea pigs are more susceptible to noise exposure than pigmented guinea pigs (Conlee et al. 1986). In birds, acoustic overexposure in the quail and budgerigar showed greater TSs and basilar papilla pathology than that in canaries and finches (Ryals et al. 1999).

The susceptibility of different species to noise exposure can, most likely, be attributed to the specific efficiency of outer, middle, and inner ear sound transmission. Peripheral ear impedance and cochlear input impedance is determined by the design of the middle and inner ears. Moreover, hair cell sensory hair bundle morphology also varies among species in their number, thickness, and height per hair cell (Saunders and Dear 1983). Similarly, the longitudinal stiffness of the basilar and tectorial membranes differs among species. These aspects, as well as others, combine in complex ways to contribute to the degree a species is susceptible to NITS (see Saunders et al. 1985b).

A crucial role may be played by the middle ear apparatus in this consideration. Two organizational principles in the configuration of the mammalian middle ear apparatus have been identified. Fleisher (1978) described a microtype middle ear in which a portion of the malleus fuses with the bony wall of the bulla. This bony fusion changes middle ear impedance and hence the efficiency of sound transmission, favoring more high-frequency transmission. A microtype middle ear is found in the mouse, rat, gerbil and hamster, for example.

The second type, which Fleisher (1978) called the freely mobile ossicular design, has the malleus suspended by ligaments within the bulla cavity. The freely mobile middle ear also creates unique impedance properties in sound conduction, but in this case, it favors lower frequency sound transmission. The freely mobile middle ear design is found in human, other primates, cat, chinchilla, and guinea pig. The distinction between these two types of middle ear morphology has recently been further refined (Mason 2013).

With regard to birds, the ossicular system is constructed of a single bony element called a columella. The columella directly connects the tympanic membrane to the oval window without any form of ligamental suspension within the middle ear cavity (see Saunders 1985). The unique mechanism of middle ear cavity air ventilation, where birds have an interconnecting air pathway between the two cavities, and the fact that they can control their Eustachian tube (Ryals et al. 1999) are thought to be aspects that contribute to the lesser susceptibly to loud sound exposure in birds compared with mammals.

Several paragraphs above, a progression of susceptibility was noted among different laboratory species, and this series parses out along the lines of the middle ear type. More susceptible ears seem to be built of the freely mobile middle ear design, whereas less susceptible ears are of the microtype design. An important consideration is that the microtype middle ear favors higher frequency sound transmission, whereas the freely mobile design favors lower frequency sound transmission. In this regard, the frequency region of best hearing sensitivity correlates among mammalian species with middle ear design.

From these observations, a rather audacious suggestion emerges. If the middle ear design of a wild species is known and if the acoustic/temporal properties of a loud sound source in the natural environment are well characterized, it may be reasonable to predict the magnitude of TS by comparing the level of NITS observed in a laboratory species with the same middle ear design after exposure to a similar loud sound. Using these relationships and based on the best available science, it should lead to a reasonable prediction about whether a particular level and duration of exposure in the natural environment is capable of causing a TTS or a PTS with accompanying damage to the auditory system.

Such theoretical estimates should be treated with caution because there are uncertainties in such predictions. Nevertheless, it provides a starting point for a discussion on whether noise levels should be a concern and whether some sort of remediation plan is important. It is essential to recognize that such predictions will only be as good as the acoustic description of the exposure (its level, spectrum, and duration) and the anatomy of the middle ear in the wild species being considered.

4.8 Summary and Conclusions

The data reviewed here come from well-controlled laboratory studies. They served as examples that might be used to predict hearing loss from loud sound exposure in animals in their natural environment. It is worth pointing out that these early animal studies were undertaken to provide a comprehensive database from which to understand and make predictions on the effects of noise exposure on human hearing. Indeed, current Occupational Safety & Health Administration (OSHA) standards regarding noise in the workplace are based, in part, on these earlier animal data. Of course, with humans, it is possible to measure hearing after noise exposures in their daily lives, their natural habitat as it were, and thus confirm the value of extrapolating from animal experiments in the laboratory. Thus, the issue of making predictions from laboratory animal research on NITS to other species is not unprecedented.

The general principles across animal studies are clear. For steady-state or intermittent noise exposures, ATS is typically reached after exposure of 18-24 hours regardless of species (Clark 1991). The amount of TTS and the occurrence of a PTS depend on the level, frequency, and duration of the exposure. Birds may be somewhat different, particularly in regard to PTS, because of their capacity for hair cell regeneration. The growth of ATS with increasing exposure SPL is generally linear over a wide range of exposure SPLs, with ATS growing slightly faster than the increasing noise level. Periodic interruptions to an exposure causes less hearing loss than the same exposure that is continuous (see Fig. 4.4). This reduction appears to be proportional to the ratio of exposure "on" and "off" durations.

There are some differences in the effects of continuous steady-state and impulsive sound exposures. Intense impulse sounds result in the growth of threshold shift that is more rapid to ATS levels than that seen with continuous noise exposures. As with continuous exposures, ATS from impulse sounds remains constant during long exposure durations (see Fig. 4.7). However, the recovery from hearing loss after impulse sound exposure appears to be nonmonotonic, with NITS increasing as much as 20 dB during the first 1-10 hours postexposure. After this "bump," a linear recovery progresses in the logarithm of time (Fig. 4.8). Linear recovery in log time is also a characteristic of continuous noise exposures (see Fig. 4.5). Compared with continuous noise exposure, the relationship between impulse noise SPL and degree of NITS has different characteristics during the initial postexposure interval or after weeks of recovery (see Fig. 4.9). Also, the growth of NITS relative to exposure level early in the recovery period is slightly steeper than that seen with continuous noise exposures (compare the 2.2-dB growth in NITS with every decibel increase in exposure level in Fig. 4.9 with the 1.7-dB growth during steady-state continuous noise in Fig. 4.4). These and other differences between continuous and impulse noise have been summarized earlier (Henderson and Hamernik 1986).

Furthermore, it was proposed that existing laboratory data on acoustic overexposure may be used to make predictions of hearing loss following a loud sound exposure in the natural environment of a wild animal. The predictions require an understanding of the acoustic properties of the exposure, the acoustic variation in the sound field, the amount of time the animal remains exposed to the loud sound, the exposure spectral properties, and an appreciation of the species middle ear conductive apparatus. In the future, scientists, regulators, and environmentalists should draw on these basic principles in addressing the susceptibility to NITS for animals in their natural environment.

Compliance with Ethics Requirements

James C. Saunders declares that he has no conflict of interest.

Robert J. Dooling declares that he has no conflict of interest.

References

Akay, A. (1978). A review of impact noise. *The Journal of the Acoustical Society of America*, 64, 977–987.

Borg, E., Canlon, B., & Engström, B. (1995). Noise-induced hearing loss, literature review and experiments in rabbits. Morphology and electrophysiological features, exposure parameters, and temporal factors and interactions. *Scandinavian Audiology Supplementum,* 40, 1–147.

Busnel, R.-G. (1978). Introduction. In J. L. Fletcher & R.-G. Busnel (Eds.), *The Effects of Noise on Wildlife* (pp. 7–22). New York: Academic Press.

Carder, H. M., & Miller, J. D. (1972). Temporary threshold shifts from prolonged exposure to noise. *Journal of Speech, Language, and Hearing Research*, 15, 603–623.
Chen, G.-D. C., Decker, B., Prakash, V., Muthaiah, K., Sheppard, A., & Salvi, R. (2014). Prolonged noise exposure-induced auditory threshold shifts in rats. *Hearing Research*, 317, 1–8.
Clark, W. W. (1991). Recent studies of temporary threshold shift (TTS) and permanent threshold shift (PTS) in animals. *The Journal of the Acoustical Society of America*, 90, 155–163.
Conlee, J. W., Abdul-Baqi, K. J., McCandless, G. A., & Creel, D. J. (1986). Differential susceptibility to noise-induced permanent threshold shift between albino and pigmented guinea pigs. *Hearing Research*, 23, 81–91.
Davis, R. R. (2006). Acoustic measurement: A tutorial for molecular biologists. *Brain Research*, 1091, 32–39.
Davis, R. R. Newlander, J. K., Ling, X. B., Cortopassi, G. A., Kreig, E. F., & Erway, L. C. (2001). Genetic basis for susceptibility to noise-induced hearing loss in mice. *Hearing Research*, 155, 82–90.
Dooling, R. J, & Saunders, J. C. (1975). Hearing in the parakeet (*Melopsittacus undulatus*): Absolute thresholds, critical ratios, frequency difference limens and vocalizations. *Journal of Comparative and Physiological Psychology*, 88, 1–20.
Dooling, R. J., West, E. W., & Leek, M. R. (2009). Conceptual and computational models of the effects of anthropogenic noise on birds. Paper presented at the 5th International Conference on Bioacoustics 2009, Holywell Park, UK, March 31 to April 2, 2009. *Proceedings of the Institute of Acoustics*, 31, 99–106.
Duan, M., Laurell, G., Qiu, J., & Borg, E. (2008). Susceptibility to impulse noise trauma in different species: Guinea pig, rat, mouse. *Acta Oto-Laryngologica*, 128, 277–283.
Finneran, J. J. (2015). Noise-induced hearing loss in marine mammals: A review of temporary threshold shift studies from 1996–2015. *The Journal of the Acoustical Society of America*, 138, 1702–1726.
Fleisher, G. (1978). Evolutionary principles of the mammalian middle ear. *Advances in Anatomy, Embryology and Cell Biology*, 55, 1–70.
Harvey, R. G., & ter Haar, G. (2016). Noise-induced hearing loss. In R. G. Harvey & G. ter Haar (Eds.), *Ear, Nose and Throat Diseases of the Dog and Cat* (pp. 214–215). Boca Raton, FL: CRC Press.
Heffner, H. E., & Koay, G. (2005). Tinnitus and hearing loss in hamsters (*Mesocricetus auratus)* exposed to loud sound. *Behavioral Neuroscience*, 119, 734–742.
Henderson, D., & Hamernik, R. P. (1986). Impulse noise: Critical review. *The Journal of the Acoustical Society of America*, 80, 569–584.
Henderson, D., Subramaniam, M., & Gratton, M. A. (1991). Impact noise: The importance of level, duration, and repetition rate. *The Journal of the Acoustical Society of America*, 89, 1350–1357.
Henderson, D., Subramaniam, M., & Boettcher, F. A. (1993). Individual susceptibility to noise-induced hearing loss: An old topic revisited. *Ear and Hearing*, 14, 152–168.
Hood, J. D. (1956). Fatigue and adaptation of hearing. *British Medical Bulletin*, 12, 125–130.
Köppl, C. (2011). Birds - same thing but different? Convergent evolution in the avian and mammalian auditory systems provides informative comparative models. *Hearing Research*, 273, 65–71.
Kujawa, S. G., & Liberman, M. C. (2009). Adding injury to insult: Cochlear nerve degeneration after "temporary" noise-induced hearing loss. *The Journal of Neuroscience*, 29, 14,077–14,085.
Mason, M. J. (2013). Of mice, moles and guinea pigs: Functional morphology of the middle ear in living mammals. *Hearing Research*, 301, 4–18.
McFadden, E. A., & Saunders, J. C. (1989). Recovery of auditory function following intense sound exposure in the neonatal chick. *Hearing Research*, 41, 205–216.
Melnick, W. (1990). Human temporary threshold shift (TTS) and damage risk. *The Journal of the Acoustical Society of America*, 90, 147–154.
Miller, J. D., Watson, C. S., & Covell, W. P. (1963). Deafening effects of noise on the cat. *Acta Oto-Laryngology, Supplement*, 176, 1–91.

Mills, J. H. (1973). Temporary and permanent threshold shifts produced by nine-day exposures to noise. *Journal of Speech and Hearing Research*, 16, 426–438.
Mills, J. H. (1976). Threshold shifts produced by a 90-day exposure to noise. In D. Henderson, R. Hammernick, D. S. Dosanjh, & J. H. Mills (Eds.), *Effects of Noise on Hearing* (pp. 265–275). New York: Raven Press.
Mills, J. H., & Talo, S. A. (1972). Temporary threshold shift produced by exposures to high frequency noise. *Journal of Speech and Hearing Research*, 15, 624–631.
Moody, D. B., Sebbins, W. C., Johnsson, L. G., & Hawkins, J. E., Jr. (1976). Noise-induced hearing loss in the monkey. In D. Henderson, R. P. Hammernick, D. S. Dosanjh, & J. H. Mills (Eds.), *Effects of Noise on Hearing* (pp. 309–325). New York: Raven Press.
Ou, H. C., Bohne, B. A., & Harding, G. W. (2000). Noise damage in the C57BL/CBA mouse. *Hearing Research*, 145, 111–122.
Pugliano, F. A., Wilcox, T. O., Rossiter, J., & Saunders, J. C. (1993). Recovery of auditory structure and function in neonatal chicks exposed to intense sound for 8 days. *Neuroscience Letters*, 151, 214–218.
Rabinowitz, P. M (2012). The public health significance of noise-induced hearing loss. In C. G. Le Prell, D. Henderson, R. R. Fay, & A. N. Popper (Eds.), *Noise-Induced Hearing Loss: Scientific Advances* (pp. 13–25). New York: Springer-Verlag.
Ramamoorthy, S, & Nuttall, A. L. (2012). Half-octave shift in mammalian hearing is an epiphenomenon of the cochlear amplifier. *PLoS ONE*, 7, e45640. doi:https://doi.org/10.1371/journal.pone.0045640.
Roberto, M, Hamernik, R. P, Salvi, R. J, Henderson, D, & Milone, R (1985). Impact noise and the equal energy hypothesis. *The Journal of the Acoustical Society of America*, 77, 1514–1520.
Ryals, B. M., Dooling, R. J., Westbrook, E., Dent, M. L., MacKenzie, A., & Larsen, O. N. (1999). Avian species differences in susceptibility to noise exposure. *Hearing Research*, 131, 71–88.
Ryan, A., & Bone, R. C. (1978). Noise-induced threshold shift and cochlear pathology in the Mongolian gerbil. *The Journal of the Acoustical Society of America*, 63, 1145–1151.
Salvi, R., Lobarinas, E., Chen, G. D., Stolzberg, D., & Ding, D. (2011). Animal models of hearing loss and tinnitus. In J. Hau & S. Y. Schapiro (Eds.), *Animal Models. Handbook of Laboratory Animal Science*, vol. II 3rd ed. (pp. 419–453). Boca Raton, FL: CRC Press.
Salvi, R. J., Henderson, D., Hamernik, R. P., & Coletti, V. (1986). *Basic and Applied Aspects of Noise-Induced Hearing Loss*. New York: Plenum Press.
Saunders, J. C. (1985). Auditory structure and function in the bird middle ear: An evaluation by SEM and capacitive probe. *Hearing Research*, 18, 253–268.
Saunders, J. C. (2010). The role of hair cell regeneration in an avian model of inner ear injury and repair from acoustic trauma. *Institute for Laboratory Animal Research Journal*, 51, 326–337.
Saunders, J. C., & Dooling, R. J. (1974). Noise-induced threshold shift in the parakeet *(Melopsittacus undulatus)*. *Proceedings of the National Academy of Sciences of the United States of America*, 71, 1962–1965.
Saunders, J. C., & Rosowski, J. J. (1979). The assessment of hearing in animals. In W. F. Rintlemann (Ed.), *The Assessment of Hearing* (pp. 487–529). Baltimore, MD: University Park Press.
Saunders, J. C., & Dear, S. P. (1983). Comparative morphology of stereocilia. In R. R. Fay and G. Gourevitch (Eds.), *Essays on Hearing in Honor of E. G. Wever* (pp. 175–197). Groton, CT: Amphora Press.
Saunders, J. C., Mills, J. H., & Miller, J. C. (1977). Threshold shift in chinchilla from daily exposure to noise for six hours *The Journal of the Acoustical Society of America*, 61, 558–570.
Saunders, J. C., Dear, S. P., & Schneider, M. (1985a). The anatomical consequences of acoustic trauma: a review and tutorial. *The Journal of the Acoustical Society of America*, 78, 833–860.
Saunders, J. C., Schneider, M. E., & Dear, S. P. (1985b). The structure and function of actin in hair cells. *The Journal of Acoustical Society of America*, 78, 299–311.

Saunders, J. C., Cohen, Y. E., & Szymko, Y. M. (1991). The structural and functional consequences of acoustic injury in the cochlea and peripheral auditory system: A five year update. *The Journal of the Acoustical Society of America*, 90, 136–146.
Saunders, J. C., Torsiglieri, A. J., & DeDio, R. M. (1993). The growth of hearing loss in neonatal chicks exposed to intense pure tones. *Hearing Research*, 69, 25–34.
Spassova, M. A., Avissar, M., Furman, A. C., Crumling, M. A., Saunders, J. C., & Parsons, T. D. (2004). Evidence that rapid vesicle replenishment of the synaptic ribbon mediates recovery from short-term adaptation at the hair cell afferent synapse. *Journal of the Association for Research in Otolaryngology*, 5, 376–390.
Syka, J., & Popelár, J. (1980). Hearing threshold shifts from prolonged exposure to noise in guinea pigs. *Hearing Research*, 3, 205–213.
Taylor, W, Pearson, J, Mair, A, & Burns, W. (1965). Study of noise and hearing in jute weaving. *The Journal of the Acoustical Society of America*, 38, 113–120.
Tontechnik-Recher-sengplielaudio. (2016). *How Does the Sound or Noise Depend on Distance from the Source?* Audio Tutorial, Tonmeister Institut. Available at http://www.sengpielaudio.com/calculator-SoundAndDistance.htm. Accessed December 22, 2016.

Chapter 5
Acoustic Conditions Affecting Sound Communication in Air and Underwater

Ole Næsbye Larsen and Craig Radford

Abstract Biodiversity across the animal kingdom is reflected in acoustic diversity, and the evolution of these signals is driven by the ability to produce and hear sounds within the complex nature of soundscapes. Signals from the sender are attenuated and their structure is changed during propagation to receivers, and other sounds contributing to the soundscape can interfere with signals intended for the receiver. Therefore, the message encoded in the sender's signal may be difficult or impossible for the potential receiver to decode unless the receiver adapts behaviorally. This chapter discusses the potential effects of sound propagation and environmental sound on communication both in air and underwater. First, the wave equation is defined; second, attenuation, absorption and scattering principles are discussed in relation to physical sound propagation effects on the sender's signal; and third, abiotic, biotic, and anthropogenic sources of environmental noise are introduced and discussed. Environmental noise is present in all habitats, and soundscapes are getting louder, in part mostly due to increased anthropogenic noise inputs. Therefore, animals that rely on sound to communicate have to adapt and evolve to their local soundscape to get their message across.

Keywords Abiotic noise · Acoustic near and far field · Biotic noise · Cylindrical attenuation · Diffraction · Ground effect · Medium absorption · Reflection · Refraction · Reverberation · Scattering · Shallow-water acoustics · Spherical attenuation · Turbulence · Wave equation

---

O. N. Larsen (✉)
Department of Biology, University of Southern Denmark, Odense M, Denmark
e-mail: onl@biology.sdu.dk

C. Radford
Leigh Marine Laboratory, Institute of Marine Science, University of Auckland, Warkworth, New Zealand
e-mail: c.radford@auckland.ac.nz

H. Slabbekoorn et al. (eds.), *Effects of Anthropogenic Noise on Animals*, Springer Handbook of Auditory Research 66,
https://doi.org/10.1007/978-1-4939-8574-6_5

5.1 Introduction

Across the animal kingdom, biodiversity is reflected in acoustic diversity or sound production. Moreover, the evolution of acoustic signals is driven by the ability to produce and hear sounds within the complexity of environmental noise. To get the message across, acoustically communicating animals face two major challenges from the environmental communication channel: degradation and ambient noise (Catchpole and Slater 2008). Emitted acoustic signals are attenuated and their structure is changed during propagation from senders to receivers. Ambient noise from other sources than the senders also interferes with the signals arriving at the receivers (Wiley and Richards 1978; Naguib and Wiley 2001). Therefore, the message encoded in an acoustic signal by the sender may be difficult or impossible for a potential receiver to decode unless the receiver adapts behaviorally. Consequently, the meaning of the signal to the receiver may be different from the encoded message (Dabelsteen and Pedersen 1988). Behaviorally, the receiver or sender (or both) may respond, for example, by moving closer together in the "here and now" (Wood and Yezerinac 2006; Halfwerk et al. 2012), by moving away from interfering sound sources (McLaughlin and Kunc 2013), or by "vocal adjustment" such as raising their voices (Brumm and Todt 2002; Patricelli and Blickley 2006).

This chapter first defines sound as a physical entity and then considers the major physical mechanisms responsible for changes in propagating acoustic signals in air and in shallow and deep water. Subsequently, two major sources of acoustic interference for animal signals, abiotic and biotic noise, are mapped and characterized in air and underwater. Man-made (anthropogenic) noise sources are not discussed here because these are dealt with at length by Dooling and Leek (Chap. 2) and Saunders and Dooling (Chap. 4).

5.2 The Nature of Sound

Sound is created by vibrating objects that produce a longitudinal mechanical wave that propagates in a compressible medium, such as water or air. For a longitudinal wave, the direction of particle vibration is parallel to the direction of sound propagation. An acoustic disturbance involves both motion of the medium (acceleration, velocity, or displacement) and changes in the state variables (pressure, temperature, or density). Therefore, the acoustic field can also be described in terms of motion and state variables, where the motion variables are vectors and the state variables are scalars.

5.2.1 Defining the Wave Equation

The wave equation is the most fundamental differential equation describing the propagation of a sound wave. The derivation is based on Medwin and Clay (1998) and on the assumption that the propagating sound wave is a plane wave, which means that it has common amplitude and phase at all points on any given plane perpendicular to the direction of wave travel. Four simple laws of physics govern the propagation of sound in any medium.

The first law, the equation of motion, governed by Newton's second law, states that the product of the mass of a particle multiplied by its acceleration is equal to the force acting on the particle (i.e., force = mass × acceleration). For a sound wave traveling in the x direction, the acceleration of a particle in any medium can be approximated by $\partial u_x/\partial t$, where u_x is the particle velocity. Therefore, the force per unit volume in the x direction (f_x) can be expressed in Eq. 5.1 as

$$f_x = \rho_0 \frac{\partial u_x}{\partial t} \tag{5.1}$$

The second law, the equation of force, states that a nonviscous medium cannot support shear stress and that a sound wave propagating in a medium will subject medium particles to compression and expansion forces. For a sound wave traveling in the x direction, the force per unit volume is related to the pressure acting on a unit volume and can be expressed as

$$f_x = -\frac{\partial p_T}{\partial x} \tag{5.2}$$

where $p_T = p_0 + p$; p_0 is the ambient pressure, which does not change over a small volume, and p is the acoustic pressure. Equating values of the force leads to

$$\frac{\partial p}{\partial x} = -\rho_0 \frac{\partial u_x}{\partial t} \tag{5.3}$$

The third law, the equation of state, describes the pressure of a fluid as a function of its density and temperature. The assumption is that as the sound wave propagates, there are slight deviations from the state of equilibrium and the small change in pressure is proportional to the small change in density, and because the changes are so small and so fast, there is no conduction of heat (i.e., no net change in temperature). Hooke's law can be applied here and states that stress is proportional to strain, where stress is the acoustic pressure (p) and strain is the relative change in density (ρ/ρ_0). The proportionality constant is the resistance of the medium to uniform compression for small amplitude sound waves. This law is expressed in Eq. 5.4 as

$$p = \left(\frac{K}{\rho_0}\right)\rho \tag{5.4}$$

where K is the bulk modulus of the medium (the resistance to uniform compression). For a homogeneous medium, the speed of sound (c) is given by

$$c = \sqrt{\left(\frac{\partial p}{\partial \rho}\right)_s} \tag{5.5}$$

where s indicates the local compression and rarefaction (the opposite of compression) caused by the sound wave at constant entropy (a measure of thermal energy not available to do work). Given this ideal scenario, a sound disturbance propagates at the speed of sound, which is related to the compressibility of the medium (Eq. 5.6). For example, the stiffer the medium, the faster the sound speed; the speed of sound in water (approximately 1,500 m s^{-1}) is about 4.5 times faster compared with that in air (approximately 340 m s^{-1})

$$c = \sqrt{\frac{K}{\rho_0}} \tag{5.6}$$

Finally, the fourth law, the conservation of mass, states that any net flow of a medium into or out of a given volume causes the density of the medium to change within it. For example, where the sound energy of a traveling wave causes a disturbance in the pressure and density of the medium, it can be expressed by the continuity of mass equation

$$\frac{\partial \rho}{\partial t} = -\rho_0 \frac{\partial u_x}{\partial x} \tag{5.7}$$

Combining the four laws, the one-dimensional wave equation can be derived by taking the $\partial p/\partial x$ of Eq. 5.3 and $\partial \rho/\partial t$ of Eq. 5.7 and canceling out the common second derivative $\partial^2\rho/\partial x\partial t$ term to get

$$\frac{\partial^2 \rho}{\partial x^2} = \frac{\rho_0}{K}\frac{\partial^2 p}{\partial t^2} \tag{5.8}$$

Because the speed of sound in a medium is independent of propagation direction and is equal to $\sqrt{K/\rho_0}$, the one-dimensional wave equation can be rewritten as

$$\frac{\partial^2 \rho}{\partial x^2} = \frac{1}{c^2}\frac{\partial^2 p}{\partial t^2} \tag{5.9}$$

In the "real world," sound does not travel in one-dimensional space; therefore, the wave equation has to be expressed in three-dimensional space by considering a sound wave traveling in a medium in the x, y, and z directions. The wave equation expressed in three-dimensional space and time can be expressed by

$$\left(\frac{\partial^2 \rho}{\partial x^2}+\frac{\partial^2 \rho}{\partial y^2}+\frac{\partial^2 \rho}{\partial z^2}\right)=\frac{1}{c^2}\frac{\partial^2 p}{\partial t^2} \tag{5.10}$$

The wave equation can be expressed in a range of variables, such as particle acceleration, velocity, and displacement. However, the most commonly used variable is pressure because the large majority of microphones (in air) and hydrophones (in water) used by researchers measure sound pressure. The wave equation is often expressed in terms of a divergence operator (∇) in vector analysis or as a measure of the magnitude of the vector's field at a given point, in terms of a signed scalar. Therefore, the three-dimensional wave equation is commonly written as

$$\nabla^2 \rho=\frac{1}{c^2}\frac{\partial^2 p}{\partial t^2} \tag{5.11}$$

5.2.2 *Near Field Versus Far Field*

Understanding the physical nature of sound starts with the ideal situation, a medium with no boundaries. Only then can the understanding be extended to more complex acoustic environments such as the shallow reefs of vocalizing fishes or a forest of vocalizing birds. In an idealized situation in unbounded media where a specific sound is produced by a source in a homogeneous infinite environment, the sound at any location can be determined. Under these conditions, sound will consist of a radial particle motion and a pressure wave that propagates radially from the source. The particle motion, or particle velocity, consists of two components. The first results from the compression of the fluid by the pressure wave and is considered the "true sound." The second is the flow component, which, for a pulsating source, decreases with the square of the distance from the source (R; i.e., proportional to R^{-2}), where the amplitude of the pressure wave is proportional to R^{-1} (see Fig. 5.1).

The flow component will dominate the sound field close to the source, which is termed the acoustic near field (Fig. 5.1). The region beyond this is termed the acoustic far field, where the pressure component will dominate the sound field and is directly related to the particle velocity component (see Eq. 5.12). This requires R to be much greater than a few wavelengths and also to be much greater than the source dimension

$$p=\rho c v \tag{5.12}$$

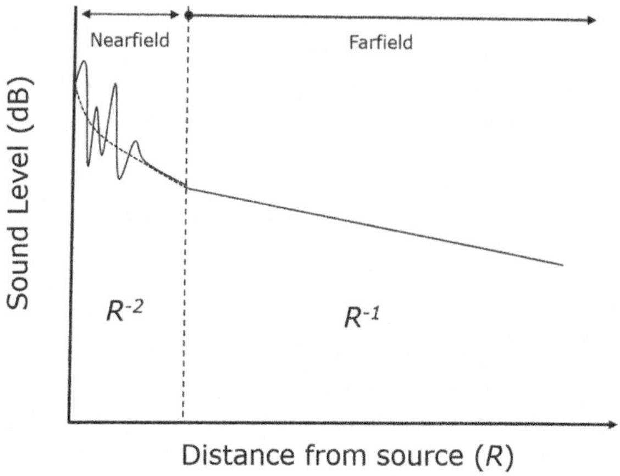

Fig. 5.1 Sound level as a function of distance (*R*) from a sound source. *Solid line* in the near field, fluctuations close to the source due to source structure and depend on direction. *Dashed line* in the near field, sound level for an ideal point source. In the far field, the amplitude of the pressure wave is proportional to R^{-1}. Modified from Higgs and Radford (2016)

where p is the sound pressure (in kg m^{-1} s^{-2}), ρ is the medium density (in kg m^{-3}), c is the speed of sound (in m s^{-1}), and v is the particle velocity (in m s^{-1}), which is defined by Eq. 5.13

$$v = \omega A \tag{5.13}$$

where ω is the angular frequency ($2\pi f$) and A is the amplitude (in m).

Near fields and far fields will exist for any kind of source. As a rough guide, the acoustic near field is confined to within one to two wavelengths from the source or a few times the source dimensions, whichever is larger. Therefore, as the wavelength of a particular sound gets smaller with increasing frequency (Eq. 5.14), the near field gets smaller too

$$\lambda = {}^{c}/_{f} \tag{5.14}$$

For example, in water at 30 Hz, the wavelength is 50.0 m and at 1,500 Hz, the wavelength is 1.0 m, whereas in air, a 30-Hz signal has a wavelength of 11.3 m and at 1,500 Hz, the wavelength is 0.2 m.

5.3 Basic Principles of Sound Propagation in Air and Water

Sound propagation in air and water is governed by a number of physical principles that limit the range and information content of the communication signal. Some of the limiting principles are relatively simple and can be expressed in simple equations that predict observations quite well. Others are very complex and can only be expressed in complicated equations or by iterative computations that predict observations with some uncertainty. Attention to their influence on animal sound communication was drawn by the pioneering studies in air by Morton (1975) and Marten and Marler (1977) and shortly after summed up in the seminal review by Wiley and Richards (1978).

Over the past 20-30 years, outdoor acousticians have made considerable progress in understanding these principles and in predicting sound propagation (e.g., Attenborough et al. 2007; Au and Hastings 2008). However, modeling of outdoor acoustics is still a developing field and software to predict propagation of outdoor sound is very expensive and complex and requires specialized understanding. In this chapter, some of the most important principles are introduced together with some considerations on their implications for sound propagation in air and underwater.

5.3.1 Geometric Attenuation

Propagating sound attenuates with distance from the source because of geometry alone. The simplest source is a point source (i.e., a theoretical source with no physical extent) that emits sound equally in all directions, (i.e., it is omnidirectional). By definition, sound energy emitted by a source per unit time is the sound power (*P*). The power only changes in magnitude when the source changes its emission. The sound power may be construed as equally distributed on the surface of an imaginary sphere that propagates away from the source with the speed of sound (*c*). By definition, sound intensity (*I*) is the sound power per unit area (*A*). This means that intensity changes as a function of distance (*R*) from the source

$$I = P/A = P/4\pi R^2 \tag{5.15}$$

where the denominator is the surface area of the imaginary sphere with radius *R* that expands as $R(t) = ct$, where *t* is time. Because sound power stays constant once it is emitted, sound intensity is reduced with the inverse square of distance (R^{-2}) from the source. Sound intensity is also proportional to the squared sound pressure

$$I = p^2 / Z_0 \tag{5.16}$$

where the proportionality factor Z_0 is the so-called acoustic impedance, which is the product of the density of the medium (ρ_0) and the speed of sound in the medium (c_0). Comparing the two expressions of intensity, it appears that sound pressure is reduced with the inverse distance (R^{-1}) from the source. This relationship is sometimes called the inverse distance law.

In acoustics, values are normally expressed in decibels (dB). The sound intensity level (L_I) is defined as

$$L_I = 10\log_{10}(I / I_0)\text{dB} \tag{5.17}$$

where I_0 is the reference intensity (1 pW). With this definition, Eq. 5.15 becomes

$$L_I(R) = L_W - 10.8 - 20\log_{10}(R) \tag{5.18}$$

where L_W is the sound power level in decibels. Using Eq. 5.16, the sound pressure level then becomes

$$L_p(R) = 10\log_{10}(p/p_0)^2 = L_W - 10.8 - 20\log_{10}(R) \tag{5.19}$$

where the reference value (p_0) is 20 µPa in air and 1 µPa in water.

In bioacoustics, and especially in underwater bioacoustics, the sound pressure level measured at a receiver-located distance (*R*) from the source [$L_p(R)$] is often referred to as the received level (*RL*), whereas the sound pressure level at a reference distance of 1 m from the center of the source [$L_p(1)$] is called the source level (*SL*). Insertion in Eq. 5.19 for sound pressure level gives the well-known relationship

$$RL - SL = -20\log_{10}(R) \tag{5.20}$$

So, at a distance of, for example, 10 m, the received level has been reduced by 20 dB relative to the sound pressure level at 1 m from the source. In general, doubling the distance from *R* to 2*R* means that the received level is reduced by

$$20\log_{10}(2R/R) = 20\log_{10}(2) = 6.021\,\text{dB} \tag{5.21}$$

Therefore, the sound pressure level is reduced by 6 dB per doubling of distance (dd) from a point source (and only in a boundless medium underwater, i.e., deep water).

This type of geometric attenuation is known as spherical spreading loss. This loss is most important up to about 100 m away from the source where *RL* is reduced by 40 dB, whereas the reduction in *RL* increases by only 6 dB from 100 to 200 m and much less for further 100-m steps. A localized sound source such as a calling elephant is far from being the theoretical point source. However, the beauty is that at sufficiently long distances, all localized sound sources behave like point sources and their sounds attenuate by 6 dB/dd. If a sound source on the ground increases its *SL* by 3 dB, then the circular area covered is doubled (within the radius at which *RL* has decreased by, for example, 40 dB), whereas the area is halved if the sound source reduces its *SL* by 3 dB.

Some localized sources are inherently directional in that the sound pressure at the same distance is not equal in all directions but is strongest in perhaps one direction. Such a source is also subject to spherical attenuation in all directions because, again, the imaginary sphere travels away from it at the speed of sound. The difference from the omnidirectional source is that the amplitude of the pressure variations in the propagating wave front of the directional source is larger in one direction than in the others.

The sound source does not need to be suspended in free space to experience spherical spreading loss because the same inverse distance law applies to sources on the ground (or, in general, next to large surfaces giving the sound source directionality). The only difference here is that the same sound energy is now emitted into a half space and therefore doubles the intensity level by 3 dB [incidentally increasing the so-called directivity index (*DI*) from 0 dB to $10\log_{10}(2) = 3$ dB]. The source could also be located, for example, where three large surfaces intersect like in the corner of a large room. Again, spherical spreading prevails, and the *DI* is now $10\log_{10}(8) = 9$ dB because the sound energy is now confined to one-eighth of the space.

If the distance to a sound source is known and the received sound pressure level is measured, then Eq. 5.20 can be used to calculate the source level. Note, however, that this back-calculated source level is theoretical and may not in practice be possible to measure with a microphone or hydrophone for a control because the theoretical position to place the transducer may be located in the near field or inside the sound-emitting elephant or sperm whale and is, by definition, located 1 m from the center of the imaginary sphere enclosing the source.

Another common sound source geometry is a line source consisting of point sources positioned close together on an infinitely long straight line. When sound is emitted from all the points of a line source, the resulting sound wave that travels away from the source at the speed of sound is shaped as a cylinder with the line source as its long axis. Only the radius of the cylinder changes with time, that is, $R(t) = ct$, whereas its height (length) does not. So, in this situation, the sound energy is evenly distributed over the surface of a larger and larger diameter cylinder, the area of which increases with the circumference $2\pi R$. Using the same logic as above, it means that the sound pressure level is reduced with $10\log_{10}(R)$. This type of geometric attenuation is known as cylindrical spreading, and the rule of thumb here is that the sound pressure level is reduced by 3 dB/dd from the source (Embleton 1996). The difference of 3 dB/dd between spherical and cylindrical spreading does not seem that important, but it does sum up. At 20 m from a spherically spreading localized sound source, for example, the sound pressure level at 1 m has been attenuated by 26 dB, but a receiver must move 400 m away from a line source to get the same attenuation.

Pure line sources probably do not exist in nature, but there are lots of sound sources that can be treated as line sources at shorter distances than their longest extent. For example, a densely trafficked motorway or the surf of an ocean coastline are reasonably well modeled as line sources and follow the 3 dB/dd attenuation rule. If such a "line-like" source is limited in extent (e.g., a flock of very vocal migrating songbirds resting on a power line), it will approach a line source with 3 dB/dd attenuation at distances shorter than its horizontal extent (e.g., 50 m), but at longer distances, it is better described as a spherical source because here the sound will attenuate by 6 dB/dd. In-between, there will be a zone where the attenuation changes from 3 dB/dd to 6 dB/dd (e.g., Embleton 1996; Attenborough 2007). Cylindrical spreading conditions may also be found when sound is trapped between two barriers and, again, transitions between spherical and cylindrical spreading may be observed.

If a localized sound source is positioned midway between the surface and the bottom of the sea, then spherical spreading prevails close to the source, whereas at longer distances, sound reflected from the surface and bottom will give cylindrical spreading conditions. So, in many cases, geometric attenuation is neither fully spherical nor fully cylindrical but is somewhere in-between.

Spherical and cylindrical attenuation, or an intermediate of the two, describe most situations in air and underwater. Geometric attenuation of a propagating sound is omnipresent and must always be taken into consideration (standing waves in enclosures are not subject to geometric attenuation, but then they do not propagate). Most importantly, geometric attenuation only attenuates the amplitude of the pressure variations in a propagating sound wave. It does not change the frequency spectrum, and it does not change the amplitude envelope of the emitted sound.

Geometric attenuation always attenuates a propagating sound wave predictably. At longer distances, however, the measured attenuation typically deviates from the one predicted by this mechanism. This deviation is called excess attenuation (EA) and is caused by a number of different physical principles such as medium absorption, refraction, reflection, diffraction, and turbulence, which call for much more complicated models than considered until now. Because all environmental mechanisms simultaneously attenuate a sound signal, it is EA that is measured in transmission measurements after geometric attenuation has been accounted for (e.g., Dabelsteen et al. 1993; Penna et al. 2012). Note, however, that close to the ground and in certain low-frequency ranges, EA may be negative by up to 6 dB, caused by constructive interference between the direct sound wave and the ground reflected wave (see Sect. 5.3.5).

5.3.2 Absorption in the Medium

Absorption (A) in the medium (air or water) is another omnipresent source of attenuation of propagating sound. In contrast to geometric attenuation, it does change the frequency spectrum and the envelope of the propagating sound. Attenuation by absorption is directly proportional to the propagation distance (*R*)

$$A_{\text{abs}}\left(R\right)=\alpha R \tag{5.22}$$

where α is the so-called absorption coefficient measured in decibels per meter. Formulas to describe α are extremely complicated because in air it depends not only on the frequency but also on the temperature, atmospheric pressure, and relative humidity (e.g., Blackstock 2000). Diagrams for absorption in air showing α as a function of frequency and relative humidity at just one meteorological condition, such as at 20 °C and 1 atm, can be found as double logarithmic graphs in the literature (e.g., Bass et al. 1995). Such graphs show that for sound frequencies below 1 kHz, α varies from about 0.005 dB/m to less than 0.0001 dB/m. Between 1 kHz

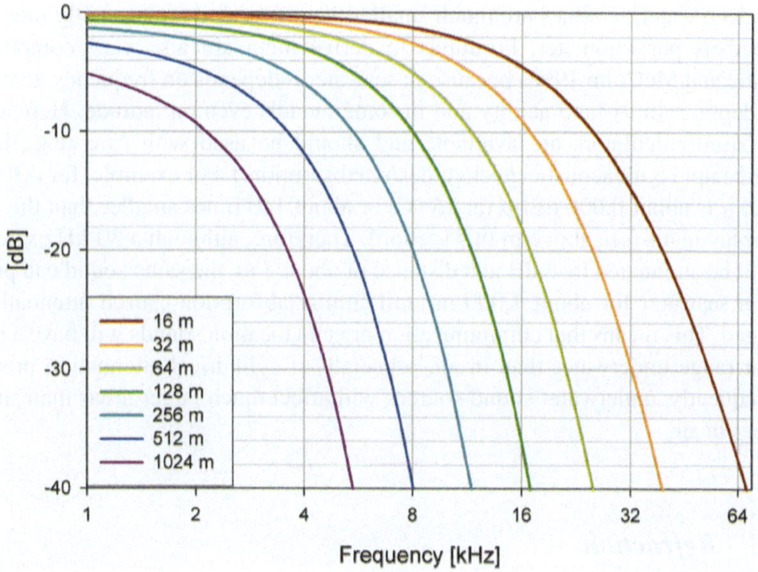

Fig. 5.2 Air absorption acts as a low-pass filter. Theoretical air absorption calculated at seven distances from a sound source should be added to the actual geometric attenuation to obtain the combined attenuation. Calculated for 20 °C, 101 kPa, and 70% relative humidity using absorption values from the Web calculator at the National Physical Laboratory, UK

and 10 kHz, it increases up to about 0.1 dB/m and from 10 kHz to 100 kHz up to about 4 dB/m. This means that air absorption is negligible close to a source at low frequencies but becomes stronger than spherical attenuation in the ultrasonic frequency range. A sound wave of 30 kHz (0.83 dB/m at 15 °C and 1 atm), for example, will be attenuated by about 50 dB at a distance of 64 m due to absorption, whereas spherical attenuation here will only attenuate by 36 dB. Only approximate α values can be read from such graphs and they are valid only for a certain value of temperature and atmospheric pressure. To give α values for optional conditions, Web-based calculators are now available. In using them, one should make sure that they are based on internationally accepted standards such as those issued by the International Standards Organization (ISO 1993) and also that they are implemented and run by internationally recognized organizations like those found on the home page of the National Physical Laboratory of the United Kingdom (http://resource.npl.co.uk/acoustics/techguides/absorption/). Because temperature, atmospheric pressure, and relative humidity vary throughout the day and season (Larsson et al. 1988), atmospheric absorption also varies, and this may be important in some studies where these Web-based calculators come in handy.

Air absorption works as a low-pass filter, which results from the dependence of the absorption coefficient on frequency (Fig. 5.2). Therefore, the higher frequency harmonics of sounds attenuate rapidly with distance and only the fundamental frequency of a vocalization will propagate any great distance (Meyer 2015).

In deep water, α values are much smaller than in air and are normally measured in decibels per kilometer. Formulas to derive them are also very complicated (Ainslie and McColm 1998) because in seawater α depends on frequency, temperature, depth, salinity, and acidity and in some models even on latitude. Here again, Web-based calculators are available and should be used with care (e.g., http://resource.npl.co.uk/acoustics/techguides/seaabsorption/). For example, for a 30-kHz sound, α is about 0.006 dB/m (at 15 °C) or about 140 times smaller than the same frequency in air (see above in this section). Therefore, although a 30-kHz sound in air will be attenuated by 6 dB at a distance of about 7 m, the same sound can propagate in seawater for about 1,000 m until similar absorption-caused attenuation is obtained. This means that communication or echo location sounds will have a much longer range underwater than in air, especially if cylindrical attenuation prevails. Consequently, underwater sound sources will affect much larger areas than similar sources in air.

5.3.3 Refraction

A sound wave propagating away from a source not only attenuates with distance, it may also change direction and thereby attenuate (or enhance) more at a given receiver position than that predicted by sender-receiver geometry alone. To illustrate this, it is practical to visualize the wave not as a progressing surface wave front but as an imaginary curve in space that describes the trajectory of a point on the wave front. This is called ray tracing (Fahy 2003). Therefore, instead of illustrating a sound wave propagating away from a point source as a number of concentric circles, it may be illustrated as a number of equally long arrows (rays) emanating from a common point and pointing in all directions.

Another tool to consider is Snell's law, which is well-known from optics (Fig. 5.3). It states that when a ray in one medium encounters a surface with a change in medium impedance, for example, from air to water, then two rays (waves) are produced, one reflected from the surface and another transmitted (refracted) through the surface. The angle of the incident ray (*i*) relative to the normal of the surface (θ_i) equals that of the reflected ray (*r*) relative to the normal (θ_r). The relationship between the angle of the incident ray and the transmitted ray (*t*) is a little more complicated because here

$$\frac{\sin\theta_i}{\sin\theta_t} = \frac{c_i}{c_t} \tag{5.23}$$

where θ_t is the angle between the transmitted ray and the normal to the surface and c_i and c_t are the sound speeds in the two media. This means that when a sound ray propagates from a medium with a lower speed of sound to one with a higher speed of sound (as from air to water), then $\theta_t > \theta_i$, whereas $\theta_t < \theta_i$, if the sound ray

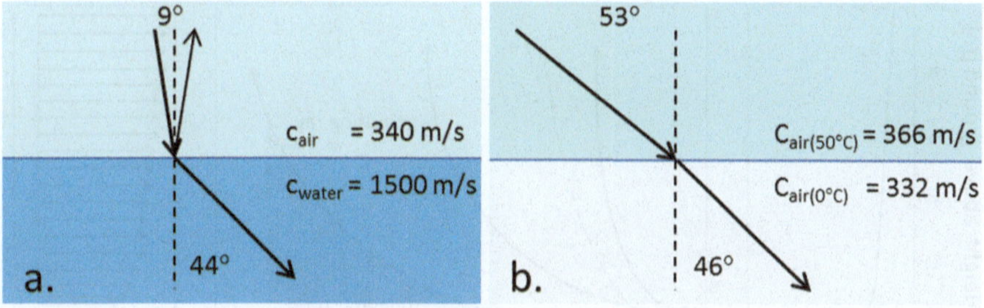

Fig. 5.3 Refraction at the boundaries is determined by the ratio between the speeds of sound in the media (cf. Eq. 5.23). **a**: At the air-water interface, there are large differences in the speed of sound (c_{air} and c_{water}). Sound from the air will be partly reflected from the water surface and partly refracted into the water when the angle of incidence (c_i; in this example 9°) is smaller than the critical angle of 13° relative to the normal. For angles of incidence larger than 13°, total reflection at the water surface will occur because the transmitted angle grows from 44° at 9° incidence to 90° already at the incidence angle of 13° and beyond. So, sound cannot penetrate the water surface at angles of incidence larger than this critical angle. **b**: In a layered medium (caused by changes in, e.g., the temperature in air or the salinity in water), only refraction will occur and the change in direction will be very modest but always in the direction of lowest speed of sound. In this example, an incident sound wave in hot air (50°C) at an angle of 53° relative to the normal is transmitted without reflection across the interface to cold air (0°C) at the lower transmission angle of 46° relative to the normal. *Dashed vertical line*, the normal to the surface

propagates in the opposite direction (Fig. 5.3). This means that the transmitted sound ray will in both cases change direction from that of the incident ray toward the medium with the lower speed of sound.

Because the sound speeds in air and water are very different, there is a considerable change in direction of the transmitted ray when sound goes from air into water (Fig. 5.3a). This means that there is a certain angle of incidence for which the angle of the transmitted ray becomes 90° or directly parallel with the air-water interface (i.e., here $\sin\theta_t = 1$). This angle is called the critical angle, and using Snell's law and the two sound speed values ($c_i = 340$ m/s and $c_t = 1{,}500$ m/s), it is easy to show that the critical angle is about $\theta_i = 13°$. If θ_i is larger than 13°, then the incident ray will be totally reflected and there will be no transmitted ray. Or, in other words, above any point on the air-water interface, there is a 26° "cone of transmission" within which sound is transmitted from air into water, whereas all sound is totally reflected for sources outside this cone at any given location. Within a medium, the change in speed of sound is usually much more gradual and an appreciable change in direction of the refracted wave requires unrealistically large differences in, for example, temperature (Fig. 5.3b).

In the case of the air-water boundary, there is only one surface between two media with different speeds of sound. The logic of proportions of reflection and transmission can be extended to a layered medium with many layers of increasing or decreasing speed of sound. Here again, the incident ray from layer to layer will keep bending toward layers with lower speeds of sound, and to a first approximation, the curved ray path is circular with the source at the periphery (Blackstock

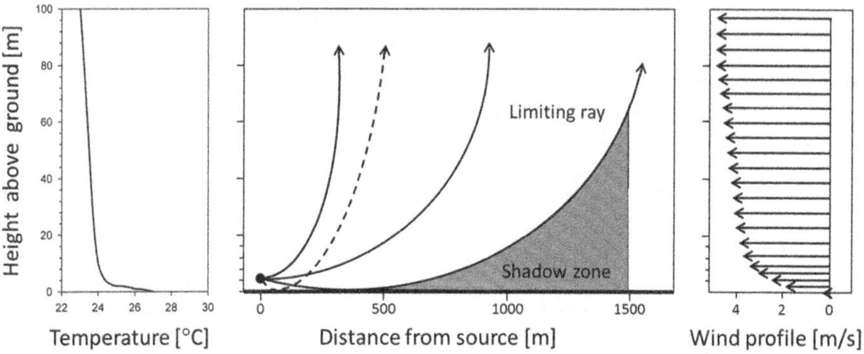

Fig. 5.4 Upward refracting conditions as shown by a ray tracing from the source (*middle*) at a temperature lapse during daytime where the curve represents air temperature as a function of height (*left*) or upwind from a source where the arrow length represents wind velocity as a function of height (*right*). In this schematic, the sound source is about 5 m above the ground and the temperature decreases by 0.01 °C/m above 10 m. *Middle, dashed curve*, reflection from the ground of downward directed rays from the source. The distance to the shadow zone defined by the limiting ray only gives the order of magnitude. For further details, see text

2000). The layer thickness can become smaller and smaller, producing a sound speed gradient at the limit. Sound speed gradients are found both in air and in water because the speed of sound in quiet air depends almost entirely on temperature, whereas in water it depends on other factors in addition to temperature. Again, Web-based calculators are available to calculate the speed of sound in both media (e.g., http://resource.npl.co.uk/acoustics/techguides/speedair/ for air and http://resource.npl.co.uk/acoustics/techguides/soundseawater/content.html for water).

Refraction in Air

In general, air temperature decreases with the height above ground by 0.01 °C/m (the so-called temperature lapse) but within the 50-100 m closest to the ground (in the surface layer), conditions may be different. Here stronger temperature gradients may arise. On a sunny afternoon with no wind, a stable temperature gradient may be established and the temperature 10 cm above the ground may be much higher, perhaps 5 °C higher, than at the meteorological reference height of 2 m above ground. When the air temperature changes by 5 °C within a normal range of ambient temperature, for example, from 0 °C to 40 °C, then the speed of sound changes by 3.2-3.6 m/s and the transmitted angle θ_t in a thin layer changes by about 0.5°-2° relative to that of the incident angle θ_i.

A decreasing temperature gradient with height produces upward refracting conditions. Such a gradient is usually found in daytime, especially on quiet sunny afternoons. Sound rays from an omnidirectional source located a few meters above the ground and propagating upward will be refracted even more upward. Sound rays propagating parallel to the ground will also be refracted upward (Fig. 5.4), whereas sound rays directed downward will hit the ground and, according to Huygens' principle, form secondary sources to bounce back up in the air according to Snell's law (Fig. 5.4, *dashed line*). It is intuitively obvious then that there must be a direction of a limiting ray that grazes the ground at a certain distance from the source, beyond which no direct sound from the source will propagate parallel to the ground. This distance marks the border of the so-called acoustic shadow zone. The distance to the border of the shadow zone is surprisingly long. If, for example, both sound source and receiver are located 1 m above the ground and there is a "normal" daytime temperature gradient decreasing with height above the ground, then the receiver must move about 300 m away from the source to get to the border of the shadow zone (Attenborough 2007). For formulas to calculate the distance, see Blackstock (2000) or Attenborough (2007). One should not be misled to think that it is impossible to hear the source within the shadow zone because there is always turbulence in the air (see Sect. 5.3.4). The turbulence will allow some sound to penetrate the zone. So, in practice, an acoustic shadow zone in air just means that sound from a source is attenuated by some 20-25 dB more than predicted from combined geometric attenuation and air absorption (Attenborough 2007).

The normal temperature lapse continues up to a height of about 10 km where the temperature starts to increase again. So, in the temperate zone in summer, the speed of sound may be 344 m/s near the ground, decreasing to about 285 m/s at a height of 10 km, and then increasing again to about 370 m/s at 40 km (Heller 2013). This means that rays of very loud sounds such as from volcanic activity or other explosions will be refracted upward but then bend downward to reach the ground again, perhaps 200 km away from the source to be heard again, whereas the volcanic activity and explosions may not be heard in a silent zone 100 km away from the source (see illustrations in Heller 2013).

So-called temperature inversions may sometimes establish stable temperature gradients in which the temperature at ground level is lower than above where the temperature increases up to 50-100 m above the ground before it returns to the "normal" 0.01 °C/m decrease with height. These are downward refracting conditions (Fig. 5.5). This happens, for example, at night and early morning when the ground and with it the layer of air immediately above it cools much faster by radiation than the air above. Here sound rays from an omnidirectional source will be bent downward to bounce off from the ground again. Therefore, no shadow zone can exist and the sound "trapped" between the ground and the warmer air above will propagate with geometric attenuation that is lower than spherical and approaching cylindrical attenuation of 3 dB/dd. This is the reason that the communication distances of low-frequency elephant calls on the savanna more than doubles to perhaps 10 km at night (Garstang et al. 1995) and that the distant railroad train never noticed in daytime suddenly is clearly heard at night. These nighttime conditions tend to persist into the morning, when dawn chorus singers benefit from the advantageous propagation properties, until the sun dissolves the inversion.

Movement of the medium itself will also change the speed of sound and therefore the direction of sound rays. In air, the wind can play a major role because its normal velocities range from 1 to 5% of the speed of sound. A constant wind will

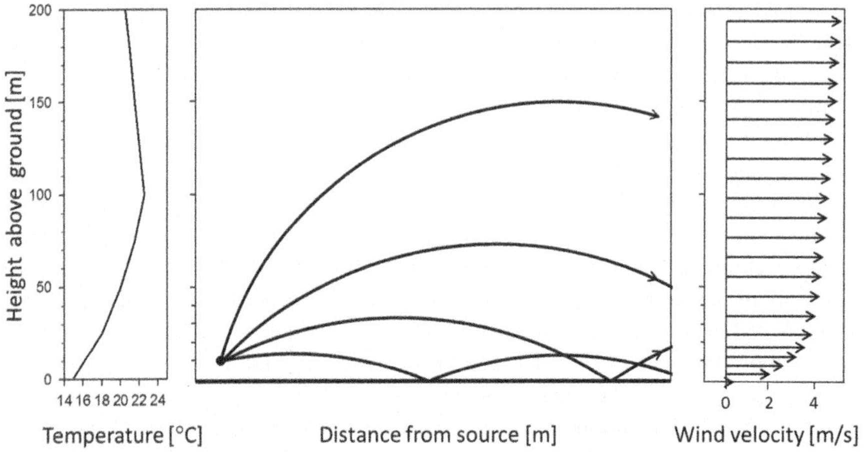

Fig. 5.5 Downward refracting conditions as shown by a ray tracing from the source (*middle*) at temperature inversion at night where the curve represents air temperature as a function of height (*left*) or during downwind conditions where the arrow length represents wind velocity as a function of height (*right*). In this schematic, the normal temperature lapse starts at a height of 100 m (*left*) and the wind is a gentle breeze (*right*). Note that a receiver located long distances from the source may receive sounds from the same source that have bounced zero, one, or two times from the ground (cf. Embleton, 1996). For further details, see text

establish a gradient with a velocity of 0 ms^{-1} at the ground layer but increasing nonlinearly up to about 10 m above the ground where it stabilizes at almost full strength but often continues to increase logarithmically with height (Attenborough et al. 2007). Upwind, the wind velocity will subtract from the speed of sound, creating upward refracting conditions with a reduction in geometric attenuation and the establishment of a shadow zone (Fig. 5.4), whereas downwind, it will add to the speed of sound, creating downward refracting conditions with a reduction in geometric attenuation (Fig. 5.5). Even a gentle breeze of 4.5 ms^{-1} will change upwind and downwind sound speeds by the same amount as changes in temperature of about 7 °C, whereas a strong breeze of 12 ms^{-1} corresponds to a temperature change of about 17 °C. Therefore, wind effects tend to dominate temperature effects when both are present and strong turbulence associated with high winds does not allow the development of marked thermal stratification (Attenborough et al. 2007).

Note that these wind effects depend on height. A downwind receiver standing on the ground may easily hear a certain sound source, whereas upwind at the same distance, it may not be heard. However, if both source and receiver are placed in two tall trees more than 10-20 m above the ground where the wind profile is stable, such an asymmetry may not exist.

Note also that due to the directional nature of the wind, refraction produced by wind is proportional to the angle between the sender-receiver direction and the downwind direction. Maximum downward refraction is observed when this angle is zero. As the angle between sender-receiver direction and the downwind direction approaches 90°, refraction approaches zero, which is reached when source and receiver are placed directly in the crosswind. As the sender-receiver direction changes toward 180°, upward refraction becomes progressively stronger and reaches its maximum when the sender-receiver direction is directly upwind.

Refraction in Water

In the ocean, temperature decreases to a depth of about 1 km from where it remains constant but then a density gradient becomes stronger with depth. This means that sounds from sources between the surface and this layer will be refracted downward while sounds from deep sources will be refracted upward and the sounds will be trapped in a layer at about 1 km where they propagate with close to cylindrical attenuation. This is the SOFAR (SOund Fixing And Ranging) channel, which may allow deep-diving whales to communicate over distances of hundreds of kilometers (Medwin and Clay 1998).

5.3.4 Turbulence

The theoretical refraction effects mentioned in Sect. 5.3.3.1 imply rather constant gradients of temperature and wind. However, in the real world, the atmosphere is in constant motion and the sound speed varies in space and time because of fluctuations in temperature and especially in wind velocity. These fluctuations are called turbulence and may be visualized as constantly circulating vertical eddies of many different diameters. Turbulence effects are very important in open environments such as coastal areas, grassland, savanna, and tundra but of limited importance in closed environments such as within temperate and tropical forests. On a hot summer day, the largest eddies extend from the ground surface to several hundred meters above. They are produced mainly by convective currents from warming over different patches of ground. The eddies are gradually broken down into smaller and smaller diameters due to instabilities in the air until they reach millimeter size and the energy within is dissipated by viscosity and thermal conduction (Piercy et al. 1977; Attenborough 2007).

A sound wave of constant frequency and amplitude propagating away from a source through a turbulent atmosphere consequently encounters a superposition of turbulent eddies of different sizes and temperatures, resulting in fluctuating speeds of sound and therefore in refraction. This means that after propagating some distance, wave fronts of constant phase no longer have the same amplitude and a signal becomes less and less correlated with the source (Piercy et al. 1977). The lack of correlation has consequences for the size of the ground effect (see Sect. 5.3.5). Under such circumstances, a stationary receiver will measure fluctuations in the sound pressure level of a constant sound source increasing with distance but stabilizing at a standard deviation of about 6 dB at a distance of a few kilometers (Piercy et al. 1977) unless the measurements are averaged over long periods, for example, 10 minutes (Attenborough 2007).

Propagation through a turbulent atmosphere has an additional effect: the envelopes of acoustic signals are irregularly amplitude modulated (Piercy et al. 1977). This turbulence-induced irregular amplitude modulation limits animals' use of amplitude modulation of acoustic signals as an information-carrying parameter. However, this only becomes a real problem for carrier frequencies above 10-15 kHz (Michelsen and Larsen 1983).

The irregular turbulence-induced fluctuation in the speed of sound also means that the direction of a sound ray changes irregularly. Sometimes this may lead to converging sound paths with higher sound pressure levels (so-called caustics that can be very intense, especially underwater) and sometimes to diverging paths with lower levels. The sound from a distant but constant sound source may consequently wax and wane depending on wind and pockets of cold or hot air in the boundary layer. In addition, turbulence effects will scatter sound waves into sound shadow zones created by upward refracting conditions in the atmosphere or by physical barriers, thus limiting the effect of their respective attenuation.

To model turbulence effects, outdoor acousticians now use the mean-squared refractive index (MSRI) that is calculated from instantaneous fluctuations in wind velocity and temperature at the receiver and typically measured with a hot-wire anemometer. MSRI depends on the squared variance of wind velocity relative to the ambient wind velocity and the squared variance in temperature fluctuations relative to ambient temperature. Typical MSRI values range from 10^{-4} for strong turbulence to 10^{-6} for quiet conditions and must be included in the calculations to successfully predict outdoor sound propagation in air (see Jensen et al. 2008).

5.3.5 *Ground and Water Surface Effects*

So far, sound propagation has been considered mainly in the medium proper, not taking into consideration intervening objects. The interaction between a propagating sound wave and an object depends on the surface properties of the object and on the ratio between the object size and the wavelength of sound. Sounds of bioacoustic importance always encounter at least one major object: the ground in air or the seafloor surface below and the air-water surface boundary above or below in water and in air, respectively. These surfaces are always much larger than the signal wavelengths, and the resultant acoustic changes are termed boundary effects.

Boundary Effects in Air

For a sound source and a receiver located within tens of meters above a flat and hard ground, the sound will first reach the receiver along a direct path (Fig. 5.6) and after a delay determined by the transmission time along the longer indirect path through

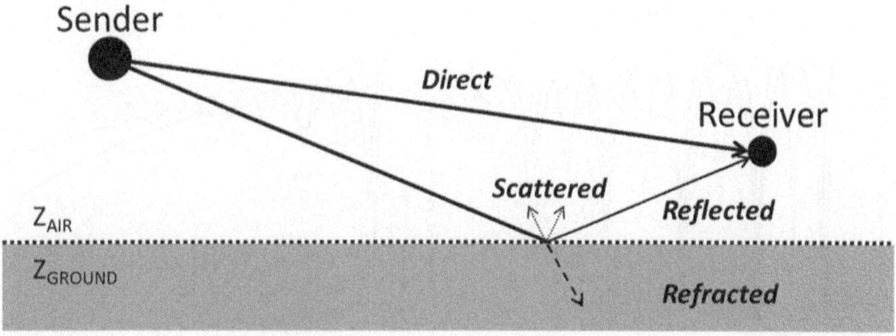

Fig. 5.6 Geometry of the ground effect. The direct wave from a sender (e.g., 4 m above the ground; *horizontal dotted line*) reaches a receiver some distance away (e.g., 2 m above the ground) along the shortest path. After a delay determined by the extra path length, the direct wave is followed by the wave reflected from the ground (suggested by the *dotted horizontal line*). Not all incident sound is reflected because it may be scattered (diffuse reflection) in other directions by a rough surface (*upward solid arrows*) and/or refracted into small air cavities ("pores"; *downward dashed arrow*) in the soft ground, where it gives rise to so-called ground and surface waves that are not shown here. The acoustic impedance of air (Z_{air}) is different from that of the ground (Z_{ground}), which is determined by ground porosity and flow resistance

reflection from the ground. If the distance between the source and receiver is short (tens of meters), the ray paths with good approximation can be considered straight, but at longer distances, refraction must be taken into consideration and the ray paths become curved by upward or downward refraction (cf. Figs. 5.4 and 5.5).

Once the two waves arrive at the receiver, they will interfere by linear superposition because most bioacoustically relevant sounds are linear (exceptions are strong explosions or shock waves from fighter jets). This means that the direct and indirect waves simply add. However, the delay between them means that for a certain low frequency, the two waves will be totally out of phase and their combined amplitude will approach zero. This is destructive interference. This will also happen at frequencies where an even multiple of their corresponding period equals the time lag between the direct and reflected waves. At another low frequency, the direct and indirect wave will be in phase and their amplitude will double (i.e., the sound pressure will increase by almost 6 dB) by constructive interference. Further amplitude doublings are located at frequencies where an uneven multiple of their corresponding period equals the time lag (Fig. 5.7a). Therefore, in the frequency domain, the presence of the ground (or air-water interface) will alternately enhance and attenuate the spectrum of the broadband sound wave, termed the ground effect. In the older literature, it was termed ground attenuation (Marten and Marler 1977), but this term should be avoided because amplitudes at low frequencies are actually enhanced and may be attenuated up to 6 dB less than those predicted by geometric attenuation alone (Fig. 5.7).

In the real world, the ground effect is most important at short distances and low frequencies up to a few kilohertz, partly because turbulence destroys the coherence between the direct and indirect waves and partly because the ground is not a perfect

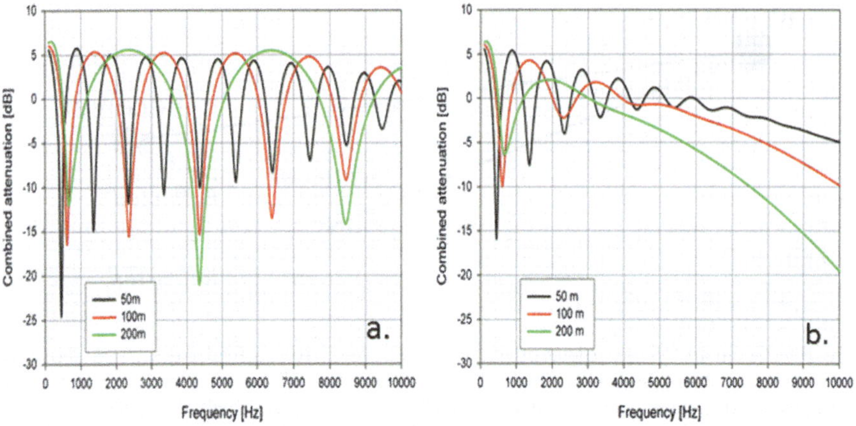

Fig. 5.7 Modeling the ground effect. **a**: The predicted ground effect at a receiver 2 m above the ground at 3 different distances from a sender located at a height of 4 m. The curves show the predicted decibel values that should be added to those resulting from geometric attenuation. The ground is a grass field (flow resistivity of 100 kPa s m^{-2}, porosity of 30%, and layer thickness of 0.01 m). **b**: Same sender-receiver geometry as in **a** but now air absorption (at 20 °C, relative humidity of 80%, and standard atmospheric pressure of 101.325 kPa) and moderate turbulence [mean-squared refractive index (MSRI) of 10^{-5}] have been included (calculated with software developed by Keith Attenborough and Shahram Taherzadeh). Effects of temperature and wind-induced refraction are not included in this model

reflector (Fig. 5.7b). When incident sound reaches the ground, it not only reflects as light from a mirror. If the surface is "rough," such as when covered by irregularities like small pebbles, then sound waves with wavelengths comparable to or smaller than the dimensions of the irregularities will be directed in different directions by diffuse reflection ("scattered"; see Fig. 5.6) and only a fraction of the incident wave will reach the receiver.

Outdoor surfaces vary not only from "acoustically rough" to "smooth" (e.g., the surface of a quiet lake) but also from "acoustically hard" to "soft" as a certain part of the incident wave is transmitted into air-filled pores in the ground (Fig. 5.6). Here sound propagation is impeded by viscous friction, resulting in a much lower sound speed than in air. This means that the transmitted wave refracts toward the normal (cf. Fig. 5.3b) and the air-ground interaction may often be considered independent of the angle of incidence. The surface is then locally reacting. Outdoor acousticians define a so-called spherical wave reflection coefficient, which takes into consideration both the acoustical softness of the ground and the shape of the wave fronts and predicts the existence of two other types of waves in addition to the reflected one, a ground wave and a surface wave (Attenborough 2007). The latter propagates parallel to, and close to, the ground and may penetrate into shadow zones (Embleton 1996). To model the ground effect, outdoor acousticians use the parameters flow resistivity, porosity, and layer depth, which together characterize the "acoustic softness" of the ground (Attenborough 2007). Flow resistivity is a measure of the ease with which air can move in and out of the pores in the ground surface. If the flow resistivity is low, then it is easy for air to flow through pores in the surface. This happens when porosity is high, such as when there is a large fraction of air space between the soil particles, roots, and other structures in the ground. Porosity is the fraction of air in a surface volume and is measured in percent. Flow resistivity is high when porosity is low as in wet compacted sand or man-made surfaces like those covered with asphalt. Ground surfaces with low porosity such as rock or asphalt can be considered "acoustically hard" and the ground effect is effective over a wide frequency range, whereas grassland, forest floors, or other vegetation-covered surfaces and new snow with high porosity can be considered "acoustically soft." In the latter, the ground effect is mainly observed at low frequencies because it becomes more and more difficult for a sound wave to penetrate pores the longer its wavelength, whereas high-frequency sound interacts with the air in the pores (cf. ISO 1996). Therefore, moving with the same sender-receiver geometry from a grass field to acoustically softer surfaces, such as newly fallen snow, will reduce the ground effect to even lower frequencies, whereas moving it to compact ground will enhance the ground effect at higher frequencies.

Boundary Effects in Water

Underwater, there are also conditions similar to the ground effect. The Lloyd's mirror effect refers to range-dependent constructive and destructive interferences caused by the interaction between the direct sound wave and the one reflected by the water surface (Medwin and Clay 1998; Au and Hastings 2008). Here sound is reflected from the interface between a medium with high impedance (water) and one with lower impedance (air), which, in contrast to conditions in air, inverts the phase of the reflected sound relative to the incident one, producing maxima and minima at frequencies opposite to those predicted in air. In principle, there should also be a "ground effect" at the seafloor, but little is known about this effect despite its potential importance for animals living on or close to the seafloor.

5.3.6 Scattering by Objects in the Sound Path

A final basic principle to consider is scattering, which is the effect of combined reflection and diffraction of a propagating sound wave by objects in its path that cause further attenuation. Here the notation *ka* is often used, where *k* is the wave number $2\pi/\lambda$ and *a* is the radius of a sphere that has the same circumference, surface area, or volume as the object.

When $ka << 1$, the sound wave will pass the object as if it were not there. As the relative size of the object grows ($ka < 1$), then fractions of the sound energy are redirected or scattered from the object in all directions (Rayleigh scattering). In the range $1 \leq ka < 30$, a new phenomenon, diffractive scattering, is observed (e.g.,

Bradbury and Vehrencamp 1998), where the reflected wave is scattered from the object in all different directions, whereas a diffracted wave (the creeping wave) bends around the object where the two waves interfere and produce a complicated sound field around the object depending on their relative phases. For $ka > 30$, very little sound is diffracted around the object and the reflected wave dominates following the laws of reflection (simple scattering).

The redirection of sound by scattering from objects in the path between the source and receiver not only reduces the sound energy in the direct wave from sender to receiver (Fig. 5.6) but also creates a large number of echoes that reach the receiver along multiple different paths. These echoes are delayed and interfere at the receiver, with the original signal outlasting its duration and creating an echo tail, which generally increases in length with the distance between sender and receiver. This distortion of the original signal by objects in the environment is known as reverberation. It is very pronounced in closed environments such as forests, where multiple scattering and reflection from surfaces like foliage, branches, and tree trunks are always present, whereas little reverberation is found in open environments.

Information encoded in the temporal structure of acoustic animal signals may be seriously distorted or lost if the reverberation tail of echoes obscures the duration of interpulse intervals by "filling out" the intervals and even masking the onset of closely following elements. On the other hand, forest-living animals may actually exploit the accumulation of reverberation effects with propagation distance for ranging (i.e., determining the distance to a vocalizing conspecific; Naguib and Wiley 2001) or with active space expansion through benefits on signal detection (Slabbekoorn et al. 2002). To date, it is very difficult to predict reverberation tail length in a closed environment because it depends on a number of factors in addition to the distance factor, but its length and amplitude may be quantified (Holland et al. 2001). In general, a much larger database is needed to understand the influence of vegetation on sound propagation (Attenborough 2007).

5.3.7 Shallow-Water Sound Propagation

Many freshwater and marine animals live in shallow water, such as temperate and tropical reefs, rivers and ponds, which makes sound propagation inherently complicated because of interaction of the propagating signal with the water surface and bottom (Rogers and Cox 1988). The transmission loss of sound propagation depends on the surface roughness, bottom properties, and water depth. In addition to the "mirror-like" reflection of the sound signal, the surface and bottom may also scatter (see Sect. 5.3.6) sound in other directions due to surface roughness. Oceans, rivers, and lake bottoms are highly variable in their composition, from hard rock to soft mud. The smoothness of the bottom can also vary remarkably, from being relatively smooth to very rough due to bottom-dwelling invertebrate communities and large boulders and outcrops. The bottom slope can also affect the propagating signal.

Therefore, bottom complexity makes modeling shallow-water sound propagation extremely difficult.

The different ways to determine shallow-water sound propagation are not discussed in detail, but the potential ways in which it can be calculated are highlighted. There are three ways in which shallow-water propagation can be modeled, which range from the relatively simple to the more complicated: spreading-law models, analytical models, and numerical models.

Spreading-law models are the most simplistic and assume that the sound levels decrease monotonically with increasing distance from the source, with the pattern of sound levels being circular. These models do not include the effect of varying bathymetry or sediment properties and so can substantially underestimate or overestimate the sound levels (Farcas et al. 2016).

Analytical models, such as those developed by Weston (1971), are intermediate in complexity, where they build on the spreading-law models to include bathymetry. Weston (1971) also pointed out the limitation of these models because they never accounted for bottom and surface interactions or for really shallow waters, such as rivers or streams.

Numerical models are arguably the most accurate (provided adequate input data are available) but also are the most complicated. These models take into account a large range of inputs, such as temperature, sound speed, bathymetry, sea surface roughness, seafloor composition, and source level. They are also based on several underlying mathematical methods, such as ray theory (Porter and Liu 1994), normal modes (Porter 1992), and parabolic equations (Collins 1993). It is also important to note that no single model is applicable to all acoustic frequencies and environments (see Farcas et al. 2016).

One such type of numerical model, normal-mode model, is a practical choice for shallow-water sound propagation modeling, as it accounts for both frequency and range dependence (Farcas et al. 2016). Therefore, sound propagation can also be treated as a value boundary problem, where the sound channel is treated as an acoustic wave guide and the solution to the wave equation (Eq. 5.11) consists of a finite sum of normal modes (Ferris 1972; Rogers and Cox 1988). Each mode is a standing value in the vertical direction that propagates in the horizontal direction with its own frequency-dependent speed. Consequently, each mode has a certain cutoff frequency below which it cannot propagate. No sound can propagate at frequencies below its cutoff frequency, which depends on propagation speed in the sediment (c_s) and is inversely proportional to the water depth (h). The absolute cutoff frequency (f_c) can be expressed by

$$f_c = \frac{c_w / 4h}{\sqrt{1 - c_w^2 / c_s^2}} \tag{5.24}$$

where c_w is the speed of sound in water. Sound propagation within the hearing range of a large majority of aquatic animals (100-2,000 Hz) will not propagate in water shallower than 10 cm (Fig. 5.8). In 1- to 2-m water depth, the useable frequency range is still very limited, especially for "slow" bottoms. Thus, fishes that have

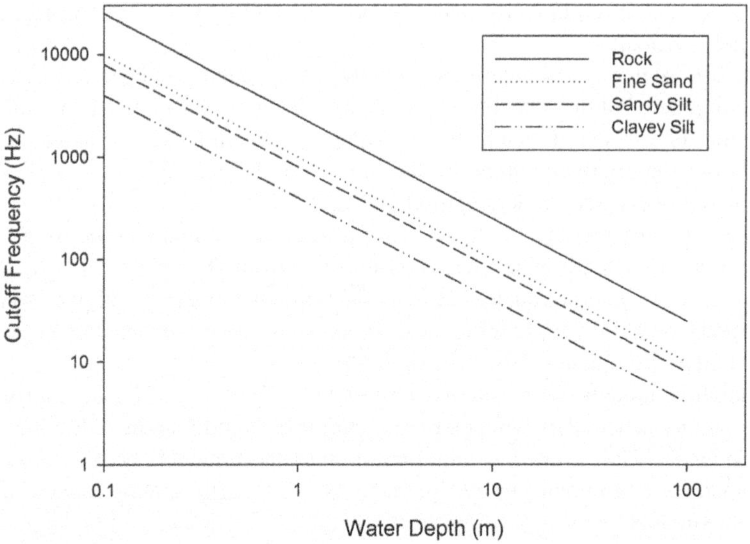

Fig. 5.8 High-pass cutoff frequency versus water depth for different types of sediment. Adapted from Rogers and Cox (1988)

ancillary hearing structures, such as Otophysans, that extend their hearing range into the higher frequencies will hear better in shallower water.

The major differences between air and underwater acoustics that influence sound propagation in the two media are summarized in Table 5.1.

5.4 Abiotic or Geophony Sound Sources

Major abiotic or geophony sources of ambient noise on land and underwater are due to weather-associated conditions such as wind and rain. These conditions not only affect the propagation of the signal to the receiver but can also increase the intensity of the natural so-called soundscape (Schafer 1993). The following sections discuss the influence of abiotic noise on the soundscape on land and underwater.

5.4.1 Abiotic Sources of Sound in Air

Sound Metrics

The influence of ambient noise on acoustic communication can only be fully appreciated when characteristics like (long-term) spectrum level, frequency bandwidth, and amplitude variation in time are available, but so far such information is sporadic

Table 5.1 Comparison of air and underwater acoustics

Variable	Air	Water
Absorption	High, dB/m	Low, dB/km
Ambient noise	Relatively low	Relatively high
Boundary effects	Strong constructive and destructive interferences (up to +6 dB and down to −15 dB) at low frequencies close to source	Effects near the surface but constructive and destructive interferences at opposite frequencies to those in air
Density, kg/m^3	1.2	1,000 (freshwater) to 1,030 (ocean)
Particle velocity at 1 Pa, m/s (cf. Eq. 5.12)	2.5×10^{-3}	6.7×10^{-7}
Propagation, km	Long-range propagation is generally not possible due to low density of air and high absorption	Long-range propagation is often possible due to high density of water and low absorption
Scattering	Air is compliant; good scatterers in closed habitats, e.g., tree trunks and rocks	Water is stiff; many good scatterers, e.g., fish swim bladders
Sound speed, m/s	340	1,500
Sound speed profile	Large changes with altitude (up to 20%) due to large changes in temperature and wind	Small changes with depth (<6%) due to temperature and pressure variation
Sound speed profile variability	Very variable (up to ±6 dB) due to winds and diurnal/seasonal temperature fluctuations	Less variable, seasonal, and generally only near the surface
Turbulence	High in open habitats, lower in closed habitats. Limits boundary effects to low frequencies	High in strong tidal currents and river flows

in the literature and either specified to precise geographic locations or generalized to major habitats such as grasslands and tropical forests. (Such information is also badly needed for water where, in addition, background noise needs to be measured as both pressure and particle motion in three dimensions to be relevant for fish and invertebrate studies.)

In addition, many terrestrial environmental studies have characterized background noise levels measured using the A-weighting filter [in dB(A); frequency range from 20 Hz to 20 kHz, but strongly attenuating frequencies below 0.7 kHz and above 8.0 kHz] because they have focused on ambient-noise influence on human life and mainly in the vicinity of human habitation. However, using the A-weighting filter seriously underestimates the low-frequency components dominating ambient noise that may be relevant for animals communicating in the frequency range below 1 kHz and above 10 kHz. A much better approach is to calculate the spectrum level of ambient noise as a function of frequency directly (Nemeth and Brumm 2010) or by measuring it in third-octave bands. Only then can, for example, the active space of an animal communicating by sound in noisy environments be predicted (Lohr et al. 2003; Jensen et al. 2008). If a measure of the overall sound level is required, then the Z-weighting filter (specified by the standard of the International Electrotechnical Commission 2013) is a better option for bioacousticians. Z stands for "zero-"frequency weighting, but, in reality, the range is 10 Hz to 20 kHz, with a maximum variation of ±1.5 dB when integrated in high-end sound level meters.

Bioacousticians have investigated a wide variety of relevant animal habitats and recorded the ambient noise using either the A-weighting filter or frequency ranges relevant for their experimental organisms but seldom calculated spectrum levels and, if so, only as power spectra with relative decibel values. Therefore, it is difficult to systematically map the background noise in major terrestrial habitats of acoustically communicating organisms. Generally, abiotic noise has the most energy below 1 kHz and the power spectrum typically decays almost linearly with frequency by 30-40 dB from 1 kHz to 10 kHz (Dingle et al. 2008). In temperate environments, the lowest level of the constant "natural quiet" typically occurs at night where it is on the order of 30-40 dB sound pressure level and limited to frequencies below a few hundred hertz (Meyer 2015).

Ambient Sound Sources

Ambient-noise sources can be divided into those that are located at specific geographic positions and subject to spherical or cylindrical attenuation and those that cannot be located but seem to be the same in all directions. Abiotic noise could, for example, be defined as the sounds that remain when sound sources of animal and human origins are removed, but there are many other definitions and terms used in the literature on this subject. In general, this background noise or "natural quiet" remains constant in all directions and positions in a given habitat. This weak background "hum" is usually dominated by low frequencies because all possible uncorrelated sound sources in a given habitat add up, but low frequencies propagate the longest (cf. Fig. 5.2) and are less attenuated by scattering.

Only under special circumstances are abiotic noise sources localizable; this applies to geological phenomena ("geophonies") and moving water. Volcanic activity, for example, produces very loud bursts of sound but like earthquakes and meteoroids exploding in the earth's atmosphere, it is episodic and usually of relatively short duration. Therefore, its influence on acoustic communication is probably negligible as are other episodic phenomena like the sounds of falling trees or rocks, landslides, breaking ice, thunder, lightning, and major forest or grassland fires.

Abiotic sound produced by moving water, on the other hand, is rather constant. This applies to fast running water in torrents and waterfalls and to constant ocean surf. These broadband sound sources remain localized and rather constant in amplitude, with intensity modulations only over long time spans (weeks to months), and, consequently, they may constitute selective forces acting on animals living in their vicinity. A famous example is the concave-eared torrent frog (*Amolops tormotus*). It lives close to fast flowing streams with high intensity, broadband noise levels peaking at about 100 Hz. Although most anurans call in the 100-5,000 Hz range, the calls of this species have surprisingly strong components above 20 kHz (Narins et al. 2004), and these components are perceived by conspecific males because the ultrasonic components are not masked by the lower frequency torrent noise (Feng et al. 2006). Small upward shifts in song frequency similar to those observed for songbirds in noisy cities (Slabbekoorn 2013) have also been observed in birds living close to waterfalls and torrents (Brumm and Slater 2006). A comparative study of anuran habitats in southeast Asia (Goutte et al. 2013) emphasizes the correlation between moving water and ambient noise because the average sound level measured at night when the frogs had stopped calling was 45 dB(A) near ponds, 52 dB(A) near rivers/lakes, and 65 dB(A) near torrents. This is less than the sound level of ocean surf, where Gough et al. (2014) found average levels of 73 dB(A) next to the shore attenuating with distance to 54 dB(A) 100 m inland or close to the predictions of cylindrical spreading (−20 dB at 100 m from the source).

Rain is another example of moving water that raises the ambient-noise level in a given habitat but often only for shorter periods of time (minutes, hours) and sometimes occurring regularly at certain times of the day or year, like the monsoon in the tropics. The rain-produced sound level is proportional to the intensity of the rainfall and its spectrum is rather broadband. Consequently, most birds and insects generally stop singing during heavy rainfall. Tawny owls (*Strix aluco*), for example, were found to vocalize considerably less during heavy rain when the sound level rose to 52 dB sound pressure level relative to dry conditions when it was only 33 dB in their perceptually relevant frequency band, 560-1,080 Hz (Lengagne and Slater 2002).

Wind probably is the most prominent source of abiotic noise in air. Wind produces the waves on lakes and oceans and thereby the sound of both the turbulence-generated air bubbles in waves and coastal surf. Wind can also induce sound by sheering noise from power lines or other objects that can shed vortices. Even moderate wind also produces sound from moving vegetation, for example, rustling leaves and twigs in the forest canopy. In general, wind speed is high (0.8-1.8 m/s) from morning (about 10 a.m.) to evening (about 6 p.m.) but much lower (<0.6 m/s) at night (McNett et al. 2010). Moderate wind will raise the ambient-noise level to 30-40 dB sound pressure level in the frequency range from 100 to 1,000 Hz, whereas strong wind may raise the level to 60-70 dB sound pressure level in this range. This is probably one of many factors that make many vocalizing animals most active in the morning and evening, with a low in the middle of the day (Luther and Gentry 2013).

5.4.2 *Abiotic Sources of Underwater Sound*

Prominent sources of abiotic underwater sound include wind and rain, where these sources dominate the frequency range (100-10,000 Hz; Fig. 5.9). The wind-generated waves and the accumulation of rain droplets on the water's surface can be very loud in extreme weather conditions. For example, during Hurricane Charley,

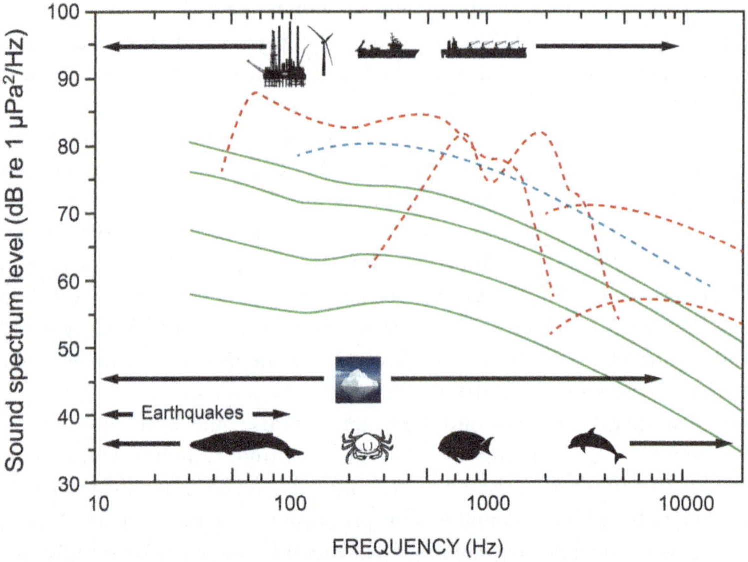

Fig. 5.9 Spectrum levels showing the contribution that abiotic and biotic noises can make to the underwater soundscape. *Red dashed lines*, biological choruses; *blue dashed line*, rain noise; *green lines*, wind-dependent noise at 30, 20, 10, and 5 knots (*top to bottom*). The pictures of anthropogenic, abiotic, and biotic sources represent their dominant energies and frequency ranges. Adapted from Cato (2014)

the mean power level in Charlotte harbor, Florida, increased to 140 dB re 1 μPa $Hz^{-1/2}$ in the bandwidth of 100-600 Hz (Locascio and Mann 2005). Depending on the weather conditions, oceans, seas, and lakes may be quiet on calm days, but fast flowing rivers and streams can be very noisy due to air bubbles caused by water turbulence (Wysocki et al. 2007). In addition to weather events, there are rarer abiotic events that can also generate loud low-frequency sounds, such as underwater earthquakes (Wilcock et al. 2014), lightning strikes (Hill 1985), and cracking or breaking ice (Lewis and Denner 1988). Earthquakes are a source of very low frequency sounds in the world's oceans and the frequencies vary depending on the magnitude of the earthquake. For example, a 5.5 magnitude earthquake will produce an approximately 1-Hz signal, where a 1.5 magnitude earthquake will produce a 100-Hz signal. Lightning strikes have been reported to be one of the loudest sources of abiotic sound in water, with the peak-to-peak source levels estimated to be 260 dB re 1 μPa (Hill 1985). Ice covered areas, such as the Arctic or Antarctica, are generally expected to have low ambient noise due to the reduced effect of wind-driven waves. However, the dynamics of ice, such as ice formation, deformation, and cracking, greatly increases ambient-noise levels over broad frequency ranges.

5.5 Biotic or Biophony Sound Sources

Biotic or biophony sound sources or animals making sound can also cause increases in the overall intensity of land and underwater soundscapes. Increased biotic background noise produced by other or competing animals can mask the signal that is intended for the receiver. Therefore, sound-generating animals are competing for acoustic space or an acoustic niche in which their signal-to-noise ratio can be optimized. This section discusses the importance of biotic noise sources in terms of the overall soundscape for both land and underwater.

5.5.1 Biotic Sources of Sound in Air

Animals produce sound mainly for communication, deterrence, and echolocation. The most vocal groups in air are found among insects, anurans, birds, and mammals. Vocalizations of solitary animals like territorial male songbirds can mask conspecific communication, but vocalizations that alternate with pauses can reduce masking when, for example, the focal bird waits for its turn to sing as demonstrated in the Eurasian wren (*Troglodytes troglodytes*; Yang et al. 2014) and in nightingales (*Luscinia megarhynchos*; Brumm 2006). However, acoustically communicating animals are often located in large aggregations (flocks of starlings, trees full of cicadas, choruses of frogs, caves full of bats) where the sound becomes almost continuous at high levels in the species-specific frequency range, effectively blocking those particular frequency bands. Consequently, acoustically communicating animals will often compete for acoustic space with both conspecifics and other species and this may lead to spectral song divergence correlated with different ambient-noise profiles, which in the end may increase the likelihood of speciation (Dingle et al. 2008).

Biotic sound sources not only have a potentially detrimental masking impact because vocalizing animals may also exploit information from heterospecific vocalizations such as alarm calls, where some species have reached an unexpected level of information extraction by eavesdropping (Templeton and Greene 2007). In addition, animals' use of heterospecific vocalizations and of soundscape characteristics for social and geographical orientation may be much more widespread than presently documented as suggested by Slabbekoorn and Bouton (2008).

Chorusing insects are the primary sources of continuous high-frequency sound in terrestrial habitats dominating the frequency bands from 4 to 8 kHz in Central African rain forests and 6 to 8 kHz in nearby gallery forests (Luther and Gentry 2013). For example, Orthopteran insects tend to produce intense sound at night by stridulation in the frequency range of 3-10 kHz but often with strong components in the ultrasonic range, in a few examples extending to 100 kHz, and at a general sound level of 60-65 dB sound pressure level in rain forests (Slabbekoorn 2004; Hartbauer et al. 2012). Some insect species omnipresent in Amazonas produce long-lasting frequency bands in the 2.5- to 3.5-kHz range during the day, giving a continuous background noise level of 50-55 dB(A) (Meyer 2015), whereas cricket species in southern India mainly call in the evening at 3.5 and 4.5 kHz (Balakrishnan 2005) and katydid species in western Australia have broadband songs with the strongest components at 6 and 12 kHz, extending up to 30 kHz (Römer et al. 1989).

Most birds vocalize at frequencies between 1 and 9 kHz, with typical source levels of 80-85 dB sound pressure level (Luther 2009). Bird song activity patterns often exhibit diurnal variation being intense during the dawn chorus (Brown and Hanford 2003), less intense during the day, and again having a small maximum at dusk, the dusk chorus (Schmidt and Belinsky 2013). Similar diurnal variations are found in the vocal activity of Amazonian frogs (Ellinger and Hödl 2003) and of mammals like howler monkeys (*Alouatta* sp.) in the frequency range of 300 to 2,000 Hz (Cornick and Markowitz 2002).

5.5.2 *Biotic Sources of Underwater Sound*

There is a range of marine and freshwater animals, from crustaceans to fishes to marine mammals, that produces sound and is responsible for different soundscapes. Production of these sounds occurs under different behavioral contexts, such as reproduction, predator avoidance, and feeding. One of the most dominant biotic sounds in the ocean is produced by an invertebrate taxon, the snapping shrimp (*Synalpheus paranemeris*), which produces peak-to-peak source levels of 183-189 dB re 1 μPa in the frequency range of 2-200 kHz (Au and Banks 1998; Fig. 5.9). Other invertebrates that produce sounds are mantis shrimp (*Hemiisquilla californiensis*; Staaterman et al. 2010), which produce a low rumbling sound to deter predators or competitors (fundamental frequency range of 20-60 Hz), and the sea urchin (*Evechinus chloroticus*; Radford et al. 2008), which produces sound at higher frequencies as a by-product of its feeding behavior (frequency range of 700-3,000 Hz). Diverse fish species produce sounds in both fresh and marine ecosystems. These sounds can range from great symphonies to synchronized spawning behavior to loud grunts to deter competitors or other threats. There are numerous species of fish that form large groups and generate sound synchronously to inform each other that they are ready to release their gametes. These great symphonies are known as choruses (Fig. 5.9) and can alter the soundscape significantly during different times of the year (Luczkovich et al. 2008). Sometimes these choruses can increase the root-mean-square intensity levels of the local ambient conditions by as much as 40 dB re 1 μPa where red drum (*Sciaenops ocellatus*) produce choruses centered around 128 Hz and weakfish (*Cynoscion regalis*) centered around 386 Hz. Other fishes, such as the plainfin midshipman fish (*Porichthys notatus*), will produce isolated sounds as warning calls (Amorim and Hawkins 2000; Amorim 2006) or nest guarding (Amorim 2006). More recently, research has shown that bigeyes (*Pempheris adspersa*) have the potential to produce sound as a contact call (van Oosterom et al. 2016).

Temperate and tropical reefs are some of the most acoustically complex habitats or soundscapes in the ocean. Research has shown that different habitats separated geographically (large distances) and also locally (small scale) have their own unique acoustic signatures, which can reflect the type or quality of the habitat (Radford et al. 2014). For example, oyster reefs in Pamlico Sound, North Carolina, have distinct underwater signatures compared with areas without reefs (Lillis et al. 2014), where reef habitats dominated by kelp forests sound different to reefs dominated by urchins (Radford et al. 2010), and two reefs separated by a mere 5 km in the Florida Keys sound completely different (Staaterman et al. 2014). In every one of these cases, the acoustic soundscape is determined by the soniferous animals present and competing for their own acoustic space.

5.6 Summary and Conclusions

Many animals both in the air and underwater depend on acoustic communication to attract mates, deter predators and conspecifics, and ultimately promote fitness. The ability of the signaler to get the encoded message to the intended receiver depends on a range of factors including the sound transmission properties of the medium and the level of environmental noise, which can impede the ability of the receiver to detect the signal. The range and information content of acoustic communication signals are limited by the physical properties of sound propagation. All signals undergo some sort of geometric sound attenuation, whether it is one of the two traditional models, spherical or cylindrical spreading, or something in-between. Sound absorption, refraction, and turbulence all alter the signal features during propagation from sender to receiver in some way or another through changing either the amplitude or frequency envelope of the signal. Due to the compressibility differences between air and water, there are differences in the propagation of sound in either medium, namely, sound travels faster in water. Therefore, sound can travel greater distances in water because for similar frequencies, attenuation is less in water than in air. As a consequence, environmental noise can propagate further and acoustic interference can occur over larger distances. However, how environmental noise masks and interferes with communication signals is similar between in air and underwater.

Environmental noise can take various forms, from natural sounds caused by weather (abiotic) to sounds produced by animals themselves (biotic). Abiotic sounds are similar between air and underwater where wind, rain, and lightning dominate the soundscape. Biotic sounds in air are dominated by arthropods, amphibians, birds, and mammals, whereas major biotic sources underwater are arthropods, amphibians, fishes, and mammals; for example, in the air, Insecta are the dominant arthropods, whereas underwater Crustacea dominate. All these are natural sounds that have been present in the environment for a long time and animals both in air and underwater have evolved acoustic niches in which they communicate.

In conclusion, every class of environmental noise produces sound that spans the entire frequency spectrum of the soundscape at various spectrum levels and has the potential to interfere or mask animal communication signals. Therefore, animals that rely on sound to communicate have to adapt and evolve to their local soundscape to get their messages across.

Acknowledgments We are grateful to the editors and an anonymous referee for improving the manuscript.

Compliance with Ethics Requirements
Ole Næsbye Larsen declares that he has no conflict of interest.
Craig Radford declares that he has no conflict of interest.

References

Ainslie, M. A., & McColm, J. G. (1998). A simplified formula for viscous and chemical absorption in sea water. *The Journal of the Acoustical Society of America*, 103(3), 1671-1672.

Amorim, M. C. P. (2006). Diversity of sound production in fish. In F. Ladich, S. P. Collin, P. Moller, & B. G. Kapoor (Eds.), *Communication in Fishes* (pp. 71-105). Enfield, NH: Science Publishers.

Amorim, M. C. P., & Hawkins, A. D. (2000). Growling for food: Acoustic emissions during competitive feeding of the streaked gurnard. *Journal of Fish Biology*, 57(4), 895-907.

Attenborough, K. (2007). Sound propagation in the atmosphere. In T. D. Rossing (Ed.), *Springer Handbook of Acoustics* (pp. 113-147). New York: Springer-Verlag.

Attenborough, K., Li, K. M., & Horoshenkov, K. (2007). *Predicting Outdoor Sound*. London, New York: Taylor and Francis.

Au, W. W. L., & Banks, K. (1998). The acoustics of the snapping shrimp *Synalpheus parneomeris* in Kaneohe Bay. *The Journal of the Acoustical Society of America*, 103, 41-47.

Au, W. W. L., & Hastings, M. C. (2008). *Principles of Marine Bioacoustics*. New York: Springer-Verlag.

Balakrishnan, R. (2005). Neurobiology and behaviour: A network of connections. *Current Science* 89, 1147-1165.

Bass, H. E., Sutherland, L. C., Zuckerwar, A. J., Blackstock, D. T., & Hester, D. M. (1995). Atmospheric absorption of sound: Further developments. *The Journal of the Acoustical Society of America*, 97(1), 680-683.

Blackstock, D. T. (2000). *Fundamentals of Physical Acoustics*. New York: John Wiley & Sons.

Bradbury, J. W., & Vehrencamp, S. L. (1998). *Principles of Animal Communication*. Sunderland, MA: Sinauer Associates.

Brown, T. J., & Hanford, P. (2003). Why birds sing at dawn: The role of consistent song transmission. *Ibis* 145, 120-129.

Brumm, H. (2006). Signalling through acoustic windows: Nightingales avoid interspecific competition by short-term adjustment of song timing. *Journal of Comparative Physiology A: Neuroethology, Sensory,Neural, and Behavioral Physiology* 192, 1279-1285.

Brumm, H., & Todt, D. (2002). Noise-dependent song amplitude regulation in a territorial songbird. *Animal Behaviour*, 63, 891-897.

Brumm, H., & Slater, P. J. B. (2006). Ambient noise, motor fatigue, and serial redundancy in chaffinch song. *Behavioural Ecology and Sociobiology*, 60, 475-481.

Catchpole, C. K., & Slater, P. J. B. (2008). *Bird Song. Biological Themes and Variations*. Cambridge, UK: Cambridge University Press.

Cato, D. H. (2014). Shipping noise impacts on marine life. In *Proceedings of the 43rd International Congress and Exposition on Noise Control Engineering (Internoise 2014): Improving the World Through Noise Control*, Melbourne, VIC, Australia, November 16-19, 2014, vol. 1, pp. 418-423.

Collins, M. D. (1993). A split-step Padé solution for the parabolic equation. *The Journal of the Acoustical Society of America*, 93, 1736-1742.

Cornick, L. A., & Markowitz, H. (2002). Diurnal vocal patterns of the black howler monkey (*Alouatta pigra*) at Lamanai, Belize. *Journal of Mammalogy*, 83(1), 159-166.

Dabelsteen, T., & Pedersen, S. B. (1988). Correspondence between messages in the full song of the blackbird *Turdus merula* and meanings to territorial males, as inferred from responses to computerized modifications of natural song. *Zeitschrift für Tierpsychologie*, 69(2), 149-165.

Dabelsteen, T., Larsen, O. N., & Pedersen, S. B. (1993). Habitat-induced degradation of sound signals: Quantifying the effects of communication sounds and bird location on blur ratio, excess attenuation, and signal-to-noise ratio in blackbird song. *The Journal of the Acoustical Society of America*, 93(4), 2206-2220.

Dingle, C., Halfwerk, W., & Slabbekoorn, H. (2008). Habitat-dependent song divergence at subspecies level in the grey-breasted wood-wren. *Journal of Evolutionary Biology*, 21, 1079-1089.

Ellinger, N., & Hödl, W. (2003). Habitat acoustics of a neotropical lowland rainforest. *Bioacoustics*, 13(3), 297-321.

Embleton, T. F. W. (1996). Tutorial on sound propagation outdoors. *The Journal of the Acoustical Society of America*, 100(1), 31-48.

Fahy, F. (2003). *Foundations of Engineering Acoustics*. London: Elsevier Academic Press.

Farcas, A., Thompson, P. M., & Merchant, N. D. (2016). Underwater noise modelling for environmental impact assessment. *Environmental Impact Assessment Review*, 57, 114-122.

Feng, A. S., Narins, P. M., Xu, C.-H., Lin, W.-Y., Yu, Z.-L., Qiu, Q., Xu, Z.-M., & Shen, J.-X. (2006). Ultrasonic communication in frogs. *Nature*, 440, 333-336.

Ferris, R. H. (1972). Comparison of measured and calculated normal-mode amplitude functions for acoustic waves in shallow water. *The Journal of the Acoustical Society of America*, 52(3), 981-988.

Garstang, M., Larom, D., Raspet, R., & Lindeque, M. (1995). Atmospheric controls on elephant communication. *Journal of Experimental Biology*, 198, 939-951.

Gough, D. C., Mennill, D. J., & Nol, E. (2014). Singing seaside: Pacific wrens (*Troglodytes pacificus*) change their songs in the presence of natural and anthropogenic noise. *Wilson Journal of Ornithology*, 126(2), 269-278.

Goutte, S., Dubois, A., & Legendre, F. (2013).The importance of ambient sound level to characterise anuran habitat. *PLoS ONE*, 8(10): e78020. doi:https://doi.org/10.1371/journal.pone.0078020.

Halfwerk, W., Bot, S., & Slabbekoorn, H. (2012). Male great tit song perch selection in response to noise-dependent female feedback. *Functional Ecology*, 26, 1339-1347.

Hartbauer, M., Siegert, M. E., Fertschai, I., & Römer, H. (2012). Acoustic signal perception in a noisy habitat: lessons from synchronising insects. *Journal of Comparative Physiology A: Neuroethology, Sensory, Neural, and Behavioral Physiology*, 198, 397-409.

Heller, E. J. (2013). *Why You Hear What You Hear. An Experimental Approach to Sound, Music, and Psychoacoustics*. Princeton, NJ: Princeton University Press.

Higgs, D. M., & Radford, C. A. (2016). The potential overlapping roles of the ear and lateral line in driving "acoustic" responses. In J. A. Sisneros (Ed.), *Fish Hearing and Bioacoustics* (pp. 255-270). New York: Springer International Publishing.

Hill, R. D. (1985). Investigation of lightning strikes to water surfaces. *The Journal of the Acoustical Society of America*, 78, 2096-2099.

Holland, J., Dabelsteen, T., Pedersen, S. B., & Paris, A. L. (2001). Potential ranging cues contained within the energetic pauses of transmitted wren song. *Bioacoustics*, 12(1), 3-20.

International Electrotechnical Commission. (2013). *Electroacoustics. Sound Level Meters, Part 2: Pattern Evaluation Tests* IEC 61672-2, International Electrotechnical Commission, Geneva, Switzerland.

International Standard Organization (ISO). (1993). *Acoustics: Attenuation of Sound During Propagation Outdoors. Part 1: Calculation of the Absorption of Sound by the Atmosphere* ISO 9613-1, International Standard Organization, Geneva, Switzerland. Available at https://www.iso.org/standard/17426.html.

Jensen, K. K., Larsen, O. N., & Attenborough, K. (2008). Measurements and predictions of hooded crow (*Corvus corone cornix*) call propagation over open field habitats. *The Journal of the Acoustical Society of America*, 123(1), 507-518.

Larsson, C., Hallberg, B., & Israelsson, S. (1988). A method to estimate meteorological effects on sound propagation near the ground. *Applied Acoustics*, 25(1), 17-31.

Lengagne, T., & Slater, P. J. B. (2002). The effects of rain on acoustic communication: tawny owls have good reason for calling in less wet weather. *Proceedings of the Royal Society B: Biological Sciences*, 269, 2121-2125.

Lewis, J. K., & Denner, W. W. (1988). Arctic ambient noise in the Beaufort Sea: Seasonal relationships to sea ice kinematics. *The Journal of the Acoustical Society of America*, 83(2), 549-565.

Lillis, A., Eggleston, D. B., & Bohnenstiehl, D. R. (2014). Estuarine soundscapes: Distinct acoustic characteristics of oyster reefs compared to soft-bottom habitats. *Marine Ecology Progress Series*, 505, 1-17.

Locascio, J. V., & Mann, D. A. (2005). Effects of Hurricane Charley on fish chorusing. *Biology Letters*, 1, 362-365.

Lohr, B., Wright, T. F., & Dooling, R. J. (2003). Detection and discrimination of natural calls in masking noise by birds: Estimating the active space of a signal. *Animal Behaviour*, 65, 763-777.

Luczkovich, J. J., Pullinger, R. C., Johnson, S. E., & Sprague, M. W. (2008). Identifying sciaenid critical spawning habitats by the use of passive acoustics. *Transactions of the American Fisheries Society*, 137, 576-605.

Luther, D. (2009). The influence of the acoustic community on songs of birds in a neotropical rain forest. *Behavioral Ecology*, 20, 864-871.

Luther, D., & Gentry, K. (2013). Sources of background noise and their influence on vertebrate acoustic communication. *Behaviour*, 150, 1045-1068.

Marten, K., & Marler, P. (1977). Sound transmission and it significance for animal vocalization. I. Temperate habitats. *Behavioral Ecology and Sociobiology*, 2, 271-290.

McLaughlin, K. E., & Kunc, H. P. (2013). Experimentally increased noise levels change spatial and singing behaviour. *Biology Letters*, 9: 20120771.

McNett, G. C., Luan, L. H., & Cocroft, R. B. (2010). Wind-induced noise alters signaler and receiver behavior in vibrational communication. *Behavioral Ecology and Sociobiology*, 64, 2043-2051.

Medwin, H., & Clay, C. S. (1998). *Fundamentals of Acoustical Oceanography*. San Diego, CA: Academic Press.

Meyer, J. (2015). Acoustic adaptation to natural environments. In J. Meyer, *Whistled Languages: A Worldwide Inquiry About Human Whistled Speech* (pp. 91-103). Berlin: Springer-Verlag.

Michelsen, A., & Larsen, O. N. (1983). *Strategies for Acoustic Communication in Complex Environments*. In F. Huber & H. Markl (Eds.), *Neuroethology and Behavioral Physiology* (pp. 321-331). Berlin Heidelberg: Springer-Verlag.

Morton, E. S. (1975). Ecological sources of selection of avian sounds. *American Naturalist*, 109, 17-34.

Naguib, M., & Wiley, R. H. (2001). Estimating the distance to a source of a sound: Mechanisms and adaptations for long-range communication. *Animal Behaviour* 62, 825-837.

Narins, P. M., Feng, A. S., Lin, W., Schnitzler, H.-U., Denzinger, A., Suthers, A., & Xu, C. (2004). Old World frog and bird vocalizations contain prominent ultrasonic harmonics. *The Journal of the Acoustical Society of America*, 115(2), 910-913.

Nemeth, E., & Brumm, H. (2010). Birds and anthropogenic noise: Are urban songs adaptive? *American Naturalist*, 176(4), 465-475.

Patricelli, G. L., & Blickley, J. L. (2006). Avian communication in urban noise: Causes and consequences of vocal adjustment. *Auk*, 123(3), 639-649.

Penna, M., Llusia, D., & Márquez, R. (2012). Propagation of natural toad calls in a Mediterranean terrestrial environment. *The Journal of the Acoustical Society of America*, 132(6), 4025-4031.

Piercy, J. E., Embleton, T. F. W., & Sutherland, L. C. (1977). Review of noise propagation in the atmosphere. *The Journal of the Acoustical Society of America*, 61(6), 1403-1418.

Porter, M. B. (1992). *The Kracken Normal Mode Program*. Technical Report NRL/MR/5120-92-6920, Naval Research Laboratory, Washington, DC.

Porter, M. B., & Liu, Y.-C. (1994). Finite-element ray tracing. *Theoretical Computing Acoustics*, 2, 947-956.

Radford, C. A., Jeffs, A. G., Tindle, C. T., & Montgomery, J. C. (2008). Temporal patterns in ambient noise of biological origin from a shallow water temperate reef. *Oecologia*, 156, 921-929.

Radford, C. A., Stanley, J. A., Tindle, C. T., Montgomery, J. C., & Jeffs, A. G. (2010). Localised coastal habitats have distinct underwater sound signatures. *Marine Ecology Progress Series*, 401, 21-29.

Radford, C. A., Stanley, J. A., & Jeffs, A. G. (2014). Adjacent coral reef habitats produce different underwater sound signatures. *Marine Ecology Progress Series*, 505, 19-28.

Rogers, P. H., & Cox, M. (1988). Underwater sound as a biological stimulus. In J. Atema, R. R. Fay, A. N. Popper, & W. N. Tavolga (Eds.), *Sensory Biology of Aquatic Animals* (pp. 131-149). New York: Springer-Verlag.

Römer, H., Bailey, W., & Dadour, I. (1989). Insect hearing in the field. III. Masking by noise. *Journal of Comparative Physiology A: Neuroethology, Sensory, Neural, and Behavioral Physiology*, 164, 609-620.

Schafer, R. M. (1993). *The Soundscape: Our Sonic Environment and the Tuning of the World*. Rochester, VT: Destiny Books.

Schmidt, K. A., & Belinsky, K. L. (2013). Voices in the dark: Predation risk by owls influences dusk singing in a diurnal passerine. *Behavioral Ecology and Sociobiology*, 67, 1837-1843.

Slabbekoorn, H. (2004). Habitat-dependent ambient noise: Consistent spectral profiles in two African forest types. *The Journal of the Acoustical Society of America*, 116(6), 3727-3733.

Slabbekoorn, H. (2013). Songs of the city: Noise-dependent spectral plasticity in the acoustic phenotype of urban birds. *Animal Behaviour*, 85, 1089-1099.

Slabbekoorn, H., & Bouton, N. (2008). Soundscape orientation: A new field in need of sound investigation. *Animal Behaviour*, 76, e5-e8.

Slabbekoorn, H., Ellers, J., & Smith, T. B. (2002). Birdsong and sound transmission: The benefits of reverberations. *Condor*, 104, 564-573.

Staaterman, E. R., Claverie, T., & Patek, S. N. (2010). Disentangling defense: the function of spiny lobster sounds. *Behaviour*, 147, 235-258.

Staaterman, E., Paris, C. B., DeFerrari, H. A., Mann, D. A., Rice, A. N., & Alessandro, E. K. (2014). Celestial patterns in marine soundscapes. *Marine Ecology Progress Series*, 508, 17-32.

Templeton, C. N., & Greene, E. (2007). Nuthatches eavesdrop on variations in heterospecific chickadee mobbing alarm calls. *Proceedings of the National Academy of Sciences of the United States of America*, 104(13), 5479-5482.

van Oosterom, L., Montgomery, J. C., Jeffs, A. G., & Radford, C. A. (2016). Evidence for contact calls in fish: Conspecific vocalisations and ambient soundscape influence group cohesion in a nocturnal species. *Scientific Reports*, 6, 19098.

Weston, D. E. (1971). Intensity-range relations in oceanographic acoustics. *Journal of Sound Vibration*, 18, 271-287.

Wilcock, W. S. D., Stafford, K. M., Andrew, R. K., & Odom, R. I. (2014). Sounds in the ocean at 1-100 Hz. *Annual Review of Marine Science*, 6, 117-140.

Wiley, R. H., & Richards, D. G. (1978). Physical constraints on acoustic communication in the atmosphere: Implications for the evolution of animal vocalizations. *Behavioral Ecology and Sociobiology*, 3, 69-94.
Wood, W. E., & Yezerinac, S. M. (2006). Song sparrow (*Melospiza melodia*) song varies with urban noise. *The Auk*, 123(3), 650-659.
Wysocki, L. E., Amoser, S., & Ladich, F. (2007). Diversity in ambient noise in European freshwater habitats: Noise levels, spectral profiles, and impact on fishes. *The Journal of the Acoustical Society of America*, 121(5), 2559-2566.
Yang, X.-J., Ma, X.-R., & Slabbekoorn, H. (2014).Timing vocal behaviour: Experimental evidence for song overlap avoidance in Eurasian wrens. *Behavioural Processes*, 103, 84-90.

Chapter 6
Effects of Man-Made Sound on Fishes

Anthony D. Hawkins and Arthur N. Popper

Abstract Sound provides animals with a means of rapid, directional, and long-distance communication. It also provides animals with a "gestalt" view of their environment by giving an acoustic image of the world that often extends far beyond what is available from other senses. Thus, sound is highly relevant for fishes, and any interference with the ability to detect sound has potential consequences for the fitness and survival of individuals, populations, and species. There is a growing body of evidence that the addition of man-made sound in the aquatic environment has the potential to affect the ability of fishes to detect and use the biologically relevant sounds that are important for their survival. Moreover, there is also evidence that especially intense sounds not only affect sound detection and behavior but also have the potential to have physiological and physical effects on fish that could result in greatly reduced fitness and, in some cases, directly to death. This chapter examines the potential effects of man-made sound on fishes. It considers the sources of such sounds, the current data on potential effects and impacts, and implications for regulation of such sounds so that the potential impact is mitigated.

Keywords Airguns · Anthropogenic · Behavior · Ear · Hearing · Noise · Pile driving · Shipping · Sonar

6.1 Introduction

Sound is important in the lives of fishes for monitoring the world around them and, in many species, for communication. Much of the early work on the significance of underwater sound to aquatic organisms was first discussed in symposium volumes edited by Tavolga (1964, 1967) and then by Tavolga et al. (1981) and Webb et al.

A. D. Hawkins (✉)
Loughine Ltd., Aberdeen, UK

A. N. Popper
Department of Biology, University of Maryland, College Park, MD, USA
e-mail: apopper@umd.edu

H. Slabbekoorn et al. (eds.), *Effects of Anthropogenic Noise on Animals*, Springer Handbook of Auditory Research 66,
https://doi.org/10.1007/978-1-4939-8574-6_6

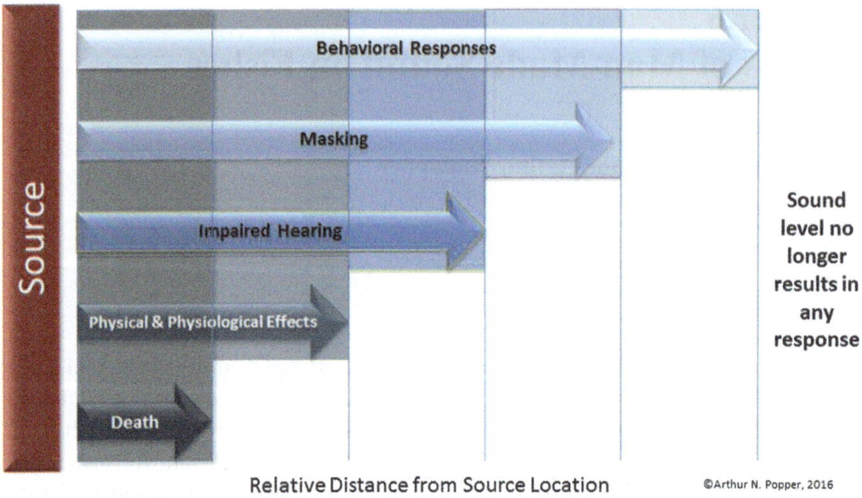

Fig. 6.1 Relationship between sound levels and potential effects on animals. See text for discussion

(2008). Most recently, the effects of sounds on fishes have been reviewed in volumes by Popper and Hawkins (2012, 2016, see also papers at http://scitation.aip.org/content/asa/journal/poma/27/1).

Fishes, depending on the species, may use sound to communicate with one another (Myrberg 1981; Ladich 2013), detect prey and predators (Remage-Healey et al. 2006), navigate from one place to another (Cotter 2008; Stanley et al. 2012), and select appropriate habitats (Simpson et al. 2004, 2005). Even though many species do not use sound to communicate with conspecifics, it is likely that all species use sound to learn about their environment by detecting and using the soundscape or "acoustic scene" (Slabbekoorn and Bouton 2008; Fay 2009). In effect, sound detection provides fishes, like other animals, with three-dimensional information from a larger space around them than is possible with vision, chemical senses, or electroreception (Bregman 1994).

Seas, lakes, and rivers are often noisy, with many sounds that are natural in origin (Wenz 1962; Martin and Popper 2016). Since the start of the industrial revolution, man-made sounds have added to the volume of background noise (Andrew et al. 2002; Hildebrand 2009). Sound travels about 4.8 times faster in water than in air and propagates substantial distances (Urick 1983), and as a consequence, the influence of underwater noise from ships, seismic exploration, offshore energy developments, and other sources can be pervasive over large areas (e.g., Popper et al. 2014; Popper and Hawkins 2016).

The effects of noise on animals can range from mild and insignificant to severe and lasting (Fig. 6.1). High sound levels can exert severe impact on individuals, affecting their fitness to survive and reproduce and, in some cases, may affect large numbers of animals, potentially damaging whole populations. It is important to underline the distinction between *effects* and *impacts* (Boehlert and Gill 2010). Effects are the broad range of potentially measurable changes that may be observed in individuals or groups of animals as a result of sound exposure. Impacts are effects that, with some certainty, rise to the level of significant risks in terms of long-term population consequences or changes to ecosystems in addition to changes in the behavior, physiology[1], or survival of individuals.

As a consequence of the substantial addition of man-made noise to the aquatic environment, there is growing concern over the impact of these sounds on aquatic animals (e.g., Hawkins and Popper 2016; National Marine Fisheries Service [NMFS] 2016). This review describes natural and man-made sounds and summarizes what is known about their effects and impact on fishes. It concludes with a discussion of the current criteria for the effects of man-made sound on fishes, including the effects on behavior and physiology, and considers whether more realistic criteria can be set to protect fishes against physical damage.

6.2 Sea Sounds

6.2.1 Natural Sounds

The sea is full of sound. The sounds of falling rain, breaking waves, cracking ice, bubbles, spray, water moving over reefs, and tidal flow (e.g., Bass and Clark 2003) provide a continuous but varying background of noise. Less frequently, earthquakes, volcanic eruptions, and lightning strikes generate intense sounds traveling over great distances. Other sounds are biological in origin, produced by marine mammals (Southall et al. 2007; Erbe et al. 2012), fishes (Hawkins and Myrberg 1983; Bass and Ladich 2008), or invertebrates (Popper et al. 2001; Morley et al. 2014). Fishes may use the acoustic environment for orientation and navigation, to facilitate foraging, and to avoid predators. Some animals, communities, and ecosystems may be especially vulnerable to changes in the soundscape and might be damaged by the imposition of man-made noise.

[1] "Physiological" effects are often used synonymously with "physical" effects. It is difficult to differentiate the two because changes in physiology (e.g., inability to produce blood cells, changes in nerve conduction) may be hard to determine, whereas it is not clear if physical effects (damage to tissues, hematoma) also result in physiological effects. This paper uses physiological to encompass all effects on the body, unless the results are clearly only one or the other.

6.2.2 Biological Sources of Sound: Fishes

There are over 32,000 species of fish, and sound production is likely to be far more extensive than the 800 species of fish from 109 families in which it is now known. Moreover, sound production seems to have evolved independently in many taxa (e.g., Kaatz 2002; Bass and Ladich 2008).

Among the vocal fishes are some of the most abundant and important commercial species, including the Atlantic cod (*Gadus morhua*) and haddock (*Melanogrammus aeglefinus*) (both Gadidae; Hawkins and Chapman 1966; Hawkins et al. 1974), as well as many croakers and drums (Sciaenidae; Ramcharitar et al. 2006). Sounds are produced when fishes are feeding, mating, or fighting, and some species also generate sounds when they are swimming (Moulton 1963; Hawkins and Myrberg 1983).

Sounds produced by spawning fishes, such as gadids and sciaenids, are sufficiently loud and characteristic for them to be used by humans to locate spawning concentrations (Luczkovich et al. 2008; Casaretto et al. 2014) and, more importantly, for females to find males. It is critically important to assess whether such spawning sites need special protection from noise generated by fishing or other sources of high-level noise.

6.2.3 Sources of Man-Made Sound

There are many sources of man-made sound in the sea, with quite different acoustical characteristics. This topic will only be briefly summarized because it has been reviewed more extensively (e.g., Popper et al. 2014; Hawkins et al. 2015). In addition to the examples discussed here, other sources of sound may include explosions (Weston 1960; Continental Shelf Associates 2004), dredging (de Jong et al. 2011; Robinson et al. 2012), and marine renewable energy devices utilizing wave and tidal forces (Thomsen et al. 2016).

Seismic Airguns

Airguns are used for seismic exploration and geological research (Gisiner 2016). They use a compressed air supply to produce a bubble that expands rapidly to create a high-level impulsive sound (e.g., Mattsson et al. 2012) typically composed of low frequencies (20 to 50 Hz, with declining energy at frequencies above 200 Hz). During a seismic survey, airguns in a large array are fired at regular intervals (e.g., every 10 to 15 seconds) as the towing vessel moves ahead (Gisiner 2016). The sound pulse enters the seabed, and the reflected sound is detected by long hydrophone arrays streamed behind the vessel. The sounds from airgun arrays may travel great distances (Nieukirk et al. 2004, 2014) and may generate seismic waves that travel along the seabed/water interface, which may affect bottom-living animals (Klages et al. 2002).

Impact Pile Driving

Impact pile driving is used for the construction of foundations for wind turbines, bridges, and other inshore and offshore structures (reviewed in Popper and Hastings 2009; Dahl et al. 2015). Sound is generated by direct contact of the pile with the water, and compression, shear and interface waves are also generated within the seabed (e.g., Hazelwood 2012; Dahl et al. 2015). Transmission of sound through the seabed and at the water/ground interface may affect bottom-living fishes and their invertebrate prey.

Of particular concern are the high-energy impulsive sounds generated by impact driving of large-diameter steel shell piles (Reyff 2016). These sounds are characterized by a relatively rapid rise time to a maximal pressure value followed by decay (Dahl et al. 2015; Reyff 2016). The peak sound pressure levels may vary substantially and depend on numerous factors such as pile diameter, hammer size, and substrate characteristics. The energy in pile impact impulses is at frequencies below 500 Hz, within the hearing range of most fishes, with much less energy above 1 kHz (Laughlin 2006; Rodkin and Reyff 2008).

Operating Wind Farms

Sounds generated by wind turbines are much lower in intensity during the operational phase than during construction (Madsen et al. 2006; Cheesman 2016). Most of the sound from the turbines themselves is at frequencies below about 700 Hz and is dominated by narrowband tones (Wahlberg and Westerberg 2005; Madsen et al. 2006). Underwater sound pressure levels within wind farms have been reported as not being significantly higher than the background noise (Nedwell et al. 2007). The highest level noted was a narrowband tone at approximately 180 Hz (Wahlberg and Westerberg 2005). There is also a particle motion component to the sounds generated by wind turbines (Sigray and Andersson 2012).

Vessel Noise

A significant part of man-made noise in the aquatic environment results from the operation of vessels, including ships and small boats (Pine et al. 2016; Rossi et al. 2016). Most vessels, but particularly large ships, produce predominantly low-frequency sound (i.e., below 1 kHz) from onboard machinery, hydrodynamic flow around the hull, and propeller cavitation, which is typically the dominant source of noise (Ross 1987, 1993). Low-frequency sounds from ships can travel hundreds of kilometers and can increase ambient-noise levels over large areas of the ocean (Southall 2005; Ellison et al. 2012).

Several studies have indicated that ambient-noise levels in busy shipping lanes have recently increased (Ross 1993; Hildebrand 2009) across much of the frequency spectrum (McKenna et al. 2012; Sertlek et al. 2016) but especially at lower frequencies (<500 Hz; Erbe et al. 2012; Bittencourt et al. 2014). One of the most serious implications of this increase in shipping noise is the effect it may have in terms of masking sounds of biological origin, affecting communication between animals (Slabbekoorn et al. 2010; Radford et al. 2014). Large numbers of smaller pleasure and recreational vessels may also result in substantial increases in sound levels in some coastal waters and lakes.

Sonar

Sonar is widely used by fishing and other vessels to detect fish schools, to survey and map the seabed, and for military purposes (Ainslie 2010). Typical sonars include echo sounders, fish-finding sonars, side-scan sonars, and a variety of sonars for mapping the topography of the seabed. Many sonars work at frequencies from 20 to 800 kHz, above the hearing ranges of most fishes but often with very high source levels. Such ultrasonic sources are detected by some fish species (Mann et al. 2001). In addition, there are some high-power, low- and midfrequency naval sonars that use signals from several hundred to several thousand hertz, within the hearing range of fishes (Popper et al. 2007; Halvorsen et al. 2012c).

6.3 Descriptions of Underwater Man-Made Sounds

A variety of metrics exist for the physical description of underwater sounds (discussed by (Ellison and Frankel 2012; Ainslie 2015). Nevertheless, measurement parameters are not well defined for underwater sounds, especially for those that may affect aquatic life. The Dutch Standards Institute TNO has published a set of standards for the measurement and monitoring of underwater sound (see Ainslie 2011), but universal agreement has not yet been reached on how such sounds should be measured in order to assess fully their effects on aquatic life.

Measurements close to sources are often in the nonlinear portion of the sound field, especially for pile drivers and explosions and to some degree for seismic sources (Sigray and Andersson 2011, 2012). It is in these regions that physical injury to fishes may occur. Information is also required on the particle motion amplitudes generated by such sources, especially close to the water surface or close to the seabed where the physics of the adjacent media must be taken into account.

A particular need to consider is which sound metrics are most appropriate for predicting the effects of sound exposure on animals (Ellison and Frankel 2012). Some sounds are more damaging than others, and for determining the effects of different sounds, it is important to describe the sounds in terms of those features that relate to the damage caused. Important features to consider are the rise time and the duration of each pulse/signal as well as the proportion of time occupied (the duty cycle) and the interpulse intervals. Weighting functions may also need to be defined and refined for fishes or fish categories, as has been done for marine mammals (Southall et al. 2007; Ellison et al. 2012).

6.4 Effects of Man-Made Sounds

There is a wide range of potential effects from sounds on fishes (and other aquatic animals as well), ranging from death (mortality) to slight changes in behavior (Table 6.1; Fig. 6.1). Very intense sounds may kill or injure animals. Extraneous sounds may impair hearing, affect orientation, and/or mask vocalizations and other sounds of biological importance (Erbe et al. 2016). Changes in behavior induced by noise may affect spawning success or disrupt foraging and feeding. Exposure to sound may also cause stress and associated physiological responses (Filiciotto et al. 2016), although not always (Wysocki et al. 2007). In some cases, sound may deny

Table 6.1 Potential effects of man-made sound on animals

Effect	Description
Death	Sound exposure leads to instantaneous or delayed mortality
Physical injury and physiological changes	Physical changes occur that temporarily or permanently impair the structure and functioning of some parts of the body. Physiological changes take place that indicate increased stress or disorientation, perhaps resulting in reduced fitness over time
Hearing changes (temporary or permanent)	Loss of hearing, temporarily or permanently, leads to inability to respond to biologically relevant sounds
Changes in behavior	Behavioral changes may be ranked, depending on their assumed severity with respect to changes in fitness. Such changes are likely to vary from species to species and in terms of context, including the location, time of day, time of year, and the condition of the animal It can be difficult to infer effects on long-term fitness of changes from behavior observed over a limited time period. Some changes in behavior, such as startle reactions, may only be transient
Masking of biologically significant sounds	Altering the ability of an animal to detect sounds that are important to it, including the sounds of predators and prey and calls from other animals of the same species There may be a reduction in "acoustically active space" as a result of a reduction in detection distance
No obvious behavioral response	If the signal level is low enough, animals may show transient or no response, even if they detect the sound. And/or habituation may take place. Moreover, even if there is no response, there is always the possibility that physical injury and physiological changes may take place without the animal showing overt changes in behavior

animals access to particular habitats, including preferred feeding grounds or spawning areas. There are, however, many gaps in understanding the effects of noise on fishes (reviewed in Hawkins et al. 2015).

The effects of sounds on fishes may depend on many factors, including the nature of the sound source and the distance from the source (which will affect the characteristics of the sounds received by the fish) as well as the condition and motivation of the animal. There is increasing recognition that sublethal effects (e.g., masking of communication and significant behavioral and physiological and/or physical responses) from chronic exposure to long-lasting sounds may be as significant for fish populations as the effects of more transient acute exposure to shorter lived sounds. Exposure to sound also has to be considered alongside exposure to other stressors such as fishing, habitat and climate change, and chemical and light pollution (Kunc et al. 2014, 2016).

Figure 6.1 sets out the relationship between sound levels and the responses from animals and makes the point that potential effects are overlapping. Thus, close to a source, where the sound is at its highest level, the effects on fishes may include death, physiological effects, temporary hearing shift, masking, and/or behavioral responses. At greater distances from the source, where the signal is still audible, the only responses may be behavioral. Indeed, even within any one class of effect, there may be different responses depending on the received sound level, what the fish is doing at the time it hears the sound, and many other factors.

Of particular interest is the effect of these extraneous man-made sounds on the ability of fishes to detect natural sounds that are biologically important to them. There are several levels of perception (Dooling et al. 2015; Dooling and Blumenrath 2016) that must be considered.

- *Detection*: the sound is just audible above the background noise.
- *Discrimination*: the sound is at a level sufficiently above background that a fish can distinguish between different sounds (e.g., sounds of conspecifics or prey animals versus those from predators).
- *Localization*: the fish is able to determine the direction from which the sounds are coming and perhaps the distance of the source, both of which may be important for responding to prey or predators and for orientation and navigation.
- *Recognition*: the fish can identify the sound and decide how significant it might be (that is, the animal can recognize the nature and context of the sound).

6.5 Hearing in Fishes

In water, sound consists of a traveling pressure wave within which the component particles of the water are alternately forced together and then apart. The sound pressure is measured by a conventional hydrophone in water. The particle motion, which can be described in terms of particle displacement, velocity, or acceleration, is less easy to measure (Martin et al. 2016; Nedelec et al. 2016). However, for most fishes, the particle motion is especially important in terms of both the detection of the sound and the localization of the sound source (Hawkins and Popper 2014; Nedelec et al. 2016).

One of the most fundamental problems in most studies of the effects of noise on fishes, and indeed on the basic studies of hearing and general bioacoustics, is that the sound fields in which laboratory experiments are done may be very complex and quite unlike the sound fields that a fish would encounter in a normal aquatic environment (see Parvulescu 1964; Rogers et al. 2016). As a result, much of the data from otherwise well-designed hearing and response studies leaves open questions as to the actual nature of the sound field to which the animals were exposed and the stimuli to which they responded (Hawkins et al. 2015).

The most valuable studies of the hearing abilities of aquatic animals are those carried out under conditions when the sound fields can be modeled and accurate measurements made of both sound pressure and particle motion. Such studies have been carried out in very specialized tanks (e.g., Hawkins and MacLennan 1976; Halvorsen et al. 2011) or in midwater in the sea (e.g., Chapman and Hawkins 1973; Hawkins and Chapman 1975). Thus, a prerequisite for studies intended to examine the hearing of fishes and their responses to sounds is that they be done under appropriate acoustic conditions where both sound pressure and particle motion can be monitored (Rogers et al. 2016).

Relatively few experiments on the hearing of fishes have been carried out under suitable acoustic conditions, and many of the measurements made in tanks and expressed solely in terms of sound pressure are unreliable. Only a few species of fish, including the Atlantic cod (Chapman and Hawkins 1973), common dab (*Limanda limanda*), European plaice (*Pleuronectes platessa*; Chapman and Sand 1974), and Atlantic salmon (*Salmo salar*; Hawkins and Johnstone 1978), and several elasmobranch species (Casper and Mann 2009), have had their hearing abilities examined under appropriate conditions. The abilities of most fish species to detect sound remain to be explored.

Figure 6.2 provides audiograms for two species of flatfish (e.g., common dab and European plaice) and for the Atlantic salmon, all obtained under field conditions. The flatfishes do not have a swim bladder or other gas bubble that would increase hearing bandwidth and provide sensitivity to sound pressure. Flatfish have a relatively narrow bandwidth of hearing (up to perhaps 300 to 500 Hz), and their sensitivity to sounds at any particular frequency is poorer than that of many fishes that have a swim bladder (Chapman and Sand 1974). Flatfish are sensitive to particle motion rather than sound pressure. The Atlantic salmon, although it has a swim bladder, is also sensitive to particle motion over the whole of its frequency range (Hawkins and Johnstone 1978; Knudsen et al. 1992).

Some fishes have adaptations that convey sensitivity to sound pressure as well as to particle motion. These adaptations include having a gas bubble or a swim bladder that is close to, or connected to, the ear. One such species is the Atlantic cod, in which the swim bladder is located close to the ear. At low frequencies (below 110 Hz), hearing in the Atlantic cod is based on the detection of particle motion, but at higher frequencies, the cod is sensitive to sound pressure (Fig. 6.3). Deflation of

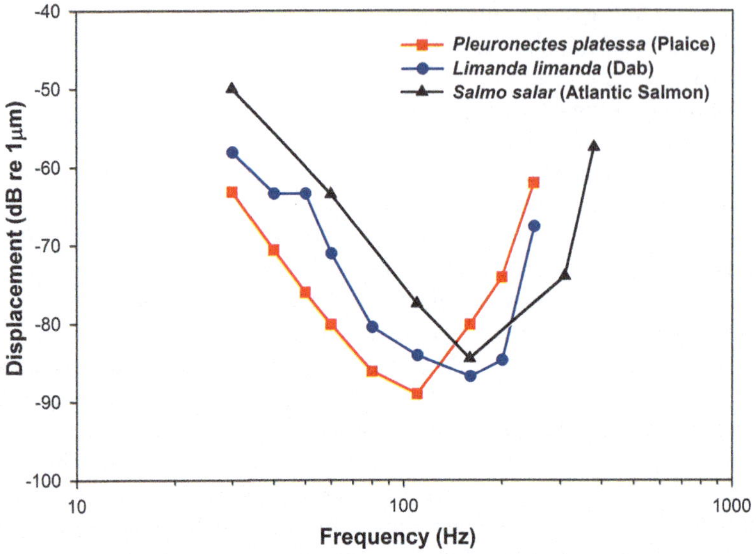

Fig. 6.2 Audiograms for European plaice (Chapman and Sand 1974), common dab (Chapman and Sand 1974), and Atlantic salmon (Hawkins and Johnstone 1978). Acoustic thresholds for all three species were obtained by cardiac conditioning to pure tones against a natural sea noise background

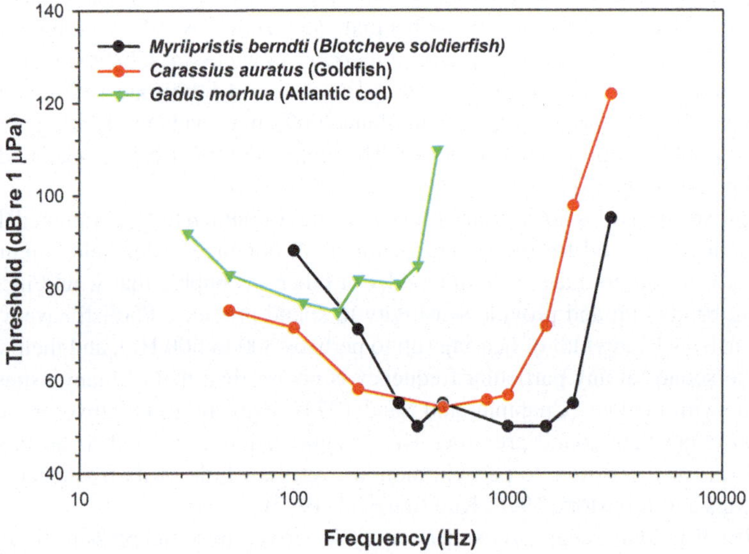

Fig. 6.3 Audiograms for blotcheye soldierfish (Coombs and Popper 1979), goldfish (Jacobs and Tavolga 1967), and Atlantic cod (Chapman and Hawkins 1973). Thresholds for Atlantic cod were obtained by cardiac conditioning to pure tones against a natural sea noise background. Thresholds for the soldierfish and goldfish were obtained using an operant conditioning paradigm in a small tank in a sound-shielded room

the gas-filled swim bladder in the cod reduces the sensitivity to sound pressure (Sand and Enger 1973). Conversely, placing a small, inflated, gas-filled rubber bladder close to the head of a common dab (which lacks a swim bladder) gives an increased sensitivity and extends the frequency range (Chapman and Sand 1974).

A number of fish species have special structures mechanically linking the swim bladder to the ear (e.g., Weberian ossicles in goldfish [*Carassius auratus*] and relatives; Popper et al. 2003; Popper and Fay 2011). In other cases, the swim bladder has extensions that come close to, or may actually contact, portions of the inner ear (Ramcharitar et al. 2006; Schulz-Mirbach et al. 2013).

In species with swim bladders or other sacs containing gas, the gas itself changes volume in response to fluctuating sound pressures. This may produce particle motion at the ears that, in turn, has the potential to cause the sensory epithelium to move relative to the otolith (Popper et al. 2003). Sand and Hawkins (1973) measured the resonance frequency and damping of the swim bladder in intact living Atlantic cod at different depths and showed that the organ served as an acoustic transformer, translating sound pressure into reradiated particle motion with little phase distortion over an extended range of frequencies.

Elasmobranchs (sharks, skates, and rays) have ears that are very similar to those in other vertebrates (Tester et al. 1972; Corwin 1977), and there is evidence that sound plays a major role in their lives (Myrberg 2001; Casper et al. 2012a). Elasmobranchs detect sounds from below 50 Hz to over 500 Hz even though they have no swim bladder or other gas sac associated with the ear. The likelihood of physiological effects from other than the most intense sounds may be substantially lower than for fishes with swim bladders (Casper et al. 2012a).

It is clear that for the more sensitive species, hearing is not limited by the fish's absolute sensitivity but by how well it is able to detect sounds against the background of ambient noise. For example, only under the quietest sea conditions do Atlantic cod show absolute thresholds (Chapman and Hawkins 1973; Fig. 6.4). Any increase in the level of ambient sea noise, either naturally as a result of an increase in wind and waves or rain or hail or from the passage of a ship, results in a raising of the auditory threshold and a decline in the ability of the fish to detect, locate, and recognize particular sounds.

The ability of fishes to detect biologically relevant signals (e.g., sounds from a predator or conspecifics) will be affected not just by variations in natural ambient noise but will be masked by any extraneous sounds that raise the level of background noise. It should be noted that many of the differences in sensitivity seen in the audiograms presented for different species might result from variable noise levels prevailing under experimental conditions. Aquarium tanks, like the sea itself, are noisy.

Not all frequency components of the background noise may contribute to the masking of a particular sound. For human subjects, it has long been known that a pure-tone signal is masked most effectively by noise components at the same and similar frequencies. Fletcher (1940) applied the term "critical band" to the frequency span of noise that is effective. It has been shown that critical bands also exist in the Atlantic cod (Hawkins and Chapman 1975) and goldfish (Fay 1974). In these

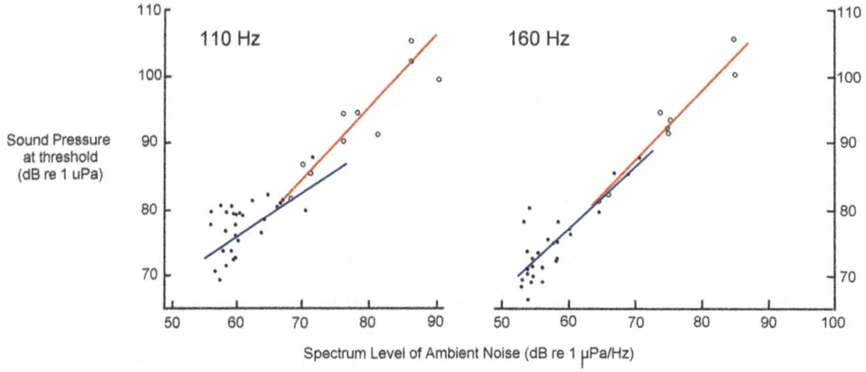

Fig. 6.4 Auditory thresholds to pure-tone signals are affected by the level of background noise in the sea (Chapman and Hawkins 1973). Thresholds determined for Atlantic cod against a natural noise background changed as the level of sea noise varied (*blue lines*). The addition of white noise from a loudspeaker also changed the thresholds (*red lines*). The results suggested that the hearing of cod at its most sensitive frequencies was limited by the level of background noise in the sea

species, only frequencies close to a pure-tone frequency will impair the detection of that tone. However, the full implications of masking have yet to be investigated fully for fishes. It is evident that it is the level of ambient sea noise that largely determines the detection distances for sounds by a fish like the Atlantic cod. The presence of extraneous man-made noise will also affect the range at which sounds can be detected. There is a need for experiments to examine the degree of masking caused by different types of noise (including both continuous and intermittent man-made noise) for different sounds of biological importance to fishes, including the calls of conspecifics, the sounds made by prey or predators, and sounds that are used for orientation and navigation. In general, but not always, the masker must also occur at the same time as the signal in order to mask it (Popper and Clarke 1979).

It cannot be concluded that the thresholds of all fish are always masked. Less sensitive species, like the common dab and the Atlantic salmon, only show masked thresholds when the background noise is raised very substantially (Hawkins and Johnstone 1978).

It is evident that some fishes can discriminate sounds from different directions (Fay 2005; Sand and Bleckmann 2008). Teleost fishes are able to discriminate between spatially separated sources under far-field conditions in both the horizontal (Chapman and Johnstone 1974; Schuijf and Buwalda 1975) and vertical (Hawkins and Sand 1977) planes. Indeed, they are able to distinguish between sources at different distances (Schuijf and Hawkins 1983). This ability not only enables fish to locate the sources of sound but may also assist them in discriminating sounds from a particular source against the general nondirectional noise background.

6.6 Effects of Sounds on Fishes

6.6.1 Behavioral Studies

Observations on the effects of sounds on fish held in tanks and even large enclosures are likely to yield equivocal results (Hawkins et al. 2015; Rogers et al. 2016). Quite apart from the difficulties of properly measuring the sound stimuli that are presented in an enclosed space (see Sect. 6.6.5), captive animals do not show the wide range of behavior observed in wild animals; they tend to behave differently when confined than when their movements are unrestricted (Holles et al. 2013), even when the enclosure is very large (e.g., Sarà et al. 2007; Mueller-Blenkle et al. 2010). Accordingly, to fully elucidate the behavior of fishes in response to sounds and assess the effects man-made sounds on them, the responses must be seen in the context of changes to natural behavior, which varies from species to species, with age, and with habitat. There have been very few studies of the effects of sound exposure on the behavior of wild (unrestrained) fishes because it is technically difficult to observe their behavior.

Although limited in number, several studies have demonstrated that man-made sounds may affect the behavior of wild teleost fishes (Nedelec et al. 2014; Voellmy et al. 2014). Studies have shown a decrease in rockfish (*Sebastes* sp.) catches when fish were exposed to sounds from a seismic airgun at a received sound pressure level of 186 to 191 dB re 1 μPa (peak; Skalski et al. 1992).

Other investigators have examined the movement of fishes in the sea during and after exposure to seismic surveys (Engås et al. 1996; Engås and Løkkeborg 2002). Rather than observe fish behavior directly, changes in catch rate of haddock and Atlantic cod by commercial trawls and long lines were taken as evidence of changes in behavior. A significant decline in catch rate occurred that lasted for several days after termination of airgun use and then returned to normal. The conclusion reached was that the decline in catch rate resulted from the fishes moving away from the area in which the seismic activity took place. Other similar experiments (Løkkeborg et al. 2012a,b) have revealed that exposure to sound could also result in an increase in fish catch as a result of changes in the behavior of the fish.

Slotte et al. (2004) used sonar to observe the behavior of schools of several pelagic species including blue whiting (*Micromesistius poutassou*) and Norwegian spring-spawning herring. The fishes appeared to swim to greater depths after airgun exposure and the abundance of animals increased 30 to 50 km away from the ensonified area, suggesting that migrating fishes were not entering the zone of seismic activity.

Wardle et al. (2001) used underwater video and an acoustic tracking system to examine the behavior of fishes on a rocky reef in response to emissions from a single seismic airgun. They observed brief startle responses and some small changes in the movement patterns of fish, but these did not last long and did not appear to affect the overall behavior of the fish. Startle responses may often be transient and without lasting effects.

The behavior of wild pelagic fish in response to sound playback was observed with a sonar/echo sounder by Hawkins et al. (2014). Schools of European sprat (*Sprattus sprattus*) and Atlantic mackerel (*Scomber scombrus*) were exposed to short sequences of repeated impulsive sounds, simulating the strikes from a pile driver, at different sound pressure levels. The incidence of behavioral responses increased with increasing sound level. Sprat schools were more likely to disperse and mackerel schools were more likely to change depth. In most instances, the sprat and mackerel schools were separate from one another, and there was little likelihood of the response of one species affecting the behavior of the other. The sound pressure levels to which the fish schools responded on 50% of presentations were about 163 dB re 1 μPa peak to peak, and the single-strike sound exposure levels (SELs) were 135.0 and 142.0 dB re 1 μPa^2 s, for sprat and mackerel, respectively. The 50% response sound levels were remarkably similar for the two species, although sprat are sensitive to sound pressure and have a much more extended frequency range than mackerel, which lack a swim bladder and are sensitive to particle motion. Strictly, the 50% response levels for mackerel should be presented as a peak-to-peak particle velocity level of −80.4 dB re 1 ms^{-1} and a single-strike particle velocity exposure level of −101.7 dB re 1 m^2 s^{-1}. Nevertheless, these results indicate that both sprat and mackerel may respond to sounds from both pile drivers and seismic airguns at similar distances, at tens of kilometers from the source. The levels of sound to which fishes respond may depend more on the context and perceived significance of the sound than the hearing abilities of the fish.

A particularly critical issue is how sound exposure ultimately affects individual fitness and the survival of individuals, with subsequent impacts on fish populations (Hawkins and Popper 2016). Behavioral effects will be specific to the species, the habitat, and even time of year and may diminish with the repetition of the sounds. Fishes of different sizes (ages) within a single species may also show differences in behavior. It is especially important to evaluate whether the exposure of organisms to man-made sounds has any detrimental effects on their populations or on the natural communities to which they belong (Francis and Barber 2013).

6.6.2 Effects of Sounds on Hearing Thresholds

Permanent threshold shift (PTS) is a permanent loss of hearing and may be a consequence of the death of sensory hair cells of the auditory epithelia of the ear. To date, there is no evidence that PTS occurs in fish as a result of sound exposure and it may not occur because fishes and sharks are able to repair or replace sensory hair cells that have been lost or damaged (e.g., Corwin 1983; Smith et al. 2006).

Temporary threshold shift (TTS) is a transient reduction in hearing sensitivity that may be caused by exposure to intense sounds or sounds of long duration. After termination of a sound causing TTS, normal-hearing ability returns over a period that may range from minutes to days depending on many factors including the intensity and duration of exposure (e.g., Amoser et al. 2004; Smith et al. 2006). TTS as a result of damage to hair cells has not been considered to legally constitute an injury under US legislation (Richardson et al. 1995; Southall et al. 2007). However, recent studies on mammals may provide insight into the likelihood that TTS is an injury. In several studies, it has been shown that even where hair cells remain intact, damage may occur to the terminals of afferent nerve cells (Kujawa and Liberman 2009; Lin et al. 2011), prolonging TTS. Moreover, during a period of TTS, animals may be at some risk to their survival in terms of communication, detecting predators or prey, and assessing their environment.

TTS has been demonstrated in a number of fish species (e.g., Smith et al. 2004; Popper et al. 2005) from a diverse array of sounds. However, in all cases, TTS was only found after multiple exposures to intense sounds (e.g., over 190 dB re 1 μPa root-mean-square [rms]) or long-term exposure (e.g., tens of minutes or hours) to somewhat less intense sounds. Even when one signal source caused TTS in some fish or some species, it did not occur in other specimens or other species (e.g., Popper et al. 2005, 2007). In most cases, normal thresholds returned within a few hours to several days. There is also evidence that given the same type and duration of sound exposure, a much louder sound will be required to produce TTS in fishes that do not hear well compared with fishes that do hear well (Smith et al. 2004; Popper et al. 2007).

It has been proposed that because TTS can arise from prolonged exposure to sound (though this is not always so), it is not likely to be of great significance for fishes that have only a brief exposure to a source (e.g., Popper et al. 2007; Halvorsen et al. 2013) because the duration of exposure would be very short. Of far greater concern is that TTS may occur when there is chronic noise exposure in an area where there is a long-term increase in sound level.

Further experiments are required on PTS and TTS in fishes. In particular, it would be appropriate to expose fish to high levels of particle motion because this may be more damaging to some fish ears than exposure to sound pressure.

6.6.3 Physical Damage to the Auditory System

Only a few studies have examined physical effects on the auditory system after exposure to intense sounds. Enger (1981) demonstrated that exposure to such sounds may result in damage to the sensory hair cells in the ears of the goldfish. Subsequently, Hastings et al. (1996) showed some damage to hair cells of the goldfish lagena (one of the three otolithic end organs of the ear) but only when the fish were kept alive for several days postexposure.

Effects of impulsive sounds on the ear give ambiguous results, very likely as a result of using different species and sources. The first study examining the effects of impulsive sounds on the ear showed that exposure to multiple airgun shots over several hours produced damage to the sensory epithelia of the saccule, the major auditory end organ of the ear, in caged pink snapper (*Pagrus auratus*; McCauley et al. 2003). A subsequent study showed no damage to the ears of three fish species in the Mackenzie River Delta: northern pike (*Esox lucius*), broad whitefish (*Coregonus nasus*), and lake chub (*Couesius plumbeus*; Popper et al. 2005), after exposure to 5 or 20 airgun shots. Although some species showed TTS, there was no apparent damage to nonauditory (Popper et al. 2005) or auditory (Song et al. 2008) tissues.

Exposure of both hybrid striped bass (*Morone* sp.) and Mozambique tilapia (*Oreochromis mossambicus*) to impulsive sounds at 210-216 dB re 1 μPa^2·s cumulative SEL (SEL_{cum}) showed that damage to sensory hair cells, if it occurs at all, only shows up at SELs that are somewhat higher than levels that will result in damage to nonauditory tissues (Casper et al. 2013b). The hypothesis is that damage as a result of exposure to impulsive sounds in the acoustic far field (which is likely to be primarily a pressure signal) results from rapid and high-amplitude motion of the walls of the swim bladder and that the tissues showing greatest damage are those closest to that organ.

The only study to date that has correlated hair cell loss in a fish ear and hearing effects was done with goldfish. This study showed that exposure to long-duration white noise resulted in extensive loss of sensory cells and that this was closely correlated with decreased hearing sensitivity (Smith et al. 2006). Over several days, hearing sensitivity returned to about normal, and this was correlated with the start of replacement of sensory cells of the ear.

6.6.4 Stress

Animals showing no overt sign of responding to an environmental stimulus may, nonetheless, show physiological changes that are often referred to as stress responses (e.g., Wysocki et al. 2006; Kight and Swaddle 2011). These may manifest as changes in heart rate, breathing rhythm, or the levels of particular hormones. Very little is known about stress effects in fishes, and the significance of such effects in response to man-made sound is even less clear (Tennessen et al. 2016). Indeed, stress is a normal part of life, integral to stimulating and maintaining healthy neuroendocrine responses and immune system activity (homeostasis). Moreover, although fishes may experience acute effects as a result of noise exposure, it is less certain that this results in long-term chronic effects (reviewed in Slabbekoorn et al. 2010). Normally, the body's stress response, essential for managing acute threats, is essential for adaptation, maintenance of homeostasis, and survival. However, repeated stress responses may damage individuals in the long term. The changes induced may introduce risks to individual fitness including loss in reproductive capacity. It is important to distinguish between normal or tolerable variations in response to environmental stress from those changes that will have consequences for survival and reproduction. At present, critical examination of these long-term changes in fishes as a result of sound exposure is lacking.

6.6.5 *Death and/or Injury*

Death and injury are probably the most easily observed responses to high levels of man-made sound for fishes. However, there are only the most limited data on mortality in fish from sound exposure, and these are when animals are very close to pile-driving sources (California Department of Transportation 2001) and explosions (Govoni et al. 2008; Popper and Hastings 2009) but not for other sound sources. Indeed, the exposure of fish to very high intensity low- and midfrequency sonars resulted in no mortality (Popper et al. 2007; Halvorsen et al. 2013) nor did exposure to seismic airguns (Popper et al. 2005, 2016).

The most commonly injured organ in fishes exposed to explosions is the swim bladder (Yelverton et al. 1975; Keevin and Hempen 1997). When pressures oscillate rapidly as they do from an explosive shock wave or other impulsive signals, the swim bladder expands and contracts rapidly and may damage the proximate organs (Halvorsen et al. 2012a,b). Of five species exposed to high-intensity simulated pile-driving signals (Halvorsen et al. 2012a,b; Casper et al. 2013a), only the hogchoker (*Trinectes maculatus*), a flatfish without a swim bladder, showed no tissue damage (Halvorsen et al. 2012b).

6.6.6 *Other Effects of Impulsive Sources*

Although the most concern has been focused on the effects of intense sounds on fishes, there are also questions as to whether such sounds may also damage other body tissues and especially tissues near the swim bladder. The sounds produced by airguns and pile drivers are often very intense, with single-strike sound exposure peak levels often exceeding 180 to 200 dB re 1 μPa^2 s and onset times that are very short.

To investigate nonauditory tissue effects, Chinook salmon (*Oncorhynchus tshawytscha*) were tested in a laboratory-based tank that could simulate very high intensity pile-driving sounds under acoustic conditions similar to those a fish would encounter if it were outside the acoustic near field of the sound source (Halvorsen et al. 2011). Results showed a close link between the extent of tissue damage and the cumulative level of sound energy to which fish were exposed. There was no tissue damage in Chinook salmon after exposure to sounds below a SEL_{cum} of 210 dB re 1 μPa^2 s. At this level, the only effects were minor hemorrhaging. At a SEL_{cum} that was a few decibels higher (but with sounds given over the same time period), internal injuries started to appear, and when the level reached 219 dB re 1 μPa^2 s, there were massive internal injuries that would likely result in death. Studies with other species showed that although there is some variation in the SEL_{cum} required for the onset of physiological effects, this is always at SEL_{cum} levels greater than 203 dB re 1 μPa^2 s (Halvorsen et al. 2012b; Casper et al. 2013a).

Studies subsequently found that Chinook salmon and hybrid striped bass recovered from all physiological effects within 10 days after exposure to sounds as high as 213 dB re 1 μPa^2 s (Casper et al. 2012b, 2013a). Hogchoker, a flatfish, showed no effect with a SEL_{cum} as high as 216 dB re 1 μPa^2 s. As described in Sect. 6.6.3, damage to ear tissues did not show up until the SEL_{cum} was 216 dB re 1 μPa^2 s (Casper et al. 2013b).

If an animal is injured, it may be more susceptible to infection because of open wounds or a compromised immune system. It is possible that minor damage will result in reducing the animal's ability to find food or making it more subject to predation, ultimately resulting in lowered fitness.

Another issue of concern is whether there is an equal trade-off between signal level and the number of impulses in terms of damage. It was originally thought that, provided the total energy of exposure was the same, the consequences were equal (the equal-energy hypothesis). However, there is recent evidence that the equal-energy hypothesis does not hold for mammals (Henderson and Hamernik 2012; Le Prell et al. 2012) or for fishes exposed to impulsive pile driving (Halvorsen et al. 2011, 2012a)

6.7 Impacts on Fish Populations and the Wider Ecosystem

The studies described up to now have largely dealt with the effects on individual animals. The effects on individuals are often of great concern for marine mammals where populations are small and relatively few offspring are produced. Some fish populations are large, however, and many offspring may be produced, often showing high levels of mortality during the juvenile stages. For these fishes, perhaps the greater concern lies with the effects on populations rather than on individual animals (Hawkins and Popper 2016). The extent to which sound affects the structure and functioning of fish populations and ecosystems is of considerable importance, although such effects have yet to be established. Currently, it is often necessary to resort to expert judgment for predicting population impacts. As discussed in Sect. 6.6.6, there is some empirical evidence of direct reactions by fish to seismic airguns, pile driving, and vessel noise, but direct impacts on populations have not been demonstrated nor have the effects of repeated exposure been reported.

In the absence of data, models may be constructed to predict the likely effects of any changes in fish behavior. Some of the models that have been utilized are based on the behavior of individuals. Thus, Rossington et al. (2013) used an individual-based model to predict the impact on Atlantic cod from noise generated during a pile-driving event. The model tracked individual "fish," each of which was represented as a particle that was subject to advection by the tides, with a set of behavioral rules governing their responses to sounds. However, there were no data available on the movements of Atlantic cod in the area to test against the model, and the assumptions made on the responses of the fish may not have reflected what would really happen if fish were actually exposed to pile-driving noise. Most importantly, examining changes in the behavior of individuals is a long way from predicting the effects on fish populations.

Attempts to model changes in population parameters were first addressed for marine mammals. The population consequences of acoustic disturbance (PCAD) approach (National Research Council 2005) recognizes that there are significant effects at individual, population, and ecosystem levels. Individual effects become significant when they result in altered reproductive success and survival. Changes in behavior that result in alterations in foraging efficiency, changes in energy expenditure, availability of preferred habitat, disruption to migrations, declines in reproductive success, or mortality through predation can be especially significant at the population and ecosystem levels.

The population consequences of disturbance (PCoD) approach (Booth et al. 2014) is a formal, mathematical version of the PCAD model that uses the opinions of experts to quantify the transfer functions that describe the relationships between the different compartments of the PCAD model. It provides a protocol that can be used by regulators and developers to examine how sound exposure might impair the ability of individual animals to survive, breed, reproduce, or rear young and to quantify how this impairment may affect the abundance of the species concerned.

The PCAD and PCoD approaches are very demanding in terms of a requirement for detailed knowledge of population dynamics. For many fishes, that knowledge is difficult to acquire and a shortage of data limits the application of such models. For data-deficient fishes and for circumstances where there is limited knowledge of ecological interactions, an alternative qualitative risk assessment tool is needed. Fisheries biologists have recently considered new risk-based approaches in assessing the effects of fishing on species for which there are only limited data on key population parameters. Such an approach attempts to evaluate the vulnerability of fish stocks to fishing based on their biological productivity and potential for resisting adverse effects. The vulnerability is compared with the susceptibility of each fish stock to the actual fisheries operating over their range of distribution. This approach has been increasingly used to identify species at risk within multispecies fisheries (e.g., Smith et al. 2007; Hobday et al. 2011) and may have wider applicability in assessing risks from noise exposure.

6.8 Setting Sound Exposure Criteria

To evaluate the effects of sounds on fishes, it is useful to set sound exposure criteria, that is, to specify those sound levels where a specified level of damage or response is likely to occur. There are extensive sets of standards and criteria to protect humans from exposure to sounds that could be detrimental (e.g., Rabinowitz 2012) and an extensive body of literature on the effects of noise on humans (see References in Le Prell et al. 2012).

For animals such as fishes, there are two approaches of importance. One is the development of criteria for behavioral effects such as changes in behavior that are perceived as being potentially harmful to fishes and fish populations in the long term. The behavior may involve animals moving from feeding sites, changing migration routes, not hearing potential predators, and other effects likely to be detrimental. The second is the effects in terms of tissue injury or changes in physiology and the onset of some kind(s) of responses (e.g., external or internal bleeding) that has the potential of harming individual animals and thereby affecting populations. These criteria for behavior and physiology are likely to be very different in terms of the sound levels that evoke a defined level of change. Developing these criteria is also problematical because there may have to be different criteria for species that differ in behavior and/or physiology and within a single species depending on several factors such as animal size, body condition, and the context in terms of life history phase and the motivational state of the fish (see Popper and Hastings 2009; Popper et al. 2014).

6.8.1 Criteria for Physiological Effects

Interim criteria for the onset of physiological effects on fishes for use on the United States West Coast were developed in 2009 (Woodbury and Stadler 2008; Stadler and Woodbury 2009). The interim criteria were a peak SPL of 206 dB re 1 μPa, a SEL_{cum} of 187 dB re 1 μPa^2 s for fishes above 2 g, and 183 dB re 1 μPa^2 s for fishes below 2 g.

Although these criteria are still in use today in some places (see Oestman et al. 2009), they were immediately criticized because they were based on very limited experimental data and did not rely on the best available data even at that time (e.g., Carlson et al. 2007; Popper and Hastings 2009). Indeed, recent studies on the effects of simulated pile driving on fishes (see Sect. 6.6.6) have demonstrated that the onset of physiological response occurs at least 16 dB above the levels in the current interim criteria and are probably over 23 dB higher (SEL_{cum}). Other factors are likely to play a major role in the potential physiological effects of impulsive sounds. For example, pulse rise time is likely to have an impact on the movement of the swim bladder and thus the way that its wall strikes the surrounding tissues. The interpulse interval is potentially of importance because a longer time between pulses may allow for some recovery from tissue damage. Other factors may include the overall duration of exposure and frequency spectrum of the source. Indeed, as recently pointed out, determining the physiological effects of impulsive sound on fishes is likely to be a substantial task (Popper et al. 2014).

Finally, the issue of physiological effects is not limited to adult animals but must also extend to eggs and larvae. Although there are fewer data for eggs and larvae from pile driving, a recent study examined the effects on flatfish larvae at life stages, including a very short period when these fishes have a swim bladder (the swim bladder is lost after the larval stage in flatfish). Using a specially designed pressure tank, (Bolle et al. 2012, 2016) found no damage to different larval stages at a SEL_{cum} of 206 dB re 1 μPa^2 s.

6.8.2 Behavioral Effects

The problem in setting criteria for behavioral effects is that there are almost no data on those sound levels that result in behavioral effects other than startle responses. The NMFS (Oestman et al. 2009) states that behavioral impact starts at a sound pressure level of 150 dB re 1 μPa in the form of startle responses, yet the origin of this criterion is unknown (Hastings 2008). Moreover, a single criterion value for behavior does not take into consideration species differences in hearing sensitivity or behavior nor does it take into consideration response changes with animal age, season, or even motivational state (see also Neo et al. 2014).

It has also been argued that strong avoidance responses by fish start at about 90 dB above the dB_{ht} (*Species*) thresholds of fish (Nedwell et al. 2006). The dB_{ht} takes account of hearing ability by referencing the sound to the species' hearing threshold. This approach has been utilized within the United Kingdom for assessing the effects of man-made sounds on fishes and it appears to have the tacit approval of some regulatory agencies. Hawkins and Popper (2014, 2016) point out, however, that substantial caution must be exercised in applying the dB_{ht} measure. Defining response criteria applicable to all species may be too simplistic an approach to evaluating behavior. Moreover, in some instances, fish are sensitive to particle motion rather than sound pressure, and the use of the dB_{ht}, which is based on sound pressure thresholds, is inappropriate.

As described in Sect. 6.6.1, experiments where two species of fish with very different hearing abilities, the sprat and the mackerel, showed that their responses to sound exposure occurred at very similar sound levels (Hawkins et al. 2014). These experiments showed that the audiogram alone or a weighted sensitivity curve does not completely predict the performance of the animal in performing complex auditory tasks. As the recent draft of the National Oceanic and Atmospheric Administration (NOAA) guidelines for marine mammals points out, auditory weighting functions best reflect an animal's ability to hear a sound (NMFS 2016). These functions may not necessarily reflect how an animal will react behaviorally to much higher levels of sound. Moreover, it is not only the level of the sound but also its frequency range, rise time, duration, repetition rate, and a number of other parameters that may be important in determining its effects, as shown by Neo et al. (2014). The context of exposure may also be highly influential in determining behavioral responses (Ellison et al. 2012).

6.8.3 Recent Guidelines

In an attempt to make better use of data on the potential effects of sound on fishes, NOAA convened an international panel to examine all of the available data (Popper et al. 2014). Interim guidelines were developed for fishes based on dividing all species into groups that were related to the way they detect sound (see Tables 6.2 and 6.3). Guidelines were also developed for sound sources that differed in terms of

Table 6.2 Interim guidelines for intermittent sources including seismic airguns and pile drivers

Type of fish	Mortality and potential mortal injury[a]	Impairment			Behavior
		Recoverable injury	TTS	Masking	
Fish: no swim bladder (particle motion detection)	>219 dB SEL_{cum}	216 dB SEL_{cum}	184 dB SEL_{cum}	Moderate Low Low	High High Moderate
Fish: swim bladder is not involved in hearing (particle motion detection)	210 dB SEL_{cum}	203 dB SEL_{cum}	184 dB SEL_{cum}	Moderate Low Low	High High Moderate
Fish: swim bladder involved in hearing (primarily pressure detection)	210 dB SEL_{cum}	203 dB SEL_{cum}	183 dB SEL_{cum}	High High Moderate	High High High
Eggs and larvae	210 dB SEL_{cum}	Moderate Low Low	Moderate Low Low	Moderate Low Low	High Moderate Moderate

Three estimates are given for the likely onset of an effect relating to three distances from the source (top to bottom within a cell); 0-100 m, 100-1,000 m, and >1,000 m

[a]960 pile strikes. See Halvorsen et al. (2011b). TTS, temporary threshold shift; SEL_{cum}, cumulative sound exposure level. From Popper et al. (2014). See text for explanation

Table 6.3 Interim guidelines for continuous sounds including shipping

Type of fish	Mortality and potential mortal injury	Impairment			Behavior
		Recoverable injury	TTS	Masking	
Fish: no swim bladder (particle motion detection)	Low Low Low	Low Low Low	Moderate Low Low	High High Moderate	High High Moderate
Fish: swim bladder is not involved in hearing (particle motion detection)	Low Low Low	Low Low Low	Moderate Low Low	High High Moderate	High High Moderate
Fish: swim bladder involved in hearing (primarily pressure detection)	Low Low Low	170 dB rms for 48 h	158 dB rms for 12 h	High High High	High High High
Eggs and larvae	Low Low Low	Low Low Low	Low Low Low	High Moderate Moderate	High Moderate Moderate

Three estimates are given for the likely onset of an effect relating to three distances from the source (top to bottom within a cell); 0-100 m, 100-1000 m, and >1,000 m rms, Root-mean-square. From Popper et al. (2014). See text for explanation their acoustic characteristics and the appropriate metrics defined for measurement of the received levels.

The resultant guidelines are presented in Tables 6.2 and 6.3. In some cases, numerical guidelines were provided, expressed in the appropriate metrics. In proposing criteria for several types of sound sources, only cases where data were available on received sound levels were considered. Where there were insufficient data to set even interim criteria, Popper et al. (2014) chose to give subjective estimates of the likelihood of there being effects at different distances from the source for different groups of fishes. The authors recognized, however, that these values were arbitrary and very much depended on the source level. New information is needed for the setting of valid criteria. The authors also pointed out that the new guidelines were still interim and that there was the expectation that as gaps in knowledge were filled, the guidelines would need to be reexamined and refined (Popper et al. 2014; Hawkins et al. 2015).

6.9 Conclusions and Future Directions

The past years have seen an increase in what is known about the potential effects of increased man-made sounds on fishes. However, the extent of data available is still very limited, and there are still very large gaps in knowledge that need to be filled (Hawkins et al. 2015; Hawkins and Popper 2016). It is recognized that filling these gaps will take many years and be costly and experimentally difficult (Popper et al. 2014), but until there are more data for different source types and for different "types" of fishes (as per Tables 6.2 and 6.3), firmer guidelines cannot be set.

It is also important to recognize that *the* most important questions relate to the behavioral responses of fishes to sounds and the potential for masking of biologically important sounds. Behavior is critical because the potential for behavioral effects could extend for hundreds to thousands of meters from a source, whereas the physiological effects are most likely to only occur in close proximity and only when a fish does not move away at the signal onset (Slabbekoorn et al. 2010).

The conduct of behavior experiments is especially difficult. It is unlikely that experiments on captive fish in the laboratory will be sufficient to establish criteria for the responses of free-living fish because captive fish will be restricted in their behavior patterns and may well be desensitized to sounds. In addition, the acoustic conditions within tanks, especially in terms of background noise levels and propagation conditions, are unlikely to reflect those in open waters. It is recognized that the view expressed here is not shared by all and that there may be some advantages in conducting laboratory studies (e.g., Slabbekoorn 2016) in terms of a more refined experimental design.

Compliance with Ethics Requirements
Anthony D. Hawkins declares that he has no conflict of interest.
Arthur N. Popper declares that he has no conflict of interest.

References

Ainslie, M. (2010). *Principles of Sonar Performance Modelling*. Berlin Heidelberg: Springer-Verlag.

Ainslie, M. (2011). *Standard for Measurement and Monitoring of Underwater Noise, Part I: Physical Quantities and Their Units*. Report TNO-DV 2011 C235, TNO, The Hague, The Netherlands.

Ainslie, M. (2015). A century of sonar: Planetary oceanography, underwater noise monitoring, and the terminology of underwater sound. *Acoustics Today*, 11(1), 12-19.

Amoser, S., Wysocki, L. E., & Ladich, F. (2004). Noise emission during the first powerboat race in an Alpine lake and potential impact on fish communities. *The Journal of the Acoustical Society of America*, 116(6), 3789-3797.

Andrew, R. K., Howe, B. M., & Mercer, J. A. (2002). Ocean ambient sound: Comparing the 1960s with the 1990s for a receiver off the California coast. *Acoustics Research Letters Online*, 3, 65-70.

Bass, A. H., & Clark, C. W. (2003). The physical acoustics of underwater sound communication. In A. M. Simmons, A. N. Popper, & R. R. Fay (Eds.), *Acoustic Communication* (pp. 15-64). New York: Springer-Verlag.

Bass, A. H., & Ladich, F. (2008). Vocal-acoustic communication: From neurons to brain. In J. F. Webb, A. N. Popper, & R. R. Fay (Eds.), *Fish Bioacoustics* (pp. 253-278). New York: Springer-Verlag.

Bittencourt, L., Carvalho, R. R., Lailson-Brito, J., & Azevedo, A. F. (2014). Underwater noise pollution in a coastal tropical environment. *Marine Pollution Bulletin*, 83(1), 331-336.

Boehlert, G. W., & Gill, A. B. (2010). Environmental and ecological effects of ocean renewable energy development: A current synthesis. *Oceanography*, 23(2), 68-81.

Bolle, L. J., de Jong, C. A., Bierman, S. M., van Beek, P. J., van Keeken, O. A., Wessels, P. W., van Damme, C. J., Winter, H. V., de Haan, D., & Dekeling, R. P. (2012). Common sole larvae survive high levels of pile-driving sound in controlled exposure experiments. *PLoS ONE*, 7(3), e33052.

Bolle, L. J., de Jong, C. A., Bierman, S. M., van Beek, P. J., Wessels, P. W., Blom, E., van Damme, C. J., Winter, H. V., & Dekeling, R. P. (2016). Effect of pile-driving sounds on the survival of larval fish. In A. N. Popper & A. Hawkins (Eds.), *The Effects of Noise on Aquatic Life II* (pp. 91-100). New York: Springer-Verlag.

Booth, C., Donovan, C., King, S., & Schick, R. (2014). *A Protocol for Implementing the Interim Population Consequences of Disturbance (PCoD) Approach: Quantifying and Assessing the Effects of UK Offshore Renewable Energy Developments on Marine Mammal Populations*. Report Number SMRUL-TCE-2013-014, Scottish Marine and Freshwater Science, 5(2). Edinburgh: Scottish Government. Available at https://goo.gl/GKu9Ek.

Bregman, A. S. (1994). *Auditory Scene Analysis: The Perceptual Organization of Sound*. Cambridge, MA: MIT Press.

California Department of Transportation. (2001). *Pile Installation Demonstration Project, Fisheries Impact Assessment*. Caltrans Contract 04A0148, San Francisco-Oakland Bay Bridge East Span Seismic Safety Project, California Department of Transportation, Sacramento.

Carlson, T. J., Hastings, M. C., & Popper, A. N. (2007). *Update on Recommendations for Revised Interim Sound Exposure Criteria for Fish During Pile Driving Activities*. Available at https://goo.gl/KRzmLh.

Casaretto, L., Picciulin, M., Olsen, K., & Hawkins, A. D. (2014). Locating spawning haddock (*Melanogrammus aeglefinus*, Linnaeus, 1758) at sea by means of sound. *Fisheries Research*, 154, 127-134.

Casper, B. M., & Mann, D. A. (2009). Field hearing measurements of the Atlantic sharpnose shark *Rhizoprionodon terraenovae*. *Journal of Fish Biology*, 75(10), 2768-2776.

Casper, B. M., Halvorsen, M. B., & Popper, A. N. (2012a). Are sharks even bothered by a noisy environment? In A. N. Popper & A. Hawkins (Eds.), *The Effects of Noise on Aquatic Life* (pp. 93-97). New York: Springer-Verlag.

Casper, B. M., Popper, A. N., Matthews, F., Carlson, T. J., & Halvorsen, M. B. (2012b). Recovery of barotrauma injuries in Chinook salmon, *Oncorhynchus tshawytscha*, from exposure to pile driving sound. *PLoS ONE, 7*(6), e39593.

Casper, B. M., Halvorsen, M. B., Matthews, F., Carlson, T. J., & Popper, A. N. (2013a). Recovery of barotrauma injuries resulting from exposure to pile driving sound in two sizes of hybrid striped bass. *PLoS ONE, 8*(9), e73844.

Casper, B. M., Smith, M. E., Halvorsen, M. B., Sun, H., Carlson, T. J., & Popper, A. N. (2013b). Effects of exposure to pile driving sounds on fish inner ear tissues. *Comparative Biochemistry and Physiology. Part A: Molecular & Integrative Physiology, 166*(2), 352-360.

Chapman, C. J., & Hawkins, A. (1973). A field study of hearing in the cod, *Gadus morhua* L. *Journal of Comparative Physiology A: Neuroethology, Sensory, Neural, and Behavioral Physiology, 85*, 147-167.

Chapman, C. J., & Johnstone, A. D. (1974). Some auditory discrimination experiments on marine fish. *Journal of Experimental Biology, 61*(2), 521-528.

Chapman, C. J., & Sand, O. (1974). Field studies of hearing in two species of flatfish *Pleuronectes platessa* (L.) and *Limanda limanda* (L.) (Family Pleuronectidae). *Comparative Biochemistry and Physiology Part A: Physiology, 47*(1), 371-385.

Cheesman, S. (2016). Measurements of operational wind turbine noise in UK waters. In A. N. Popper & A. Hawkins (Eds.), *The Effects of Noise on Aquatic Life II* (pp. 153-160). New York: Springer-Verlag.

Continental Shelf Associates. (2004). *Explosive Removal of Offshore Structures - Information Synthesis Report*. Outer Continental Shelf (OCS) Study MMS 2003-070 prepared by Continental Shelf Associates for the Minerals Management Service,, Gulf of Mexico OCS Region, US Department of the Interior, New Orleans, LA.

Coombs, S., & Popper, A. N. (1979). Hearing differences among Hawaiian squirrelfish (family Holocentridae) related to differences in the peripheral auditory system. *Journal of Comparative Physiology A: Neuroethology, Sensory, Neural, and Behavioral Physiology, 132*(2), 203-207.

Corwin, J. T. (1977). Morphology of the macula neglecta in sharks of the genus *Carcharhinus*. *Journal of Morphology, 152*(3), 341-362.

Corwin, J. T. (1983). Postembryonic growth of the macula neglecta auditory detector in the ray, *Raja clavata*: Continual increases in hair cell number, neural convergence, and physiological sensitivity. *Journal of Comparative Neurology, 217*(3), 345-356.

Cotter, A. J. R. (2008). The "soundscape" of the sea, underwater navigation, and why we should be listening more. In A. Payne, J. Cotter and T. Potter (Eds.), *Advances in Fisheries Science: 50 Years on from Beverton and Holt* (pp. 451-471). Oxford, UK: Blackwell Publishing.

Dahl, P. H., de Jong, C. A. F., & Popper, A. N. (2015). The underwater sound field from impact pile driving and its potential effects on marine life. *Acoustics Today, 11*(2), 18-25.

de Jong, C., Ainslie, M., & Blacquière, G. (2011). *Standard for Measurement and Monitoring of Underwater Noise, Part II: Procedures for Measuring Underwater Noise in Connection with Offshore Wind Farm Licensing*. Report TNO-DV 2011 C235, TNO, The Hague, The Nertherlands. Available at https://goo.gl/hzQrGV

Dooling, R. J., & Blumenrath, S. H. (2016). Masking experiments in humans and birds using anthropogenic noises. In A. N. Popper & A. Hawkins (Eds.), *The Effects of Noise on Aquatic Life II* (pp. 239-243). New York: Springer-Verlag.

Dooling, R. J., Leek, M. R., & Popper, A. N. (2015). Effects of noise on fishes: What we can learn from humans and birds. *Integrative Zoology, 10*(1), 29-37.

Ellison, W. T., & Frankel, A. S. (2012). A common sense approach to source metrics. In A. N. Popper & A. Hawkins (Eds.), *The Effects of Noise on Aquatic Life* (pp. 433-438). New York: Springer-Verlag.

Ellison, W. T., Southall, B. L., Clark, C. W., & Frankel, A. S. (2012). A new context-based approach to assess marine mammal behavioral responses to anthropogenic sounds. *Conservation Biology, 26*(1), 21-28.

Engås, A., & Løkkeborg, S. (2002). Effects of seismic shooting and vessel-generated noise on fish behaviour and catch rates. *Bioacoustics,* 2(3), 313-316.

Engås, A., Løkkeborg, S., Ona, E., & Soldal, A. V. (1996). Effects of seismic shooting on local abundance and catch rates of cod (*Gadus morhua*) and haddock (*Melanogrammus aeglefinus*). *Canadian Journal of Fisheries and Aquatic Sciences,* 53, 2238-2249.

Enger, P. S. (1981). Frequency discrimination in teleosts—Central or peripheral? In W. N. Tavolga, A. N. Popper, & R. R. Fay (Eds.), *Hearing and Sound Communication in Fishes* (pp. 243-255). New York: Springer-Verlag.

Erbe, C., MacGillivray, A., & Williams, R. (2012). Mapping cumulative noise from shipping to inform marine spatial planning. *The Journal of the Acoustical Society of America,* 132(5), EL423-EL438.

Erbe, C., Reichmuth, C., Cunningham, K., Lucke, K., & Dooling, R. (2016). Communication masking in marine mammals: A review and research strategy. *Marine Pollution Bulletin,* 103(1-2), 15-38.

Fay, R. R. (1974). Masking of tones by noise for the goldfish (*Carassius auratus*). *Journal of Comparative and Physiological Psychology,* 87(4), 708-716.

Fay, R. R. (2005). Sound source localization by fishes. In A. N. Popper & R. R. Fay (Eds.), *Sound Source Localization* (pp. 36-66). New York: Springer-Verlag.

Fay, R. R. (2009). Sound source segregation by goldfish: Two simultaneous tones. *The Journal of the Acoustical Society of America,* 125(6), 4053-4059.

Filiciotto, F., Cecchini, S., Buscaino, G., Maccarrone, V., Piccione, G., & Fazio, F. (2016). Impact of aquatic acoustic noise on oxidative status and some immune parameters in gilthead sea bream *Sparus aurata* (Linnaeus, 1758) juveniles. *Aquaculture Research,* 48, 1895-1903.

Fletcher, H. (1940). Auditory patterns. *Reviews of Modern Physics,* 12(1), 47.

Francis, C. D., & Barber, J. R. (2013). A framework for understanding noise impacts on wildlife: An urgent conservation priority. *Frontiers in Ecology and the Environment,* 11(6), 305-313.

Gisiner, R. (2016). Sound and marine seismic surveys. *Acoustics Today,* 12(4), 10-18.

Govoni, J. J., West, M. A., Settle, L., Lynch, R. T., & Greene, M. D. (2008). Effects of underwater explosions on larval fish: Implications for a coastal engineering project. *Journal of Coastal Research,* 24, 228-233.

Halvorsen, M. B., Casper, B. M., Woodley, C. M., Carlson, T. J., & Popper, A. N. (2011). *Hydroacoustic Impacts on Fish from Pile Installation.* NCHRP Research Results Digest 363, Project 25-28, National Cooperative Highway Research Program, Transportation Research Board, National Academy of Sciences, Washington, DC. Available at http://www.trb.org/Publications/Blurbs/166159.aspx.

Halvorsen, M. B., Casper, B. M., Woodley, C. M., Carlson, T. J., & Popper, A. N. (2012a). Threshold for onset of injury in Chinook salmon from exposure to impulsive pile driving sounds. *PLoS ONE,* 7(6), e38968.

Halvorsen, M. B., Casper, B. M., Matthews, F., Carlson, T. J., & Popper, A. N. (2012b). Effects of exposure to pile-driving sounds on the lake sturgeon, Nile tilapia and hogchoker. *Proceedings of the Royal Society B: Biological Sciences,* 279(1748), 4705-4714.

Halvorsen, M. B., Zeddies, D. G., Ellison, W. T., Chicoine, D. R., & Popper, A. N. (2012c). Effects of mid-frequency active sonar on hearing in fish. *The Journal of the Acoustical Society of America,* 131(1), 599-607.

Halvorsen, M. B., Zeddies, D. G., Chicoine, D., & Popper, A. N. (2013). Effects of low-frequency naval sonar exposure on three species of fish. *The Journal of the Acoustical Society of America,* 134(2), EL205-EL210.

Hastings, M. C. (2008). Coming to terms with the effects of ocean noise on marine animals. *Acoustics Today,* 4(2), 22-34.

Hastings, M. C., Popper, A. N., Finneran, J. J., & Lanford, P. J. (1996). Effects of low-frequency underwater sound on hair cells of the inner ear and lateral line of the teleost fish *Astronotus ocellatus*. *The Journal of the Acoustical Society of America,* 99(3), 1759-1766.

Hawkins, A. D., & Chapman, C. J. (1966). Underwater sounds of the haddock, *Melanogrammus aeglifinus*. *Journal of the Marine Biological Association of the United Kingdom*, 46, 241-247.
Hawkins, A. D., & Chapman, C. J. (1975). Masked auditory thresholds in the cod, *Gadus morhua* L. *Journal of Comparative Physiology A: Neuroethology, Sensory, Neural, and Behavioral Physiology*, 103(2), 209-226.
Hawkins, A. D., & MacLennan, D. N. (1976). An acoustic tank for hearing studies on fish. In A. Schuijf & A. D. Hawkins (Eds.), *Sound Reception in Fish* (149-169). Amsterdam: Elsevier.
Hawkins, A. D., & Sand, O. (1977). Directional hearing in the median vertical plane by the cod. *Journal of Comparative Physiology A: Neuroethology, Sensory, Neural, and Behavioral Physiology*, 122(1), 1-8.
Hawkins, A. D., & Johnstone, A. D. F. (1978). The hearing of the Atlantic salmon, *Salmo salar*. *Journal of Fish Biology*, 13, 655-673.
Hawkins, A. D., & Myrberg, A. A., Jr. (1983). Hearing and sound communication underwater. In B. Lewis (Ed.), *Bioacoustics: A Comparative Approach* (pp. 347-405). London: Academic Press.
Hawkins, A. D., & Popper, A. N. (2014). Assessing the impacts of underwater sounds on fishes and other forms of marine life. *Acoustics Today*, 10(2), 30-41.
Hawkins, A. D., & Popper, A. N. (2016). Developing sound exposure criteria for fishes. In A. N. Popper & A. Hawkins (Eds.), *The Effects of Noise on Aquatic Life II* (pp. 431-439). New York: Springer-Verlag.
Hawkins, A. D., MacLennan, D. N., Urquhart, G. G., & Robb, C. (1974). Tracking cod *Gadus morhua* L. in a Scottish sea loch. *Journal of Fish Biology*, 6(3), 225-236.
Hawkins, A. D., Roberts, L., & Cheesman, S. (2014). Responses of free-living coastal pelagic fish to impulsive sounds. *The Journal of the Acoustical Society of America*, 135(5), 3101-3116.
Hawkins, A. D., Pembroke, A., & Popper, A. N. (2015). Information gaps in understanding the effects of noise on fishes and invertebrates. *Reviews in Fish Biology and Fisheries*, 25, 39-64.
Hazelwood, R. A. (2012). Ground roll waves as a potential influence on fish: Measurement and analysis techniques. In A. N. Popper & A. Hawkins (Eds.), *The Effects of Noise on Aquatic Life* (pp. 449-452). New York: Springer-Verlag.
Henderson, D., & Hamernik, R. P. (2012). The use of kurtosis measurement in the assessment of potential noise trauma. In C. G. Le Prell, D. Henderson, R. R. Fay, & A. N. Popper (Eds.), *Noise-Induced Hearing Loss: Scientific Advances* (pp. 41-55). New York: Springer-Verlag.
Hildebrand, J. A. (2009). Anthropogenic and natural sources of ambient noise in the ocean. *Marine Ecology Progress Series*, 395, 5-20.
Hobday, A., Smith, A., Stobutzki, I., Bulman, C., Daley, R., Dambacher, J., Deng, R., Dowdney, J., Fuller, M., & Furlani, D. (2011). Ecological risk assessment for the effects of fishing. *Fisheries Research*, 108(2), 372-384.
Holles, S., Simpson, S. D., Radford, A. N., Berten, L., & Lecchini, D. (2013). Boat noise disrupts orientation behaviour in a coral reef fish. *Marine Ecology Progress Series*, 485, 295-300.
Jacobs, D. W., & Tavolga, W. N. (1967). Acoustic intensity limens in the goldfish. *Animal Behaviour*, 15(2), 324-335.
Kaatz, I. M. (2002). Multiple sound-producing mechanisms in teleost fishes and hypotheses regarding their behavioural significance. *Bioacoustics*, 12(2-3), 230-233.
Keevin, T. M., & Hempen, G. L. (1997). *The Environmental Effects of Underwater Explosions with Methods to Mitigate Impacts*. SDMS Doc ID 550560, US Army Corps of Engineers, St. Louis District, St. Louis, MO. Available at https://semspub.epa.gov/work/01/550560.pdf.
Kight, C. R., & Swaddle, J. P. (2011). How and why environmental noise impacts animals: An integrative, mechanistic review. *Ecology Letters*, 14(10), 1052-1061.
Klages, M., Muyakshin, S., Soltwedel, T., & Arntz, W. E. (2002). Mechanoreception, a possible mechanism for food fall detection in deep-sea scavengers. *Deep-Sea Research Part I: Oceanographic Research Papers*, 49(1), 143-155.
Knudsen, F. R., Enger, P. S., & Sand, O. (1992). Awareness reactions and avoidance responses to sound in juvenile Atlantic salmon, *Salmo salar* L. *Journal of Fish Biology*, 40, 523-534.

Kujawa, S. G., & Liberman, M. C. (2009). Adding insult to injury: Cochlear nerve degeneration after "temporary" noise-induced hearing loss. *The Journal of Neuroscience,* 29(45), 14077-14085.

Kunc, H. P., Lyons, G. N., Sigwart, J. D., McLaughlin, K. E., & Houghton, J. D. R. (2014). Anthropogenic noise affects behavior across sensory modalities. *The American Naturalist,* 184(4), E93-E100.

Kunc, H. P., McLaughlin, K. E., & Schmidt, R. (2016). Aquatic noise pollution: Implications for individuals, populations, and ecosystems. *Proceedings of the Royal Society B: Biological Sciences,* 283(1836), 20160839. doi:https://doi.org/10.1098/rspb.2016.0839.

Ladich, F. (2013). Effects of noise on sound detection and acoustic communication in fishes. In H. Brumm (Ed.), *Animal Communication and Noise* (pp. 65-90). Berlin Heidelberg: Springer-Verlag.

Laughlin, J. (2006). *Underwater Sound Levels Associated with Pile Driving at the Cape Disappointment Boat Launch Facility, Wave Barrier Project.* Report prepared by the Washington State Department of Transportation, Seattle.

Le Prell, C. G., Henderson, D., Fay, R. R., & Popper, A. N. (Eds.). (2012). *Noise-Induced Hearing Loss: Scientific Advances.* New York: Springer-Verlag.

Lin, H., Furman, A., Kujawa, S., & Liberman, M. C. (2011). Primary neural degeneration in the guinea pig cochlea after reversible noise-induced threshold shift. *Journal of the Association for Research in Otolaryngology,* 12(5), 605-616.

Løkkeborg, S., Ona, E., Vold, A., & Salthaug, A. (2012a). Effects of sounds from seismic air guns on fish behavior and catch rates. In A. N. Popper & A. Hawkins (Eds.), *The Effects of Noise on Aquatic Life* (pp. 415-419). New York: Springer-Verlag.

Løkkeborg, S., Ona, E., Vold, A., Salthaug, A., & Jech, J. M. (2012b). Sounds from seismic air guns: Gear- and species-specific effects on catch rates and fish distribution. *Canadian Journal of Fisheries and Aquatic Sciences,* 69(8), 1278-1291.

Luczkovich, J. J., Pullinger, R. C., Johnson, S. E., & Sprague, M. W. (2008). Identifying sciaenid critical spawning habitats by the use of passive acoustics. *Transactions of the American Fisheries Society,* 137(2), 576-605.

Madsen, P. T., Wahlberg, M., Tougaard, J., Lucke, K., & Tyack, P. (2006). Wind turbine underwater noise and marine mammals: Implications of current knowledge and data needs. *Marine Ecology Progress Series,* 309, 279-295.

Mann, D. A., Higgs, D. M., Tavolga, W. N., Souza, M. J., & Popper, A. N. (2001). Ultrasound detection by clupeiform fishes. *The Journal of the Acoustical Society of America,* 109(6), 3048-3054.

Martin, B., Zeddies, D. G., Gaudet, B., & Richard, J. (2016). Evaluation of three sensor types for particle motion measurement. In A. N. Popper & A. Hawkins (Eds.), *The Effects of Noise on Aquatic Life II* (pp. 679-686). New York: Springer-Verlag.

Martin, S. B., & Popper, A. N. (2016). Short-and long-term monitoring of underwater sound levels in the Hudson River (New York, USA). *The Journal of the Acoustical Society of America,* 139(4), 1886-1897.

Mattsson, A., Parkes, G., & Hedgeland, D. (2012). Svein Vaage broadband air gun study. In A. N. Popper & A. Hawkins (Eds.), *The Effects of Noise on Aquatic Life II* (pp. 469-471). New York: Springer-Verlag.

McCauley, R. D., Fewtrell, J., & Popper, A. N. (2003). High intensity anthropogenic sound damages fish ears. *The Journal of the Acoustical Society of America,* 113(1), 638-642.

McKenna, M. F., Ross, D., Wiggins, S. M., & Hildebrand, J. A. (2012). Underwater radiated noise from modern commercial ships. *The Journal of the Acoustical Society of America,* 131(1), 92-103.

Morley, E. L., Jones, G., & Radford, A. N. (2014). The importance of invertebrates when considering the impacts of anthropogenic noise. *Proceedings of the Royal Society B: Biological Sciences,* 281(1776), 20132683.

Moulton, J. M. (1963). Acoustic behaviour of fishes. In R.-G. Busnel (Ed.), *Acoustic Behaviour of Animals* (pp. 655-693). Amsterdam: Elsevier.

Mueller-Blenkle, C., McGregor, P. K., Gill, A. B., Andersson, M. H., Metcalfe, J., Bendall, V., Sigray, P., Wood, D. T., & Thomsen, F. (2010). *Effects of Pile-Driving Noise on the Behaviour of Marine Fish*. Cowrie Ref: Fish 06-08, Technical Report 31st March 2010. Available at https://goo.gl/YXDC8i.

Myrberg, A. A., Jr. (1981). Sound communication and interception in fishes. In W. N. Tavolga, A. N. Popper, & R. R. Fay (Eds.), *Hearing and Sound Communication in Fishes* (pp. 395-426). New York: Springer-Verlag.

Myrberg, A. A., Jr. (2001). The acoustical biology of elasmobranchs. *Environmental Biology of Fishes*, 60(1-3), 31-46.

National Marine Fisheries Service (NMFS). (2016). *Technical Guidance for Assessing the Effects of Anthropogenic Sound on Marine Mammal Hearing: Underwater Acoustic Thresholds for Onset of Permanent and Temporary Threshold Shifts*. NOAA Technical Memorandum NMFS-OPR-55, National Oceanic and Atmospheric Administration (NOAA), US Department of Commerce, Washington, DC. Available at https://goo.gl/F2VPU6.

National Research Council. (2005). *Marine Mammal Populations and Ocean Noise: Determining When Noise Causes Biologically Significant Effects*. Washington, DC: National Academies Press.

Nedelec, S. L., Radford, A. N., Simpson, S. D., Nedelec, B., Lecchini, D., & Mills, S. C. (2014). Anthropogenic noise playback impairs embryonic development and increases mortality in a marine invertebrate. *Science Reports*, 4, 5891. doi:https://doi.org/10.1038/srep05891.

Nedelec, S. L., Campbell, J., Radford, A. N., Simpson, S. D., & Merchant, N. D. (2016). Particle motion: The missing link in underwater acoustic ecology. *Methods in Ecology and Evolution*, 7, 836-842.

Nedwell, J. R., Turnpenny, A. W. H., Lovell, J. M., & Edwards, B. (2006). An investigation into the effects of underwater piling noise on salmonids. *The Journal of the Acoustical Society of America*, 120(5), 2550-2554.

Nedwell, J. R., Parvin, S. J., Edwards, B., Workman, R., Brooker, A. G., & Kynoch, J. E. (2007). *Measurement and Interpretation of Underwater Noise During Construction and Operation of Offshore Windfarms in UK Waters*. Subacoustch Report No. 544R0738 for Cowrie Ltd., UK. Available at https://goo.gl/cCJyfK.

Neo, Y. Y., Seitz, J., Kastelein, R. A., Winter, H. V., ten Cate, C., & Slabbekoorn, H. (2014). Temporal structure of sound affects behavioural recovery from noise impact in European seabass. *Biological Conservation*, 178, 65-73.

Nieukirk, S. L., Stafford, K. M., Mellinger, D. K., Dziak, R. P., & Fox, C. G. (2004). Low-frequency whale and seismic airgun sounds recorded in the mid-Atlantic Ocean. *The Journal of Acoustical Society of America*, 115, 1832-1843.

Nieukirk, S. L., Klinck, H., Mellinger, D. K., Klinck, K., & Dziak, R. P. (2014). Seismic airgun surveys and vessel traffic in the Fram Strait and their contribution to the polar soundscape. *The Journal of the Acoustical Society of America*, 136(4), 2154.

Oestman, R., Buehler, D., Reyff, J., & Rodkin, R. (2009). *Technical Guidance for Assessment and Mitigation of the Hydroacoustic Effects of Pile Driving on Fish*. Report by ICF International and Illingworth and Rodkin Inc. prepared for the California Department of Transportation, Sacramento.

Parvulescu, A. (Ed.). (1964). *Problems of Propagation and Processing*. Oxford, UK: Pergamon Press.

Pine, M. K., Jeffs, A. G., Wang, D., & Radford, C. A. (2016). The potential for vessel noise to mask biologically important sounds within ecologically significant embayments. *Ocean & Coastal Management*, 127, 63-73.

Popper, A. N., & Clarke, N. L. (1979). Non-simultaneous auditory masking in the goldfish, *Carassius auratus*. *Journal of Experimental Biology*, 83, 145-158.

Popper, A. N., & Hastings, M. C. (2009). The effects of anthropogenic sources of sound on fishes. *Journal of Fish Biology,* 75(3), 455-489.

Popper, A. N., & Fay, R. R. (2011). Rethinking sound detection by fishes. *Hearing Research,* 273(1), 25-36.

Popper, A. N., & Hawkins, A. (Eds.). (2012). *The Effects of Noise on Aquatic Life*. New York: Springer-Verlag.

Popper, A. N., & Hawkins, A. (Eds.). (2016). *The Effects of Noise on Aquatic Life II*. New York: Springer-Verlag.

Popper, A. N., Salmon, M., & Horch, K. W. (2001). Acoustic detection and communication by decapod crustaceans. *Journal of Comparative Physiology A: Neuroethology, Sensory, Neural, and Behavioral Physiology,* 187(2), 83-89.

Popper, A. N., Fay, R. R., Platt, C., & Sand, O. (2003). Sound detection mechanisms and capabilities of teleost fishes. In S. P. Collin & N. J. Marshall (Eds.), *Sensory Processing in Aquatic Environments* (pp. 3-38). New York: Springer-Verlag.

Popper, A. N., Smith, M. E., Cott, P. A., Hanna, B. W., MacGillivray, A. O., Austin, M. E., & Mann, D. A. (2005). Effects of exposure to seismic airgun use on hearing of three fish species. *The Journal of the Acoustical Society of America,* 117(6), 3958-3971.

Popper, A. N., Halvorsen, M. B., Kane, A. S., Miller, D. L., Smith, M. E., Song, J., Stein, P., & Wysocki, L. E. (2007). The effects of high-intensity, low-frequency active sonar on rainbow trout. *The Journal of the Acoustical Society of America,* 122(1), 623-635.

Popper, A. N., Hawkins, A. D., Fay, R. R., Mann, D. A., Bartol, S., Carlson, T. J., Coombs, S., Ellison, W. T., Gentry, R. L., Halvorsen, M. B., Lokkeborg, S., Rogers, P., Southall, B. L., Zeddies, D. G., & Tavolga, W. N. (2014). Sound exposure guidelines. In *ASA S3/SC1. 4 TR-2014 Sound Exposure Guidelines for Fishes and Sea Turtles. A Technical Report Prepared by ANSI-Accredited Standards Committee S3/SC1 and Registered with ANSI* (pp. 33–51). New York: Springer International Publishing.

Popper, A. N., Gross, J. A., Carlson, T. J., Skalski, J., Young, J. V., Hawkins, A. D., & Zeddies, D. (2016). Effects of exposure to the sound from seismic airguns on pallid sturgeon and paddlefish. *PLoS ONE,* 11(8), e0159486.

Rabinowitz, P. M. (2012). The public health significance of noise-induced hearing loss. In C. G. Le Prell, D. Henderson, R. R. Fay, & A. N. Popper (Eds.), *Noise-Induced Hearing Loss: Scientific Advances* (pp. 13-26). New York: Springer-Verlag.

Radford, A. N., Kerridge, E., & Simpson, S. D. (2014). Acoustic communication in a noisy world: Can fish compete with anthropogenic noise? *Behavioral Ecology,* 25, 1022-1030.

Ramcharitar, J., Gannon, D. P., & Popper, A. N. (2006). Bioacoustics of the family Sciaenidae (croakers and drumfishes). *Transactions of the American Fisheries Society,* 135, 1409-1431.

Remage-Healey, L., Nowacek, D. P., & Bass, A. H. (2006). Dolphin foraging sounds suppress calling and elevate stress hormone levels in a prey species, the Gulf toadfish. *Journal of Experimental Biology,* 209, 4444-4451.

Reyff, J. A. (2016). Underwater sound propagation from marine pile driving. In A. N. Popper & A. Hawkins (Eds.), *The Effects of Noise on Aquatic Life II* (pp. 909-915). New York: Springer-Verlag.

Richardson, W. J., Greene, C. R., Jr., Malme, C. I., & Thomson, D. H. (1995). *Marine Mammals and Noise*. New York: Academic Press.

Robinson, S. P., Theobald, P. D., Lepper, P. A., Hayman, G., Humphrey, V. F., Wang, L.-S., & Mumford, S. (2012). Measurement of underwater noise arising from marine aggregate operations. In A. N. Popper & A. Hawkins (Eds.), *The Effects of Noise on Aquatic Life* (pp. 465-468). New York: Springer-Verlag.

Rodkin, R. B., & Reyff, J. A. (2008). Underwater sound from marine pile driving. *Bioacoustics,* 17(1-3), 138-140.

Rogers, P. H., Hawkins, A. D., Popper, A. N., Fay, R. R., & Gray, M. D. (2016). Parvulescu revisited: Small tank acoustics for bioacousticians. In A. N. Popper & A. Hawkins (Eds.), *The Effects of Noise on Aquatic Life II* (pp. 933-941). New York: Springer-Verlag.

Ross, D. (1987). *Mechanics of Underwater Noise*. Los Altos, CA: Peninsula Publishing.
Ross, D. (1993). On ocean underwater ambient noise. *Acoustics Bulletin,* 18, 5-8.
Rossi, E., Licitra, G., Iacoponi, A., & Taburni, D. (2016). Assessing the underwater ship noise levels in the North Tyrrhenian Sea. In A. N. Popper & A. Hawkins (Eds.), *The Effects of Noise on Aquatic Life II* (pp. 943-949). New York: Springer-Verlag.
Rossington, K., Benson, T., Lepper, P., & Jones, D. (2013). Eco-hydro-acoustic modeling and its use as an EIA tool. *Marine Pollution Bulletin,* 75(1-2), 235-243.
Sand, O., & Enger, P. S. (1973). Function of the swimbladder in fish hearing. In A. Moller (Ed.), *Basic Mechanisms of Hearing* (pp. 893-908). New York: Academic Press.
Sand, O., & Hawkins, A. D. (1973). Acoustic properties of the cod swim bladder. *Journal of Experimental Biology,* 58, 797-820.
Sand, O., & Bleckmann, H. (2008). Orientation to auditory and lateral line stimuli. In J. F. Webb, R. R. Fay, & A. N. Popper (Eds.), *Fish Bioacoustics* (pp. 183-222). New York: Springer-Verlag.
Sarà, G., Dean, J. M., D'Amato, D., Buscaino, G., Oliveri, A., Genovese, S., Ferro, S., Buffa, G., Lo Martire, M., & Mazzola, S. (2007). Effect of boat noise on the behaviour of bluefin tuna *Thunnus thynnus* in the Mediterranean Sea. *Marine Ecology Progress Series,* 33, 243-253.
Schuijf, A., & Buwalda, R. (1975). On the mechanism of directional hearing in cod (*Gadus morhua* L.). *Journal of Comparative Physiology A: Neuroethology, Sensory, Neural, and Behavioral Physiology,* 98(4), 333-343.
Schuijf, A., & Hawkins, A. (1983). Acoustic distance discrimination by the cod. *Nature.* 302, 143-144.
Schulz-Mirbach, T., Hess, M., Metscher, B. D., & Ladich, F. (2013). A unique swim bladder-inner ear connection in a teleost fish revealed by a combined high-resolution microtomographic and three-dimensional histological study. *BMC Biology,* 11, 1-13.
Sertlek, H. Ö., Aarts, G., Brasseur, S., Slabbekoorn, H., ten Cate, C., von Benda-Beckmann, A. M., & Ainslie, M. A. (2016). Mapping underwater sound in the Dutch part of the North Sea. In A. N. Popper & A. Hawkins (Eds.), *The Effects of Noise on Aquatic Life II* (pp. 1001-1006). New York: Springer-Verlag.
Sigray, P., & Andersson, M. H. (2011). Particle motion measured at an operational wind turbine in relation to hearing sensitivity in fish. *The Journal of the Acoustical Society of America,* 130(1), 200-207.
Sigray, P., & Andersson, M. H. (2012). Underwater particle acceleration induced by a wind turbine in the Baltic Sea. In A. N. Popper & A. Hawkins (Eds.), *The Effects of Noise on Aquatic Life* (pp. 489-492). New York: Springer-Verlag.
Simpson, S., Meekan, M., McCauley, R., & Jeffs, A. (2004). Attraction of settlement-stage coral reef fishes to reef noise. *Marine Ecology Progress Series,* 276(1), 263-268.
Simpson, S. D., Meekan, M., Montgomery, J., McCauley, R., & Jeffs, A. (2005). Homeward sound. *Science,* 308(5719), 221.
Skalski, J. R., Pearson, W. H., & Malme, C. I. (1992). Effects of sounds from a geophysical survey device on catch-per-unit-effort in a hook-and-line fishery for rockfish (*Sebastes* spp.). *Canadian Journal of Fisheries and Aquatic Sciences,* 49, 1357-1365.
Slabbekoorn, H. (2016). Aiming for progress in understanding underwater noise impact on fish: Complementary need for indoor and outdoor studies. In A. N. Popper & A. Hawkins (Eds.), *The Effects of Noise on Aquatic Life II* (pp. 1057-1065). New York: Springer-Verlag.
Slabbekoorn, H., & Bouton, N. (2008). Soundscape orientation: A new field in need of sound investigation. *Animal Behaviour,* 76, e5-e8.
Slabbekoorn, H., Bouton, N., van Opzeeland, I., Coers, A., ten Cate, C., & Popper, A. N. (2010). A noisy spring: the impact of globally rising underwater sound levels on fish. *Trends in Ecology and Evolution,* 25(7), 419-427.
Slotte, A., Hansen, K., Dalen, J., & Ona, E. (2004). Acoustic mapping of pelagic fish distribution and abundance in relation to a seismic shooting area off the Norwegian west coast. *Fisheries Research,* 67(2), 143-150.

Smith, A., Fulton, E., Hobday, A., Smith, D., & Shoulder, P. (2007). Scientific tools to support the practical implementation of ecosystem-based fisheries management. *ICES Journal of Marine Science,* 64(4), 633-639.

Smith, M. E., Kane, A. S., & Popper, A. N. (2004). Acoustical stress and hearing sensitivity in fishes: Does the linear threshold shift hypothesis hold water? *Journal of Experimental Biology,* 207, 3591-3602.

Smith, M. E., Coffin, A. B., Miller, D. L., & Popper, A. N. (2006). Anatomical and functional recovery of the goldfish (*Carassius auratus*) ear following noise exposure. *Journal of Experimental Biology,* 209, 4193-4202.

Song, J., Mann, D. A., Cott, P. A., Hanna, B. W., & Popper, A. N. (2008). The inner ears of Northern Canadian freshwater fishes following exposure to seismic air gun sounds. *The Journal of the Acoustical Society of America,* 124(2), 1360-1366.

Southall, B. L. (2005). *Shipping Noise and Marine Mammals: A Forum for Science, Technology, and Management.* Final Report of the National Oceanic and Atmospheric Administration (NOAA) International Symposium, Arlington, VA, May 18–19, 2004.

Southall, B. L., Bowles, A. E., Ellison, W. T., Finneran, J. J., Gentry, R. L., Greene, C. R., Jr., Kastak, D., Ketten, D. R., Miller, J. H., Nachtigall, P. E., Richardson, W. J., Thomas, J. A., & Tyack, P. L. (2007). Marine mammal noise exposure criteria: Initial scientific recommendations. *Aquatic Mammals,* 33, 411–521.

Stadler, J. H., & Woodbury, D. P. (2009). Assessing the effects to fishes from pile driving: Application of new hydroacoustic criteria. *Proceedings of the 38th International Congress and Exposition on Noise Control Engineering 2009 (Inter-Noise 2009)*, Ottawa, ON, Canada, August 23–25, 2009.

Stanley, J. A., Radford, C. A., & Jeffs, A. G. (2012). Location, location, location: Finding a suitable home among the noise. *Proceedings of the Royal Society of London B: Biological Sciences,* 279(1742), 3622-3631.

Tavolga, W. N. (1964). *Marine Bio-Acoustics*. Oxford, UK: Pergamon Press.

Tavolga, W. N. (1967). *Marine Bio-Acoustics II*. Oxford, UK: Pergamon Press.

Tavolga, W. N., Popper, A. N., & Fay, R. R. (1981). *Hearing and Sound Communication in Fishes*. New York: Springer-Verlag.

Tennessen, J. B., Parks, S. E., & Langkilde, T. L. (2016). Anthropogenic noise and physiological stress in wildlife. In A. N. Popper & A. Hawkins (Eds.), *The Effects of Noise on Aquatic Life II* (pp. 1145-1148). New York: Springer-Verlag.

Tester, A. L., Kendall, J. I., & Milisen, W. B. (1972). Morphology of the ear of the shark genus *Carcharhinus*, with particular reference to the macula neglecta. *Pacific Science,* 26, 264-274.

Thomsen, F., Gill, A., Kosecka, M., Andersson, M., Andre, M., Degraer, S., Folegot, T., Gabriel, J., Judd, A., Neumann, N., Norro, A., Risch, D., Sigray, P., Wood, D., & Wilson, B. (2016). *MaRVEN—Environmental Impacts of Noise, Vibrations and Electromagnetic Emissions from Marine Renewable Energy*. Final Study Report, Directorate General for Research and Innovation, European Commission, Brussels. Available at https://goo.gl/wzQSyc.

Urick, R. J. (1983). *Principles of Underwater Sound,* 3rd ed. New York: McGraw-Hill.

Voellmy, I. K., Purser, J., Flynn, D., Kennedy, P., Simpson, S. D., & Radford, A. N. (2014). Acoustic noise reduces foraging success in two sympatric fish species via different mechanisms. *Animal Behaviour,* 89, 191-198.

Wahlberg, M., & Westerberg, H. (2005). Hearing in fish and their reactions to sound from offshore wind farms. *Marine Ecology Progress Series,* 288, 298-309.

Wardle, C. S., Carter, T. J., Urquhart, G. G., Johnstone, A. D. F., Ziolkowski, A. M., Hampson, G., & Mackie, D. (2001). Effects of seismic air guns on marine fish. *Continental Shelf Research,* 21, 1005-1027.

Webb, J. F., Fay, R. R., & Popper, A. N. (Eds.). (2008). *Fish Bioacoustics*. New York: Springer-Verlag.

Wenz, G. M. (1962). Acoustic ambient noise in the ocean: Spectra and sources. *The Journal of the Acoustical Society of America,* 34, 1936-1956.

Weston, D. E. (1960). Underwater explosions as acoustic sources. *Proceedings of the Physical Society,* 76, 233-249.

Woodbury, D., & Stadler, J. (2008). A proposed method to assess physical Injury to fishes from underwater sound produced during pile driving. *Bioacoustics,* 17, 289-297.

Wysocki, L. E., Dittami, J. P., & Ladich, F. (2006). Ship noise and cortisol secretion in European freshwater fishes. *Biological Conservation,* 128(4), 501-508.

Wysocki, L. E., Davidson, J. W., III, Smith, M. E., Frankel, A. S., Ellison, W. T., Mazik, P. M., Popper, A. N., & Bebak, J. (2007). Effects of aquaculture production noise on hearing, growth, and disease resistance of rainbow trout *Oncorhynchus mykiss*. *Aquaculture,* 272(1-4), 687-697.

Yelverton, J. T., Richmond, D. R., Hicks, W., Saunders, H., & Fletcher, E. R. (1975). *The Relationship Between Fish Size and Their Response to Underwater Blast.* Report DNA 3677T prepared for the Defense Nuclear Agency by the Lovelace Foundation For Medical Education and Research, Albuquerque, NM. Available at http://www.dtic.mil/dtic/tr/fulltext/u2/a015970.pdf.

Chapter 7
Effects of Anthropogenic Noise on Amphibians and Reptiles

Andrea Megela Simmons and Peter M. Narins

Abstract Anurans are highly vocal species that rely on acoustic communication for social behaviors. The advertisement (mating) calls of many anurans contain considerable energy within the predominant spectral range of traffic and other anthropogenic-noise sources. Whether and how these noise sources affect reproductive success and species viability is unclear. Data that address how anthropogenic sources affect the spatial distribution of breeding ponds, production and propagation of males' vocal signals, and detection and discrimination of these signals by females are inconsistent. Anurans may respond to anthropogenic noise using many of the same strategies that they use to deal with biotic and abiotic noise. But there are considerable differences between species in their responses to noise, related to habitat and other variables. Interpretation of data is hampered by the small numbers of species that have been studied; moreover, experiments to date focus only on the perception of advertisement calls and do not address how other biologically important vocalizations, such as aggressive and courtship calls, might be affected by anthropogenic noise. Some species of reptiles are also vocal, but data on the effects of anthropogenic noise on reptile social behaviors are severely lacking. Extensive research is needed to determine the impact of acoustic habitat degradation on these classes of animals.

A. M. Simmons (✉)
Department of Cognitive, Linguistic and Psychological Sciences, Brown University, Providence, RI, USA

Department of Neuroscience, Brown University, Providence, RI, USA
e-mail: andrea_simmons@brown.edu

P. M. Narins
Department of Integrative Biology and Physiology, University of California, Los Angeles, Los Angeles, CA, USA

Department of Ecology and Evolutionary Biology, University of California, Los Angeles, Los Angeles, CA, USA
e-mail: pnarins@ucla.edu

H. Slabbekoorn et al. (eds.), *Effects of Anthropogenic Noise on Animals*, Springer Handbook of Auditory Research 66,
https://doi.org/10.1007/978-1-4939-8574-6_7

Keywords Advertisement calls · Anurans · Calling surveys · Evoked calling · Mate choice · Noise rejection · Phonotaxis · Species distribution

7.1 Introduction

The biological classes Amphibia and Reptilia are composed of tetrapod (four-legged) animals that are either anamniotes (eggs developing without an amniotic membrane; amphibians) or amniotes (reptiles). Amphibians include the anurans (tailless amphibians: frogs and toads), the caudates (tailed amphibians: salamanders and newts), and the caecilians (limbless amphibians). The anurans are the most conspicuously vocal amphibians; as a result, much more is known about their hearing and vocal production abilities than for any other amphibians. Even though caudates can hear, vocalizations are limited in this group. Caecilians, on the other hand, are not known to hear airborne sounds or to vocalize, but they likely detect ground-borne vibrations. Reptiles (the testudines [turtles and tortoises], the crocodilians [crocodiles, alligators, and caimans], the squamates [snakes and lizards], and the rhynchocephalids [tuatara]) are sensitive to sounds. Some squamates, including geckos and crocodiles, have extensive vocal repertoires, whereas others, including the rhynchocephalids and other squamates, do not appear to use sounds for conspecific communication. Studies on the impact of anthropogenic-noise sources have been limited largely to anurans; thus, this chapter focuses on the analysis of anthropogenic noise on vocal communication within this order. All species names used in this chapter are consistent with those listed in AmphibiaWeb (2017).

7.1.1 The Acoustic World of Anurans

Various anthropogenic activities contributing to climate change, habitat loss, pollution, and disease are related to ongoing declines in amphibian populations (Pechmann et al. 1991; Hof et al. 2011). It is not at all clear, however, to what extent amphibians have been affected by acoustic habitat degradation specifically. Anthropogenic-noise sources might be expected to impact breeding and social behaviors in anurans because these behaviors in these groups are so heavily dependent on acoustic cues. Anuran species differ considerably in the acoustic structure of their advertisement calls, in their chorus structure, and in their habitats. All of these variables need to be understood in order to construct models of the potential effects of anthropogenic noise on anuran behavior.

Most anurans are chorusing species in which males form dense calling assemblages, often containing both conspecifics and heterospecifics, to attract females for mating. Within these choruses, males need to vocalize while other males are calling simultaneously and in the midst of abiotic sound sources such as wind, flowing water, and rain. A large body of literature (reviewed by Narins and Zelick 1988; Schwartz and Bee 2013) has described strategies by which chorusing males alleviate masking of their own calls within these acoustically cluttered conditions. Behavioral adaptations that have been identified in several species include (1) changing calling patterns (in rate, duration, frequency, or amplitude), (2) shifting the timing of calls with respect to calls of competitors, (3) calling at times of day when other species are not active, (4) varying the complexity or the type of call produced, and (5) maintaining specific spatial locations within choruses to minimize call overlap with neighbors. Different species are capable of modulating the rate of their vocal output in different directions and to different degrees; for example, some species increase calling rate in response to acoustic interference from conspecifics, whereas others decrease their calling rate. Even accounting for species variability, acoustic adaptations such as these also emerge when chorusing males are exposed to heterospecific vocalizations or abiotic sources (Penna et al. 2005; Penna and Zúñiga 2014). Of significance is whether these or other behavioral adaptations also operate to alleviate the impact of anthropogenic noise. If anthropogenic noise is treated as just another interfering source, then male anurans should be able to adapt their vocal output in its presence and thus remain immune to acoustic habitat degradation. There are only a few experiments directly addressing this question. And if behavioral adaptations do occur, they may carry costs that limit their effectiveness in a noisy soundscape with high human impact. The small number of species in which the impact of anthropogenic compared with biotic or abiotic noise has been explicitly examined limits the generalizability of the available data.

Mating behavior in anurans is by and large a female choice system, so females must be able to locate male choruses and to discriminate conspecific from heterospecific vocalizations within these choruses. Compared with the extensive literature on female responses to biotic sound sources (Gerhardt and Huber 2002; Velez et al. 2013), much less is known about the effects of abiotic or anthropogenic noise on female choice behavior. Wollerman and Wiley (2002) showed that even moderate levels of background chorus sound reduced the preferences of female hourglass treefrogs (*Dendropsophus ebraccattus*) for conspecific advertisement calls. They suggested that noise changed the female's decision tactics, thus potentially increasing mating errors (i.e., in noisy conditions, choosing a male who might be less fit and who would not have been chosen in quiet conditions). If such effects occur in the presence of anthropogenic sound sources, then overall population fitness might be affected, notwithstanding any adaptive modifications of a male's vocalizations under these conditions.

Hearing and Vocalizations in Anurans in Relation to Anthropogenic Sound

Except for highly specialized species with ultrasonic sensitivity (Narins et al. 2004, 2014; Feng et al. 2006; Shen et al. 2011), most anurans produce and hear sounds within the frequency range of 50 to 6000 Hz (Fay and Simmons 1999). Some species, including the white-lipped frog (*Leptodactylus albilabris*; Lewis and Narins 1985) and the red-eyed treefrog (*Agalychnis callidryas*; Caldwell et al. 2010), also produce and detect very low frequency (10- to 40-Hz) seismic signals that propagate through vegetation or the ground (for reviews, see Gridi-Papp and Narins 2010; Narins et al. 2016). Still other species, including the bullfrog (*Rana catesbeiana*) and the túngara frog (*Engystomops pustulosus*), vocalize while partially submerged in shallow water. Movements of their vocal sacs produce waterborne vibrations (ripples) that emanate in concentric circles into the water away from the male. These ripples are detectable by both conspecifics and predators (Boatright-Horowitz et al. 1999; Halfwerk et al. 2014). Seismic and waterborne signals are low frequency (<200 Hz) and propagate either through the solid substrate or through the water; thus, they would be less subject to masking by air-propagated vehicular or engine noise than are airborne components of advertisement calls. This raises the possibility that anurans communicating in habitats that are heavily impacted by anthropogenic-noise sources may increase reliance on seismic or waterborne signals for communication.

The acoustic structure of anuran advertisement calls is highly diverse. Some vocalizations contain tonal call elements, some are trill-like, some are made up of a series of repeated short pulses, some are harmonically structured with a clear periodicity, and some are noisy or buzz-like. Males typically emit their advertisement calls in repeated series or bouts. Because male anurans attract mates over relatively long distances compared with their body length, in general, advertisement calls are structured to resist masking from extraneous ambient sounds and to propagate effectively in the species' habitat (Ey and Fischer 2009). For example, the ultrasonic harmonics in the advertisement calls of the concave-eared torrent frog (*Odorrana tormota*) are high enough in frequency to avoid being masked by the intense, low-frequency broadband noise of the streams and waterfalls in their natural habitat (Feng et al. 2006). The ambient sound pressure level (SPL), with contributions from both biotic and abiotic factors, was shown to be a strong predictor of microhabitat selection in 34 Asian species (Goutte et al. 2013; see also Larsen and Radford, Chap. 5). This relationship implies that changes in ambient-noise levels, such as increased anthropogenic noise, have the capacity to radically alter calling site selection and communication behavior in chorusing species.

The spectral range of common anthropogenic-noise sources such as road traffic, aircraft noise, and construction noise extends from about 50 Hz up to 7000 Hz (Sun and Narins 2005; Cunnington and Fahrig 2010) but is concentrated at frequencies <2000 Hz, within the region of best auditory sensitivity and of peak spectral energy in the advertisement calls of many anuran species (Fig. 7.1). This suggests that anthropogenic noise, if sufficiently intense, can mask vocalizations and thus hinder species-specific communication. Even for species with vocalizations that lie outside the major spectral energy band of traffic noise, perception of communication sounds might still be impaired because of the upward spread of masking. Low-frequency sounds are effective in masking higher frequency sounds, and so masking grows nonlinearly on the high-frequency side (see Dooling and Leek, Chap. 2). The extent of this upward spread depends on the intensities of the masking sound, the frequency separation between the masker and the stimulus, and the sharpness of the frog's internal auditory filters.

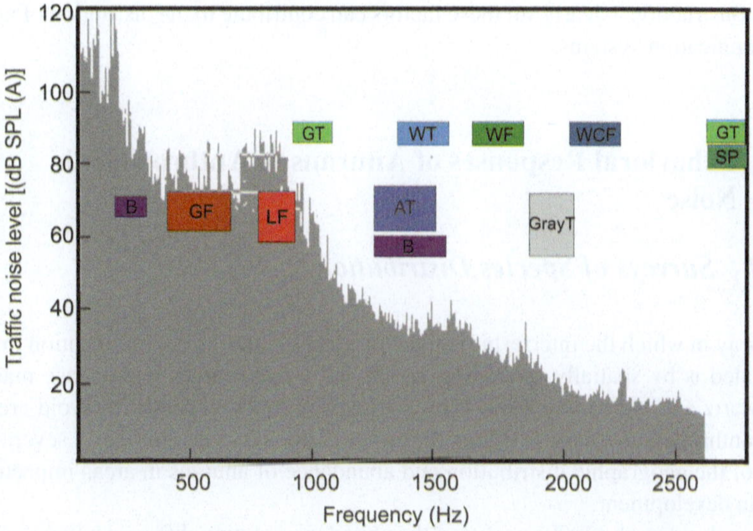

Fig. 7.1 Spectrum of highway traffic noise in relation to the spectral composition of advertisement calls of 10 North American anuran species. The power spectrum (in dB sound pressure level [SPL] A- weighting) of the traffic was measured 5 m from the highway. *Boxes* show the peak frequencies and estimated amplitudes (at a 5-m distance, calculated using the inverse square law from values recorded at 1 m; Gerhardt 1975; Megela-Simmons 1984) of advertisement calls. The low-frequency components of the advertisement calls of the green frog, the leopard frog, and the American bullfrog show the most overlap with the spectrum of traffic noise. *AT* American toad, *B* bullfrog (the two dominant spectral peaks are shown separately), *GrayT* gray treefrog, *GF* green frog, *GT* green treefrog (the two dominant spectral peaks shown separately), *LF* leopard frog, *SP* spring peeper, *WCF* western chorus frog, *WF* wood frog, *WT* Woodhouse's toad. Modified from Cunnington and Fahrig (2010), with permission from Elsevier Masson

Anuran choruses can be intense, with aggregate sound levels measured at a 1-m distance from the source ranging from 80 to over 110 dB SPL re 20 μPa depending on the species and its particular habitat (Gerhardt 1975; Megela-Simmons 1984). Aggregate sound levels of traffic noise can vary widely depending on the location, time of day, and the metric used to characterize it. Cunnington and Fahrig (2010) reported A-weighted traffic-noise levels around 78 dB SPL, a measurement that gradually attenuates frequencies <800 Hz at breeding ponds 5 m away from a major highway in Canada. Bee and Swanson (2007) measured an average peak level of traffic noise of 97 dB SPL (C-weighting) at 10 m from an active chorusing site of Cope's gray treefrogs (*Hyla chrysoscelis*) in Minnesota. In many locations, traffic noise is more intense during the day when nocturnal anurans are not vocalizing. Nevertheless, if the noise is more intense than species' vocalizations, its intermittent character may preclude its interference with transmission or perception of these calls. Eigenbrod et al. (2009) and Cunnington and Fahrig (2013) suggested that the real danger of traffic for anuran populations is, in fact, not noise masking of vocalizations but direct mortality by vehicles and the elimination of breeding sites due to road construction. Clearly, all these factors can contribute to the disruption of anuran communication systems.

7.2 Behavioral Responses of Anurans to Anthropogenic Noise

7.2.1 Surveys of Species Distribution

One way in which the interfering impact of noise on anuran communication may be mitigated is by spatially separating signal and noise sources to decrease masking (Schwartz and Gerhardt 1989). Thus, anurans might be expected to avoid areas of high anthropogenic noise, such as highways. This has been studied in several surveys of the geographic distribution and abundance of anurans in areas impacted by human development.

Eigenbrod et al. (2009) analyzed the abundance of nine different anuran species (American toad [*Anaxyrus americanus*], bullfrog, gray treefrog [*Hyla versicolor*], green frog [*Rana clamitans*], northern leopard frog [*Rana pipiens*], mink frog [*Rana septentrionalis*], spring peeper [*Pseudacris crucifer*], western chorus frog [*Pseudacris triseriata*], and wood frog [*Rana sylvatica*]) at 34 ponds located at various distances from a major, heavily trafficked highway (average traffic volume 18,200 vehicles/day) in eastern Ontario, Canada. A representative spectrum of the traffic noise recorded at a distance of 5 m from this highway is plotted in Fig. 7.1 along with the dominant spectral energy and typical calling levels (at a 5-m distance from the male) in these species' advertisement calls. Eigenbrod et al. (2009) assessed species population size by 5-min-long nighttime (quantifying numbers of calling males) and daytime (quantifying numbers of visible frogs) surveys of 34 ponds in 2006 and in surveys of 22 ponds repeated in 2007. The results showed the presence of "road effect zones," extending as far as 1000 m away from the highway, in which species abundance (number of animals) and species richness (number of species; Fig. 7.2A) were reduced compared with levels observed farther away from the highway. Overall, the distance from the highway explained 51% of the variance in species richness, with five species (gray treefrog, leopard frog, spring peeper, western chorus frog, and wood frog) most strongly negatively affected by the presence of the road. Of these five species, only the leopard frog emits an advertisement call with spectral components within the dominant spectral energy in the recorded traffic noise. Bullfrogs, whose advertisement calls also fall within the dominant energy of traffic noise, did not vary in abundance with distance to the highway. The analysis of movements of leopard frogs in the same area showed that during the spring migration from overwintering sites to breeding ponds, only 72% of the surveyed frogs were able to cross roads in high-traffic areas whereas 94% were able to cross roads in low-traffic areas (Bouchard et al. 2009). The authors suggested that

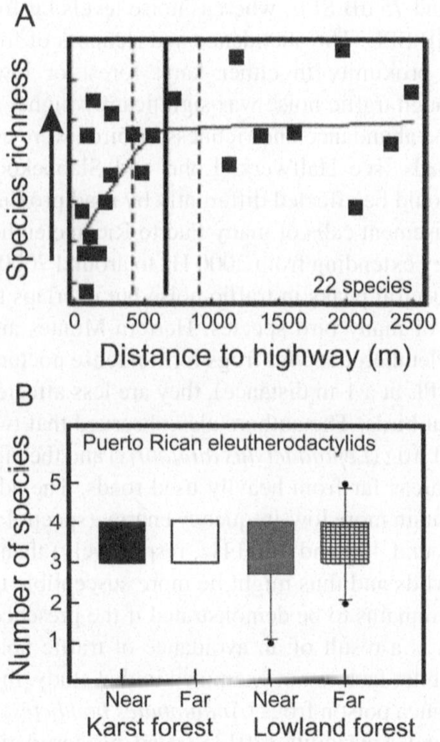

Fig. 7.2 Spatial distribution of anuran species in relation to the presence of roadways. (A) Relationship between species richness, defined as the relative abundance of 22 species in 34 ponds (*solid squares*), and the distance to a highway in Ontario, Canada, assessed in 2007. Species richness increases with distance from the highway, up to a breakpoint at 750 m (*solid vertical line*; *dashed vertical lines*, 1 SD). The breakpoint indicates the boundary of the "road effect zone." The curved line is the LOESS regression through the data. Modified from Eigenbrod et al. (2009), under a Creative Commons Attribution License. (**B**) Box plots showing the richness of four species in the genus *Eleutherodactylus* near the San Juan, Puerto Rico, metropolitan area relative to the distance from roads (near and far) and type of forest habitat (karst and lowland). The number of species does not differ significantly between the four sampled sites. *Boxes,* 25th to 75th percentiles; *vertical lines,* 10th to 90th percentiles; *solid circles,* values less than the 10th and greater than the 90th percentiles. Reproduced from Herrera-Montes and Aide (2011), with permission from Springer

highways constitute physical barriers to breeding site migration and in this way limit the numbers of animals at these sites and, consequently, reproductive activity.

Studies of population distribution in other anurans found no significant impact on species abundance related to proximity to roads. Herrera-Montes and Aide (2011) conducted acoustical surveys of the distribution and richness of ten species of birds and four species of frogs in the genus *Eleutherodactylus* at forest sites located either near (<100 m) or far from (>300 m) roads in the San Juan, Puerto Rico, metropolitan area. A-weighted traffic-noise levels at sites near these roads ranged between 65 and 75 dB SPL, whereas noise levels far from the roads ranged between 50 and 60 dB SPL. The abundance and richness of frogs did not vary statistically with road proximity in either karst forest or lowland forest habitat (Fig. 7.2B), even though traffic noise was significantly higher at locations near the roads. Conversely, the abundance and richness of birds were considerably reduced at sites closer to roads (see Halfwerk, Lohr, and Slabbekoorn, Chap. 8). Why anurans and birds should be affected differently by road proximity is an interesting question. The advertisement calls of many Puerto Rican eleutherodactylids contain high-frequency energy extending from 2000 Hz to around 7000 Hz, which is higher than the dominant spectral peaks in traffic noise but overlaps the range of spectral energy in the songs of many bird species. Herrera-Montes and Aide (2011) suggested that because eleutherodactylid frogs form intense nocturnal choruses (at levels reaching 95 dB SPL at a 1-m distance), they are less affected by daytime traffic noise than are diurnal birds. The authors also observed that two other anuran species, the white-lipped frog (*Leptodactylus albilabris*) and the pig frog (*Rana grylio*), were found only in areas far from heavily used roads. The advertisement calls of these two species contain more low-frequency energy (spectral peaks between 1100 and 2200 Hz and around 400 and 1900 Hz, respectively) than those of the Puerto Rican eleutherodactylids and thus might be more susceptible to direct masking by traffic noise. Still, it remains to be demonstrated if the presence of these species in areas far from roads is a result of an avoidance of traffic noise or represents the influence of other habitat factors not examined in that study. In a survey of the spatial distribution of Cauca poison frogs (*Andinobates bombetes*) in a Columbian forest, Vargas-Salinas and Amézquita (2013) found no correlation between species abundance and distance (15–300 m) from a heavily trafficked road. Yet the availability of bromeliads as breeding sites was highly correlated with species abundance, suggesting that in this species, the effects of traffic are secondary to other ecological effects.

A study of the distribution of two species, the green treefrog (*Hyla cinerea*) and Woodhouse's toad (*Anaxyrus woodhousii*), within a nature reserve in Texas found no significant differences in either frog abundance or chorus density related to road proximity (Barrass 1985). The advertisement calls of these two species contain maximal energy outside the dominant low-frequency energy in traffic noise (Fig. 7.1). The author observed, however, that intermale spacing within breeding ponds differed in relation to the type of roadway near these ponds. In both species, males calling in ponds near an interstate highway were spaced farther apart from each other than expected by chance. Conversely, in quiet ponds far from any roads, intermale spacing was closer, indicating a more compact and aggregated chorus. These data suggest that in areas with high traffic noise, males may choose more dispersed calling sites so as to lessen acoustic interference from neighbors and also to be able to better detect a neighbor's vocalizations against the noisier background levels. Because data on reproductive success of callers at the various ponds were not collected in this study, the effectiveness of this strategy could not be determined.

Together, these population surveys do not pinpoint clearly a specific effect of traffic noise on anuran distribution or population size that can be separated from biotic variables or from nonacoustic impacts of habitat destruction, habitat loss, or direct mortality due to the presence of roads.

7.2.2 *Variations in Spontaneous Calling Behaviors*

Aside from modifying spatial distribution, the question of how the spontaneous calling of male anurans is affected by anthropogenic noise has been addressed in studies of 13 different species in which variations in total calling and in advertisement call parameters during natural chorusing were investigated for correlations with noise levels at breeding ponds. As with their effects on spatial distribution, the effects of traffic and other anthropogenic-noise sources on male vocal behavior are inconsistent across species; the calling of some species was substantially altered and showed evidence of adaptations similar to those observed in studies of male-male interactions in natural choruses (Gerhardt and Huber 2002; Schwartz and Bee 2013), whereas the calling of other species was unaffected (Table 7.1). Interpretations are limited by the few species studied and the different call parameters assessed in different studies. Whether any acoustic modifications that did occur are adaptive in maintaining reproductive success and species isolation under conditions of high anthropogenic noise remains unclear.

Barrass (1985) examined the vocal behaviors of male green treefrogs and Woodhouse's toads at ponds in a nature reserve at different locations from nearby roadways. The trill-like, pulsed advertisement calls of Woodhouse's toad were emitted at lower call rates with shorter intercall intervals and were longer in duration at ponds close to an interstate highway compared with those at ponds with less traffic noise. Call dominant frequencies (around 1500–1800 Hz) did not differ with pond location. In contrast, the rate and duration of the advertisement calls of the green treefrog did not vary with distance to the roads and intercall intervals increased rather than decreased at the quieter ponds. The green treefrog's advertisement call is a harmonically-structured sound with two dominant spectral peaks around 900–1100 Hz and around 2000–3000 Hz. The high-frequency peak in this call was higher by a mean of 415 Hz in males living closer to compared with those living farther from traffic-noise sources. Barrass (1985) did not report any changes in the low-frequency spectral peak in the advertisement call, so it is not clear if the entire call shifted upward in frequency or if only the higher frequency components were affected. It is also unknown if males at the different calling sites varied in body size, which is related to the frequency composition of vocal signals (smaller males tend to produce higher frequencies).

Species differences in the impact of traffic noise on advertisement call parameters were also observed in a study of four species (American toad, gray treefrog, green frog, and leopard frog) living in eastern Ontario at different distances from a major highway (Cunnington and Fahrig 2010; these are the same species and the same locations sampled in Eigenbrod et al. 2009). Decreases in call rate at noisier ponds closer to the highway (mean traffic noise of 73 dB SPL A-weighting compared

Table 7.1 Spontaneous calling surveys show species-dependent changes in advertisement call parameters in the presence of anthropogenic noise

Species	Total calling[a]	Call rate	Call duration	Call amplitude	Dominant frequency	Reference
American toad	No change	No change		No change	No change	Cunnington and Fahrig (2010); Vargas-Salinas et al. (2014)
Bullfrog	Decrease					Vargas-Salinas et al. (2014)
Gray treefrog	No change	Decrease		No change	No change	Cunnington and Fahrig (2010); Vargas-Salinas et al. (2014)
Green frog	Decrease	Decrease		Decrease	Increase	Cunnington and Fahrig (2010); Vargas-Salinas et al. (2014)
Leopard frog		Decrease		No change	Increase	Cunnington and Fahrig (2010)
Green treefrog		No change	No change		Increase	Barrass (1985)
Woodhouse's toad		Decrease	Increase		No change	Barrass (1985)
Cauca poison frog	Decrease					Vargas-Salinas and Amézquita (2013)
Common eastern froglet					Increase	Parris et al. (2009)
Ewing's tree frog					Increase	Parris et al. (2009)
Painted chorus frog		Decrease				Sun and Narins (2005)
Sapgreen stream frog		Decrease				Sun and Narins (2005)
Two-striped grass frog		Increase			New call	Sun and Narins (2005)
Pacific chorus frog		No change	No change	No change	Decrease	Nelson et al. (2017)

Entries show the direction of change from call parameters measured under low-noise conditions. Blank cells mean no data

[a]Total calls in conditions where anthropogenic noise was present compared with no-noise conditions

with 44 dB SPL at quieter ponds farther from the highway) were observed for three (gray treefrog, green frog, and leopard frog) of those four species. Male green frogs and leopard frogs inhabiting ponds closer to the highway produced calls with significantly higher dominant frequencies (200–400 Hz higher) than those produced by conspecifics living farther from the highway. The predominant spectral peaks in the advertisement calls of these two species are within the major energy band of traffic noise (Fig. 7.1), suggesting that the upward shift in frequency composition is an adaptation to reduce masking by that noise. There were no differences in call dominant frequency related to pond location for the other two species, whose advertisement calls have spectral peaks outside the predominant energy band in traffic noise. Green frog calls also were at a lower amplitude at the ponds closer to the highway, but a similar effect was not observed in the other three species sampled. Cunnington and Fahrig (2010) suggested that these modifications in call parameters reflected the males' immediate responses to ambient-noise levels and that the species differences they observed reflected breeding strategy. American toads and gray treefrogs are explosive breeders, meaning that males actively search for mates, whereas green frogs and leopard frogs are prolonged breeders where females actively search for mates. Cunnington and Fahrig (2010) hypothesized that because explosive breeders do not typically form choruses, these species may be subject to less selective pressure than are chorusing species for adapting their vocalizations in response to noise. This hypothesis was not directly tested nor was reproductive success quantified in any of these species. In addition, the sample sizes were small (the vocalizations of only one male of each species was recorded at each pond) and the body sizes of the target males were not measured.

In a follow-up study in the same locations, Vargas-Salinas et al. (2014) tested the hypothesis that males decreased their total calling during times when traffic-noise levels were higher. This hypothesis was supported for the bullfrog and the green frog but not for the American toad and gray treefrog (Fig. 7.3A). Although the difference in traffic-noise intensity between calling and noncalling times was significant for both the bullfrog and the green frog, the absolute differences were small (1.48 dB and 0.74 dB, respectively). The advertisement calls of the bullfrog and the green frog both have frequency components within the dominant spectral range of traffic noise, so their calls might be more susceptible to masking than those of the American toad or the gray treefrog. Vargas-Salinas and Amézquita (2013) observed that the Cauca poison frog, whose high-frequency advertisement call contains maximal energy outside the spectral range of traffic noise, also called less when background traffic noise was higher (difference in traffic-noise intensity of 3.4 dB). These data suggest that a variable other than the spectral composition of vocalizations may modify vocal production in the presence of anthropogenic noise.

Similar to the results observed in the green treefrog (Barrass 1985), two species of Australian treefrogs, the common eastern froglet (*Crinia signifera*) and Ewing's tree frog (*Litoria ewingii*), shifted the spectral content of their advertisement calls upward under conditions of high levels of anthropogenic noise (Parris et al. 2009). These workers quantified the calling activity (5-min sampling period) of these two species at 47 sites in southern Australia that had undergone over 20 years of anthropogenic activity. The advertisement call of Ewing's tree frog is a narrowband note with peak energy from 2200 to 2600 Hz. This high-frequency peak was higher by a mean of 123 Hz at sites with higher levels of traffic noise (mean levels of 77 dB SPL A-weighting at 40 m; Fig. 7.3B). This shift is greater than the upward frequency shifts seen in the common eastern froglet (63 Hz; Parris et al. 2009) but smaller than the shifts observed in the green treefrog (415 Hz; Barrass 1985) and in the green frog and the leopard frog (200–400 Hz; Cunnington and Fahrig 2010). Parris et al. (2009) suggested, based on the mathematical modeling of sound propagation, that

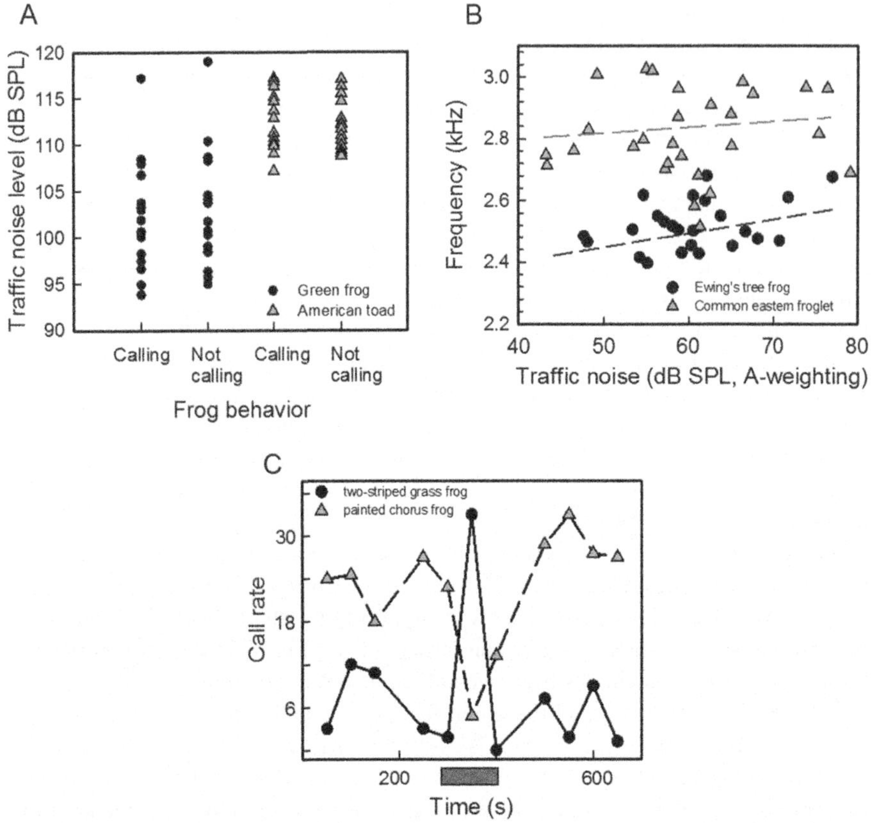

Fig. 7.3 Traffic noise affects spontaneous calling rate, time spent calling, and call dominant frequency in some male anurans. (**A**) Background traffic-noise level (in dB SPL, linear weighting) during calling and noncalling times for the green frog (*solid circles*) and the American toad (*gray triangles*). Each symbol is the data point from one animal, showing the amplitude of the traffic noise when that animal was calling and was not calling. Green frogs called significantly more when traffic-noise levels were lower than when they were higher. American toads did not vary their calling in relation to traffic-noise levels, but traffic noise was higher overall at their breeding ponds. Replotted from Vargas-Salinas et al. (2014), with permission from Springer. (**B**) Call dominant frequency shifts upward with increasing traffic-noise levels (in dB SPL A-weighting) in two Australian species. The dominant frequency shifts upward by an average of 125 Hz in Ewing's tree frog (*solid circles*; *black dashed line,* Bayesian regression through the data). Frequency shifts by a mean of 63 Hz in the common eastern froglet (*gray triangles; gray dashed line*, regression). Measurements of call frequency were corrected for the ambient temperature. Modified from Parris et al. (2009), under a Creative Commons Attribution License. (**C**) call rate of two species recorded at the same chorus site in Thailand at times before, during (*gray box*), and after airplane overflights. The two-striped grass frog (*solid circles*) significantly increased its call rate during the overflights, whereas the heterospecific painted chorus frog (*gray triangles*) significantly decreased its call rate from levels before and after the overflights. Reproduced and modified from Sun and Narins (2005), with permission from Elsevier

the upward frequency shifts in the calls of Ewing's tree frog would increase this species' active space for communication. Direct measurements of active space under conditions of anthropogenic noise are needed to confirm these predictions. Call parameters other than dominant frequency were not reported in this study, so it is unclear if a shift in call spectral frequency is the only adaptation made by the two Australian species in response to traffic noise.

Nelson et al. (2017) quantified advertisement call parameters of the Pacific chorus frog (*Pseudacris regilla*) at eight breeding sites subjected to different amounts of road noise (unweighted levels ranging from 26 to 58 dB SPL). Recordings lasted either 4 or 8 h and were repeated during two breeding seasons. Call amplitude, call duration, and call rate did not vary significantly with road-noise level. Call dominant frequency (2000–3000 Hz in this species) shifted slightly but significantly downward with increases in road noise. The authors concluded that Pacific chorus frogs did not adjust their vocalizations to mitigate the effects of masking from road noise, even though at the highest noise levels the transmission radius of vocalizations decreased (by a mean of 0.498 m per 1-dB increase in noise).

Many anurans vocalize in mixed-species choruses, and the presence of heterospecifics in these choruses might affect a particular species' responses to anthropogenic noise. Sun and Narins (2005) recorded advertisement calls from a mixed-species chorus in a lowland wet tropical forest in Thailand before, during, and after exposure to airplane overflights. During overflights, the overall sound level of the chorus decreased by as much as 14 dB (from a peak of 86 dB SPL to an average peak [$n = 4$] of 72 dB SPL, measured at a 1-m distance). This decrease in chorus level reflected a reduction in calling rate by males of two species, the painted chorus frog (*Microhyla butleri*) and the sapgreen stream frog (*Sylvirana nigrovittata*), and an increase in calling rate by males of a third species, the two-striped grass frog (*Hylarana taipehensis*; Fig. 7.3C). Sun and Narins (2005) proposed that the increased calling by the two-striped grass frog was a response to the cessation of calling by the other two species rather than a direct response to the airplane noise. The two-striped grass frog also produced a different kind of call during periods of lower chorus noise, shifting from a high-amplitude, long-duration, multiple-harmonic call to a series of lower amplitude, "squeak-like" notes with lower spectral energy. The function of this new vocalization is unknown. These results emphasize that the entire soundscape of the breeding site, including the biotic-noise level, must be taken into consideration when interpreting variations in calling parameters that may occur in the presence of anthropogenic noise.

7.2.3 Experimental Exposure Studies

The short-term effects of anthropogenic noise on the vocalizations of male anurans have been assessed in playback experiments conducted in the field and in the laboratory. Results from these experiments, like those from field surveys, highlight species diversity in the degree of call modification due to the presence of background noise

Table 7.2 Results of playback experiments assessing changes in advertisement call parameters in the presence of anthropogenic noise

Species	Total calling[a]	Call rate	Call duration	Call amplitude	Dominant frequency	Reference
American toad		No change		No change	No change	Cunnington and Fahrig (2010)
Green frog		Decrease		Decrease	Increase	Cunnington and Fahrig (2010)
Gray treefrog		Decrease		No change	No change	Cunnington and Fahrig (2010)
Leopard frog		Decrease		Decrease	Increase	Cunnington and Fahrig (2010)
Painted chorus frog	Decrease	Decrease				Sun and Narins (2005)
Sapgreen stream frog	Decrease	Decrease				Sun and Narins (2005)
Two-striped grass frog	Increase	Increase				Sun and Narins (2005)
Asian painted frog	Decrease					Sun and Narins (2005)
Gulf Coast toad		No change	Increase			Kaiser et al. (2011)
Hourglass treefrog		No change				Kaiser et al. (2011)
Loquacious treefrog		Increase				Kaiser et al. (2011)
Morelet's treefrog		No change				Kaiser et al. (2011)
Painted treefrog		No change	No change			Kaiser et al. (2011)
Red-eyed treefrog		Increase				Kaiser et al. (2011)
Yellow treefrog		Increase				Kaiser et al. (2011)
Triangle treefrog	Increase	Increase				Kaiser and Hammers (2009)
Common tree frog	Decrease	No change	Decrease		No change	Lengagne (2008)

Entries show the direction of change from no-noise conditions. Blank cells mean no data
[a]Total calls refers to changes in the total number of vocalizations during playbacks of anthropogenic noise

(Table 7.2). Interpretations are constrained by the small number of call parameters and the lack of consistency in the parameters quantified in different studies. Adaptations to the presence of anthropogenic noise are similar to those seen in response to biotic- and abiotic-noise sources.

Field Studies

Cunnington and Fahrig (2010) broadcast traffic noise (mean level of 76 dB SPL A-weighting, 5-m distance) to males of four species (American toad, gray treefrog, green frog, and leopard frog) living in ponds in eastern Ontario at sites categorized as low (mean level of 44 dB SPL A-weighting) traffic-noise areas. Compared with spontaneous calling, male green frogs and leopard frogs decreased their call rates and call amplitudes and increased call dominant frequency in response to these playbacks. Male gray treefrogs decreased their call rates but did not significantly alter either call amplitude or call dominant frequency. These vocal modifications made the advertisement calls of these three species more similar to those recorded from conspecifics living in ponds categorized as high traffic-noise areas. The authors suggested that these modifications minimize attenuation of the call and maintain its propagation range and active space, but this has yet to be tested. They also proposed that differences in call characteristics between breeding ponds exposed to different levels of anthropogenic noise represent immediate, adaptive responses that function to allow tolerance to changes in ambient acoustic conditions.

Cunnington and Fahrig (2010) did not examine any calling interactions between different species located in the same breeding ponds. Sun and Narins (2005) showed that the calling behavior of heterospecifics can affect a species' responsiveness to anthropogenic noise. These authors quantified the responsiveness of frogs in a mixed species chorus in Thailand to playbacks of low-frequency (principal spectral peak <2000 Hz, 30-s duration, 1-m distance) motorcycle engine noise. Male Asian painted frogs (*Kaloula pulchra*), painted chorus frogs, and sapgreen stream frogs all decreased their calling rates both during and after playbacks, whereas male two-striped grass frogs significantly increased their calling rate after the cessation of playbacks. As also observed during spontaneous airplane overflights, two-striped grass frogs increased their vocal output during lulls in calling by heterospecifics. These data show the importance of monitoring all anuran activity at a given breeding site in order to develop a full understanding of the impact of anthropogenic noise on vocal communication.

The influence of microhabitat on anuran responses to anthropogenic noise was investigated in seven species of frogs inhabiting areas around Las Cuevas Research Station, Belize (Kaiser et al. 2011). Three of these species (the hourglass treefrog, red-eyed treefrog, and Morelet's treefrog [*Agalychnis moreletii*]) inhabited ponds in "undisturbed" forest sites and four of these species (the yellow treefrog [*Dendropsophus microcephalus*]; Gulf Coast toad [*Incilius valliceps*], loquacious treefrog [*Tlalocohyla loquax*], and painted treefrog [*Tlalocohyla picta*]) inhabited ponds in more open, "disturbed" areas closer to a logging road. During chorusing bouts, background noise levels (around 60 dB SPL C-weighting) did not vary between forest and open sites. Males' responses to playbacks of automobile noise were compared with those to playbacks of white noise, both of which were broadcast at levels of 90 dB SPL (at 1 m from the source). One of the three forest species (the red-eyed tree frog) and two of the four open-habitat species (the loquacious tree frog and the yellow treefrog) increased call rates in response to playbacks of both automobile and white noise, with no significant difference between these noise types. Three other species did not modify their calling rates in response to playbacks (Fig. 7.4A; Table 7.2). Changes in call duration were analyzed in the Gulf Coast toad, which increased call duration in response to playbacks, and the painted treefrog, which did not change call duration. Overall, there was no significant relationship between habitat (forest or open) and responses to anthropogenic noise. Kaiser et al. (2011) also monitored two different choruses of the yellow treefrog to determine whether chorus tenure, defined as the number of nights a male participated in the chorus, was affected by noise. Chorus tenure at the pond exposed to playbacks of engine noise was significantly shorter than that at a second pond not exposed to these playbacks. In addition, chorus length (duration of the chorus on a given night) and total chorusing time over the season were significantly shorter at the pond exposed to noise. These data suggest that anthropogenic noise may reduce reproductive success by limiting chorus tenure and duration. Replications of these effects at more sites, with greater numbers of frogs and in other species, are needed.

Kaiser and Hammers (2009) compared the rates of advertisement calling of the triangle treefrog (*Dendropsophus triangulum*) in response to playbacks of motorcycle noise, recorded music, and chorus noise. Males of this species increased their calling rates during playbacks of all of these stimuli, and there were no significant differences between the responses to motorcycle and to chorus noise. Because motorcycle noise was presented at only one level (60 dB SPL) while the music and chorus playbacks were presented at a higher level (75 dB SPL), the results may not be directly comparable. These researchers also observed that in response to short bursts of engine noise, males called more during the noise itself than in the silent intervals between the bursts. That is, there was no evidence of gap calling or the alternation of calls with silent periods between noise bursts.

Laboratory Playbacks

Laboratory experiments provide more control over the entire auditory soundscape than is possible with field experiments, but they also provide a more restricted view of the vocal communication system. Lengagne (2008) recorded the evoked vocal responses of male common tree frogs (*Hyla arborea*) housed in a tank to playbacks of traffic noise (spectral energy <2000, peak at about 1300 Hz, 3-h duration, mean levels between 71 and 81 dB SPL A-weighting at 1 m). The advertisement call of this species contains its predominant spectral energy at about 2400–2600 Hz. Males did not move away from the source of either the low- (72 dB SPL) or the high- (88 dB SPL) level playbacks, suggesting that they were not disturbed by the noise. Males that were housed alone lowered their call rate after, but not during, low-level playbacks (Fig. 7.4B) and also decreased the duration of their calling bouts. In response to high-level playbacks, they reduced both call rate and call bout duration. Males that were housed in groups called at higher rates than males that were housed alone, and they also reduced their calling rates during high-level playbacks. Call duration and call spectral content did not vary with playback level. The author suggested that traffic noise influences chorus dynamics, but this was not explicitly tested.

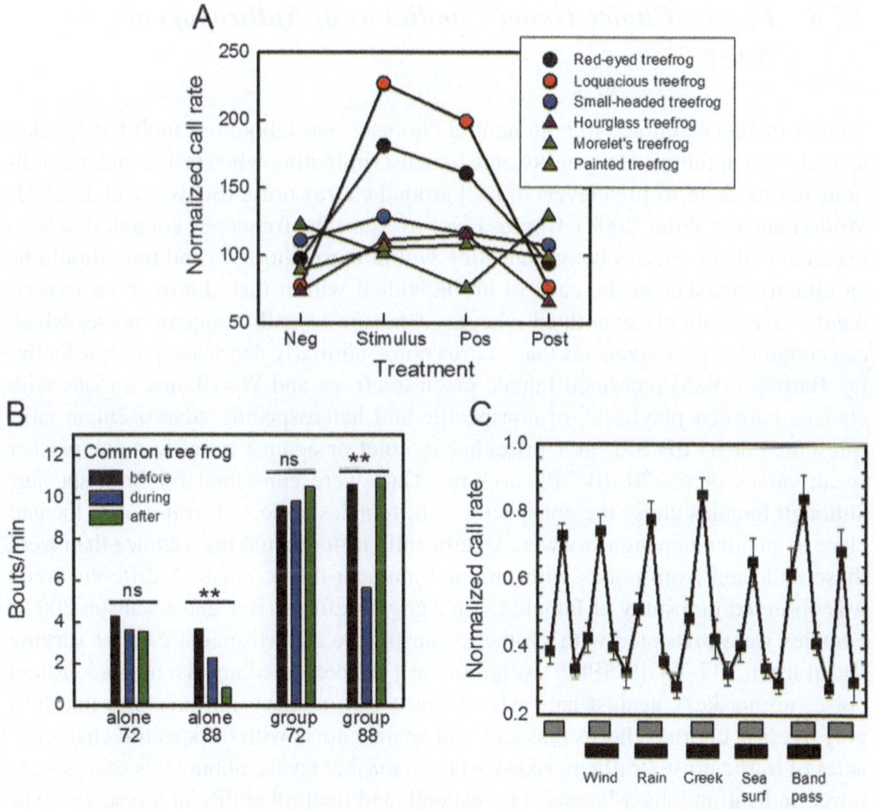

Fig. 7.4 Evoked calling of male anurans is affected by noise. (**A**) Call rate, normalized to pretreatment baseline levels, in six species of sympatric tropical frogs varied with noise treatment. Males of three species (*circles*) increased their call rate during playbacks of engine noise and white noise. Males of three other species (*triangles*) did not change their call rates. Neg, silent control; Stimulus, automobile engine noise (90 dB SPL at 1 m; 60–70 dB SPL at the position of the male); Pos, white noise (90 dB SPL at 1 m); Post, poststimulus spontaneous calling. Data were calculated and replotted from values presented in Kaiser et al. (2011). (**B**) Playbacks of traffic noise affected call rate (plotted as call bouts/min) in male common tree frogs tested in a laboratory tank. The number of calls (bouts/min) was recorded in response to two levels of traffic noise playback (72 and 88 dB SPL A-weighting at 1 m) for males housed alone (alone 72 and alone 88) and for males housed in a group (group 72 and group 88). *Black bars*, call rate before playbacks; *light gray bars*, call rate during playbacks; *dark gray bars*, call rate after playbacks. There were no significant differences (ns) in call rate in response to noise playback levels of 72 dB SPL in either the alone 72 or group 72 condition. Call rates were significantly reduced (**) after 88 dB SPL playbacks for males housed alone and during playbacks for males housed in a group. Modified from Lengagne (2008), with permission from Elsevier. (**C**) Call rate of male wood frogs, normalized to the maximal rate, in response to playbacks of different categories of abiotic noise (67 dB SPL, linear weighting at a mean distance of 0.73 m). *Gray boxes*, presentation time of synthetic advertisement calls (70 dB SPL at 1 m); *black boxes*, presentation time of the different noises; *vertical gray lines*, standard deviations. The first and the last presentations of advertisement calls were made in no-noise conditions. Males vocalized significantly more when creek, rain, and band-pass noise were presented with the synthetic advertisement calls. Modified from Penna and Zúñiga (2014), with permission from Springer

7.2.4 Female Choice Under Conditions of Anthropogenic Noise

Studies of female choice in both natural choruses and laboratory multiple-speaker tests show that female anurans become less discriminating when tested under conditions of moderate-to-high levels of background chorus noise (Schwartz et al. 2001; Wollerman and Wiley 2002). Chorus noise overlaps the frequency content of advertisement calls of species communicating within these choruses and thus should be an effective masker of the calls of an individual within that chorus. Four experiments have explicitly examined whether exposure to anthropogenic noise, which can contain lower frequencies than chorus noise, similarly degrades female selectivity. Barrass (1985) presented female green treefrogs and Woodhouse's toads with choices between playbacks of conspecific and heterospecific advertisement calls (presented at 65 dB SPL at 1 m) either in quiet or against a traffic-noise masker (peak values of 65–70 dB SPL at 1 m). Data were consistent for both species; although females chose the conspecific call, females collected from ponds located close to an interstate highway were significantly faster in making a choice than were those collected from ponds with minimal ambient-traffic noise. A different trend was observed in a study of female Cope's gray treefrogs (Bee and Swanson 2007). Females were presented with synthetic conspecific advertisement calls at varying signal levels (37–85 dB SPL C-weighting at 1 m) presented against no background noise (no masker), against narrowband noise with a spectrum matching that of a gray treefrog chorus (chorus masker), and against noise with the spectral characteristics of traffic noise (traffic masker). At most masker levels, phonotaxis scores were lower, indicating longer latencies to respond, and the probability of a response was lower when females were tested under chorus or traffic-noise conditions than when tested in a no-noise condition (Fig. 7.5A). Only at high signal levels (79–85 dB SPL) were the phonotaxis scores similar in all three conditions. Bee and Swanson concluded that background noise, either from a chorus or from an anthropogenic source, interferes with female choice by decreasing the active space of advertisement calls. In another experiment (Caldwell and Bee 2014), female Cope's gray treefrogs were tested for their ability to localize advertisement calls (79–85 dB SPL) presented from an array of loudspeakers both under no-noise conditions and against a background of band-limited noise (500–4500 Hz, overlapping spectral components in traffic noise; 73–76 dB SPL C-weighting at 1 m). Females responded with longer latencies and more localization errors when tested in a noisy background. These data suggest that low-frequency traffic noise can render females less discriminating to conspecific vocalizations by impairing localization accuracy. Further studies of localization accuracy under different types and levels of background noise are needed.

Even if males modify their calls in the presence of anthropogenic noise, females may still be attracted to these modified calls and remain able to select appropriate mates despite increases in decision latency. As shown in Table 7.1, male green frogs shifted the frequency composition of their advertisement calls upward in the pres-

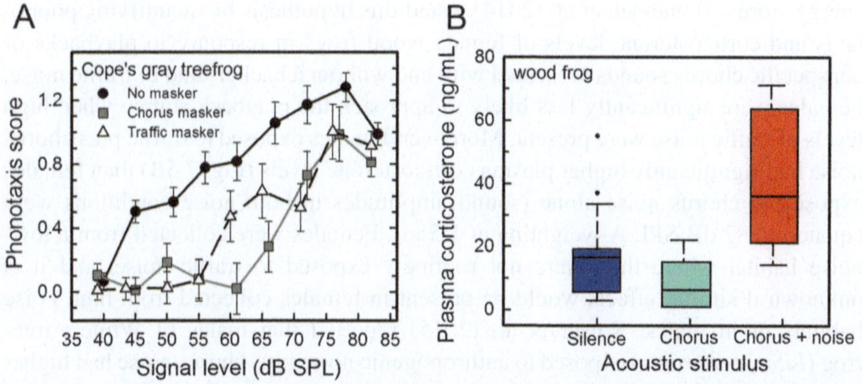

Fig. 7.5 Performance and hormone levels of female frogs in phonotaxis tests are affected by the presence of traffic noise. (**A**) Phonotaxis of female Cope's gray treefrogs to playbacks of conspecific advertisement calls is influenced by the presence of maskers. Phonotaxis is quantified as a score related to response latency. A phonotaxis score of 0 indicates no response and a score of 1.5 indicates a short latency to respond to the stimulus. Advertisement calls were broadcast at different signal levels (in dB SPL C-weighting) in a no-masker condition, against a background of narrowband noise mimicking chorus noise, and against a background of traffic noise. Background noise levels were 70 dB SPL at the female's release site. Scores are higher, indicating faster choices, in no-masker conditions and lower in both chorus-masker and traffic-masker conditions between signal levels of 43 to 73 dB SPL. There are no significant differences in phonotaxis scores between chorus and traffic maskers. Modified from Bee and Swanson (2007), with permission from Elsevier. (**B**) Plasma corticosterone levels of female wood frogs increased in response to playbacks of chorus + traffic noise. Stimuli were presented at a level of 87 dB SPL A-weighting at 1 m. Silence, no acoustic stimulus; chorus, male wood frog chorus; chorus + noise, male chorus and traffic noise. *Colored boxes*, interquartile ranges; *horizontal lines in boxes*, median values; *dashed vertical lines*, standard deviations (1.5 times the height of the box); *circles*, outliers. Modified from Tennessen et al. (2014), under a Creative Commons Attribution License

ence of traffic noise (Cunnington and Fahrig 2010). Cunnington and Fahrig (2013) quantified phonotactic responses of female green frogs to conspecific calls that were either unmodified (recorded under no-noise conditions) or modified (recorded in the presence of traffic noise). Calls were broadcast to the females from traps, with and without an added background of traffic noise (level of 76 dB SPL A-weighting at 5 m), and the rate of capture of females at these traps was used as an index of mate attraction. Unmodified calls broadcast in the presence of traffic noise attracted fewer females than unmodified calls broadcast without a background of traffic noise, but modified calls attracted the same number of females with or without traffic noise. Cunnington and Fahrig (2013) interpreted these data as suggesting that the presence of traffic noise does not negatively impact mate attraction because males can modify their calls to compensate for any masking effects. They suggested that the major influence of traffic on anuran populations might be due to direct mortality rather than to indirect effects on vocal communication and mate choice.

Another possible effect of traffic noise is increased stress, which, in turn, would impact reproduction by its influence on hormone levels, immune function, and energy stores. Tennessen et al. (2014) tested this hypothesis by quantifying phonotaxis and corticosterone levels of female wood frogs in response to playbacks of conspecific chorus sounds presented with and without a background of traffic noise. Females were significantly less likely to approach the playback source when high levels of traffic noise were present. Moreover, females exposed to traffic plus chorus noise had significantly higher plasma corticosterone levels (Fig. 7.5B) than females exposed to chorus noise alone (sound amplitudes in both noise conditions were equated to 87 dB SPL A-weighting at 1.5 m). Females were collected from a low-noise habitat where they were not routinely exposed to traffic noise, and it is unknown if similar effects would be present in females collected from high-noise habitats or in males. Kaiser et al. (2015) reported that males of White's treefrog (*Litoria caerulea*) exposed to anthropogenic noise plus chorus noise had higher corticosterone levels and lower sperm counts than males exposed to chorus noise alone (sound amplitudes in both conditions were 70 dB SPL at 1 m). What is unclear from these studies is whether the results are attributable specifically to traffic noise or to high sound levels. Both experiments suggest that acoustic habitat degradation has physiological consequences that can impair anuran reproduction and thus species survival.

7.2.5 Comparisons of Anuran Vocal Communication in Biotic, Abiotic, and Anthropogenic Noise

Anurans have evolved to vocalize and to choose mates within dense and often multispecies choruses in a variety of habitats. A number of behavioral adaptations have been identified that facilitate vocal communication and mate choice in these acoustically complex environments (Narins and Zelick 1988; Schwartz and Bee 2013). The question remains whether the adaptations that have been shown to operate in the presence of biotic and abiotic noise also operate under conditions of anthropogenic-noise exposure. Direct comparisons of male vocal modifications and female approach behaviors in different types and levels of background noise are needed for isolating any specific effects of anthropogenic noise on communication. Further work should include testing more species and larger sample sizes at more locations and should account for the effects of body size on call parameters.

Schwartz and Bee (2013) argued that modifications in the timing of vocal signals are the most significant means by which males act to reduce acoustic interference from other males. As one example, in mixed-species choruses of yellow treefrogs and hourglass treefrogs, males alternated their calls such that hourglass treefrogs called preferentially in silent intervals between bouts of calling by yellow treefrogs (Schwartz and Wells 1983a,b). In effect, yellow treefrogs inhibited calling by hourglass treefrogs. More fine-scale, note-by-note adjustments in call timing have also been observed. In a set of acoustic playback experiments, male coqui frogs (*Eleutherodactylus coqui*) interleaved their call notes with those of the stimulus and inserted their notes in the silent gaps both between periodic (Zelick and Narins 1983) and aperiodic (Zelick and Narins 1985) interfering tonal stimuli. Thus, in this species, the temporal intermittency and aperiodic structure of vehicle noise may not limit the degree to which males can predict the occurrence of silent gaps in the noise within which to insert their vocalizations. Surveys comparing vocal activity between conditions of exposure to traffic noise have not documented the precise insertion of male calls into silent periods between bouts of vehicular noise (Kaiser and Hammers 2009). Males of some species have been shown to increase call rate and call duration in dense choruses. Such a strategy is adaptive because females of some species prefer higher calling rates and longer call durations (Gerhardt and Huber 2002; Schwartz and Bee 2013). Some species have been shown to increase call rate to abiotic-noise sources as well. Penna and Zúñiga (2014) compared the evoked vocal responses of male banded wood frogs (*Batrachyla taeniata*) to playbacks of synthetic advertisement calls (levels of 70 dB SPL linear weighting at the position of the male) presented against abiotic noises (wind, creek, and rain) and band-pass noise (2000-Hz center frequency, overlapping the high-frequency peak in the species call); all noise stimuli were equated to a level of 67 dB SPL measured at a mean distance of 0.73 m. Normalized calling rates increased significantly in response to all playbacks (Fig. 7.4C). But another species, Emilio's ground frog (*Eupsophus emiliopugini*), did not exhibit comparable changes in its calling behavior to the same noise stimuli (Penna et al. 2005). The authors proposed that different species have evolved different strategies for dealing with abiotic noise.

The available data are inconsistent with respect to the effects of anthropogenic noise on call rate and duration. Of ten species in which the call rate was compared between high traffic-noise and low traffic-noise conditions (Table 7.1), one species showed an increase in call rate, six species showed a decrease in call rate, and three species showed no change. Call duration was measured in three of these species; one showed an increase under high traffic noise while the other two showed no change. Of the 16 species in which the call rate during playbacks of anthropogenic noise has been measured (Table 7.2), five species decreased call rate, five species increased call rate, and six species showed no change. Call duration was measured in three of these species; one species decreased call duration, one species increased call duration, and one species showed no change. Even within the same breeding pond or local area, anthropogenic noise can affect vocal parameters in different directions and at different magnitudes (Sun and Narins 2005; Kaiser et al. 2011). Species differences in the extent of any vocal modifications in response to traffic noise are not clearly related to the spectral structure of the conspecific advertisement call or to the degree of spectral overlap between the call and the spectrum of the noise. These species differences may reflect, instead, microhabitat differences, differences in hearing sensitivity or selectivity, variability in male-male interactions within a chorus, or aspects of female choice that are as yet unidentified. Identifying these species differences will be important for conservation purposes.

Within the same species, anurans living closer to traffic sources can have higher dominant frequencies in their advertisement calls than those living farther from these sources. This has been documented in five of nine species tested (Table 7.1), but across these species, the low-frequency spectral peak in the conspecific call did not overlap the spectrum of traffic noise consistently. Playback experiments show that in five of the nine species tested, there was an upward shift in the high-frequency component of the advertisement call under high traffic-noise conditions (Table 7.2), and even in these species, the magnitude of the shift differed. In contrast, upward shifts in the high frequencies of bird songs in noise environments are more consistently observed (see Halfwerk, Lohr, and Slabbekoorn, Chap. 8). Even during natural male-male interactions in anuran choruses, shifts in the spectral content of advertisement calls have been observed only rarely and may reflect a switch in strategy from mate attraction to intermale aggression (Bee et al. 2000). Although some anurans have evolved advertisement calls with spectral content outside the range of ambient abiotic masking noise (Narins et al. 2004; Feng et al. 2006; Arch et al. 2008), in other species, playbacks of abiotic-noise sources did not result in changes in call spectral content (Penna et al. 2005; Penna and Zúñiga 2014). Short-term shifts in the spectral content of advertisement calls may not be adaptive for anurans if these shifts move the dominant frequency of the calls outside the female's most sensitive hearing range (the matched filter hypothesis; Capranica and Moffat 1983; Narins and Clark 2016) or recognition space (Amézquita et al. 2011). This may limit the effectiveness of this type of short-term calling strategy for these animals.

Other vertebrates vocally respond to increases in noise levels by increasing the amplitude of their own calls. This is called the Lombard effect (Brumm and Zollinger 2011), and its existence in anurans is controversial (Love and Bee 2010). To demonstrate a Lombard effect, one must document an increasing amplitude of a frog's call in the presence of increasing amplitudes of interfering noise. Because males typically vocalize at very high levels (Gerhardt 1975) and because advertisement calls are so energetically costly to produce, it may not be adaptive or even possible for these animals to increase the amplitude of their calls in response to increases in background noise (Parris et al. 2009). Increases in call amplitude in response to playbacks of conspecific calls at increasing levels were observed by Lopez et al. (1988) in the white-lipped frog, but because broadband noise was not used in those experiments, the results cannot be considered an example of the Lombard effect. Penna et al. (2005) observed increases in call amplitude in two Chilean species (Emilio's ground frog and the Chiloe Island ground frog [*Eupsophus calcaratus*]) in response to increasing playback levels of band-pass noise, whereas Love and Bee (2010) found no evidence of a Lombard effect in gray treefrogs in response to playbacks of chorus-shaped noise. Halfwerk et al. (2016) reported the existence of the Lombard effect in the túngara frog, although call amplitudes in that study did not increase proportionally with masking-noise levels. Most of the males increased the amplitude of their calls by only 1–3 dB with every 10-dB increase in noise, a much smaller effect than that observed in other vertebrates. Halfwerk et al. (2016) suggested that the apparent Lombard effect functions to communicate a male's readiness to fight rather than as a mechanism to increase signal-to-noise ratios. In experiments examining call amplitude in response to playbacks of vehicular noise specifically, two species decreased the amplitude of their calls and two other species made no change (Table 7.2), thus showing no Lombard effect.

7.2.6 *Auditory Filters and Susceptibility to Noise*

Knowing how well an animal's auditory system can reject noise that interferes with signal detection, that is, the width of internal auditory filters, is important for interpreting and predicting how anthropogenic sound sources affect perception. Filter width is typically measured using two techniques, critical ratios and critical bands (see Dooling and Leek, Chap. 2). The critical ratio (CR) is the ratio of the intensity of a signal at its masked threshold to the spectrum level of the masking noise and is an indirect measure of filter bandwidth. Large values of the CR indicate very broad filters that pass considerable noise and thus have high susceptibility to masking, whereas small CR values indicate narrow filters that pass less noise and have a lower susceptibility to masking. Behavioral estimates of CRs are available for three anuran species based on the techniques of selective phonotaxis (green treefrog: Ehret and Gerhardt 1980), evoked calling (coqui frog: Narins 1982) and reflex modification (bullfrog and green treefrog: Moss and Simmons 1986; Simmons 1988). Estimates of the lowest CRs from these experiments, all measured at signal frequencies at or around 1000 Hz, vary from 17 to 22 dB in the green treefrog (Ehret and Gerhardt 1980; Moss and Simmons 1986) and the bullfrog (Simmons 1988) to 31 dB in the coqui frog (Narins 1982). CRs of 25 dB and below are within the range of those measured psychophysically at similar frequencies in other nonmammalian vertebrates (Fay and Simmons 1999) and indicate good filtering of signals from background noise. CRs of 30 or above are so high that they suggest that these auditory filters are poor at rejecting noise and that frequencies within these filters will be easily masked. Narins (1982) estimated that the critical band around 1000 Hz in the coqui frog is 500 Hz. Such broad filters may explain the decreased selectivity to advertisement calls of female anurans in dense choruses (Schwartz et al. 2001; Wollerman and Wiley 2002). Male anurans may be able to mitigate the impact of broad auditory filters by vocalizing at high-amplitude levels or at times when background noise is less prominent or by shifting their call frequencies to other, more narrowly tuned filters. Similarly, female anurans may be able to shift their hearing sensitivity to narrower filters in the presence of noise. These are all attractive questions for future research.

7.3 Physiological Coding of Noise

Although various experiments have examined the effects of noise on the neural representation of sounds in the anuran's auditory system (Narins and Zelick 1988), the interpretation of results has been confined largely to noting the changes in the parameters of the physiological responses due to the presence or absence of noise. The integrative mechanisms of hearing that lead to perception of a target sound, either in quiet or in noise, are largely unknown in anurans. This section touches briefly on those studies that may be most relevant to understanding the neural impacts of anthropogenic-noise sources.

Broadband noise affects both the thresholds (measured by changes in spike rates) and the temporal synchrony (measured by the cycle-by-cycle phase locking to the stimulus envelope) of individual auditory nerve fibers to single-tone stimuli (coqui frog: Narins 1987; Narins and Wagner 1989; bullfrog: Freedman et al. 1988). In response to noise, thresholds to tones increased and phase locking was reduced, suggesting less efficient neural representation of that signal. Phase locking remained significant at noise levels where the firing rate had already saturated, also suggesting that this temporal coding scheme is available at high-noise levels. In addition, synchronization to the first harmonic period of a synthetic advertisement call was maintained at noise levels where rate responses to that stimulus had already saturated (Simmons et al. 1992). In the auditory midbrain, synchronization to the period of amplitude-modulated sounds remained robust in the presence of low levels of narrowband noise (European marsh frog [*Rana ridibunda*]; European common frog [*Rana temporaria*]; Bibikov 2002). These data suggest that signal periodicity, conveyed as the period of a single tone, the repetition rate of a short pulse or click, or the first harmonic period of a multiple-harmonic signal, remains useful as a cue for detection or discrimination in even high levels of background noise. Reliance on a synchrony code could help mitigate the effects of broad internal filters. Preservation of neural synchronization could explain why many species increased their call repetition rate under conditions of background noise, including traffic noise. Not only can these changes increase the redundancy of the signal but also they could maintain the representation of signal periodicity or repetition rate in the auditory system.

In an experiment directly comparing behavioral performance with neural responses, Schwartz and Gerhardt (1998) showed that the ability of female spring peepers to discriminate between synthetic advertisement calls improved when these calls were presented against low-to-moderate levels of modulated chorus noise. This behavioral improvement was reflected in the shapes of neural audiograms recorded from the auditory midbrain. When high-intensity calls were presented in quiet conditions, audiograms were flat over the entire audible range, with no differences in sensitivity to individual frequencies in that call. Conversely, when high-intensity calls were presented against a noise background, the audiogram showed peaks in response at the dominant frequencies in the advertisement call. These data suggest that the addition of modulated noise can improve neural discrimination. They are consistent with an interpretation based on a comodulation masking release: the combination of a signal with a modulated noise produces a combined stimulus with a modulated envelope. A comodulated signal is more easily discriminated than a nonmodulated signal and so the result is less masking (see Dooling and Leek, Chap. 2). Goense and Feng (2012) showed that comodulation of signals with noise decreased the masking of responses of groups of neurons in the auditory midbrain. If anthropogenic-noise sources are modulated, either in frequency or over time, then their impact on the processing of important biological signals may not be as severe as might be expected.

7.4 The Acoustic Worlds of Reptiles in Relation to Anthropogenic Noise

Although most reptiles hear to some extent (Dooling et al. 2000), only a few groups use sounds for communication. The most vocal reptiles include the geckos (squamata) and the crocodilians. Tokay geckos (*Gekko gecko*) are sensitive to sounds in the frequency range from about 200 to 5000 Hz, with the best sensitivity in the range around 2000 Hz. The spectra of their vocalizations match their best auditory sensitivity measured using auditory brainstem responses (Brittan-Powell et al. 2010). Both vocalizations and best hearing sensitivity lie within the major spectral energy in traffic and other anthropogenic-noise sources, suggesting that anthropogenic noise will affect the auditory communication of geckos. It is not known if these animals exhibit compensatory strategies similar to those shown by male anurans in the presence of high background noise levels.

Mancera et al. (2017) analyzed the reactions of the blue-tongued lizard (*Tiliqua scincoides*) to playbacks of synthetic mining machinery noise (bulldozer, coal truck, and drill, filtered to comprise frequencies less than or greater than 2000 Hz) at mean levels of 74 dB SPL and 63 dB SPL A-weighting. Lizards exposed to high-frequency high-amplitude noise spent more time freezing, with their heads oriented downward compared with lizards exposed to other noise combinations. The authors interpreted these reactions as indicative of fear or stress.

Alligators and crocodiles employ relatively large vocal repertoires for both long-distance and short-distance communication. Juvenile crocodilians produce a variety of harmonically-structured sounds with energy extending up to 5000 Hz for communicating contact, distress, or threat. Adults produce intense bellows in the context of courtship and mating, with most of the energy concentrated below 250 Hz (Vergne et al. 2009). These animals are amphibious and so could be impacted by anthropogenic noise both on land (e.g., vehicular traffic) and in the water (e.g., boat engines). As is the case with geckos, direct behavioral measurements of the impact of anthropogenic noise on these vocal animals are lacking.

Hearing has also been assessed in several species of turtles (testudines). The first study in which the auditory system of a turtle was studied in detail revealed that auditory nerve fibers of the red-eared slider (*Pseudemys scripta elegans*) were tuned to the range of 70–700 Hz (Crawford and Fettiplace 1980). Sea turtles (green sea turtle [*Chelonia mydas*] and loggerhead turtles [*Caretta caretta*]) can detect sounds in the frequency range from <100 Hz to about 2000 Hz, with the best sensitivity between 100 and 400 Hz (Martin et al. 2012; Popper et al. 2014). Given this range of hearing, sea turtles might be able to detect noise from shipping, recreational boating, underwater explosions, pile driving, or low-frequency sonar soundings. The widths of auditory filters in the turtle's ear for rejecting these sources of background noise have not been measured. Observations suggest that sea turtles avoid boats and firing as opposed to silent airguns (Popper et al. 2014), but there are no quantitative data to support these observations.

Ferrara et al. (2014) reported that the Amazonian river turtle (*Podocnemis expansa*) produces a series of distinct vocalizations with frequencies in the range from about 95 to 460 Hz. The authors speculated that these animals use sounds in social contexts and that sound plays an important role in the synchronization of group activities during the nesting season. These provocative results open up an entire order to the potential deleterious effects of anthropogenic noise on their acoustic biology.

7.5 Summary

As vertebrate animals first adjusted to terrestrial life, there was a strong selection pressure to develop airborne acoustic communication signals that were at once well suited for the exchange of biologically significant information between conspecifics, inconspicuous to or poorly localizable by potential predators, and robust in the face of high-level background noise. More recently, selection pressure has been exerted on animals to mitigate the deleterious effects of anthropogenic noise on acoustic communication systems. This chapter has focused on the Amphibia, a highly vocal vertebrate taxon, and the changes to their vocal signals that may have arisen as a response to environmental noise. Sufficient data are not yet available to quantify the effects of anthropogenic noise on either the sound production or reception systems of amphibians or their courtship and reproductive behavior. From the available literature, it is clear that anthropogenic noise affects social and reproductive behavior in complex ways that differ between species and habitats. Many of the behaviors anurans have evolved to cope with biotic- and abiotic-noise sources also appear to operate to mitigate the impact of anthropogenic noise. To date, however, the number of species in which the effects of anthropogenic noise on behavior and physiology have been explicitly studied is small, and results vary across species. Moreover, even less is known about how anthropogenic noise impacts the acoustic behaviors of reptiles. Knowing how some species have adapted to noise is important for understanding and modeling potential population and ecosystem consequences. It is clear that additional research is needed to document and quantify the long-term ecological consequences of anthropogenic-noise exposure in both terrestrial and underwater environments.

Compliance with Ethics Requirements

Andrea M. Simmons declares that she has no conflict of interest.

Peter M. Narins declares that he has no conflict of interest.

References

Amézquita, A., Flechas, S. V., Lima, A. P., Gasser, H., & Hödl, W. (2011). Acoustic interference and recognition space within a complex assemblage of dendrobatid frogs. *Proceedings of the National Academy of Sciences of the United States of America*, 108, 17058-17063.

AmphibiaWeb (2017). Available at http://www.amphibiaweb.org/.

Arch, V. S., Grafe, T. U., & Narins, P. M. (2008). Ultrasonic signaling by a Bornean frog. *Biology Letters*, 4, 19–22.

Barrass, A. N. (1985). The effects of highway traffic noise on the phonotactic and associated reproductive behavior of selected anurans. Unpublished PhD Dissertation, Vanderbilt University, Nashville, TN.

Bee, M. A., & Swanson, E. M. (2007). Auditory masking of anuran advertisement calls by road traffic noise. *Animal Behaviour*, 74, 1765–1776.

Bee, M. A., Perrill, S. A., & Owen, P. C. (2000). Male green frogs lower the pitch of acoustic signals in defense of territories: A possible dishonest signal of size? *Behavioral Ecology*, 11, 169–177.

Bibikov, N. G. (2002). Addition of noise enhances neural synchrony to amplitude-modulated sounds in the frog's midbrain. *Hearing Research*, 173, 21–28.

Boatright-Horowitz, S. S., Cheney, C. A., & Simmons, A. M. (1999). Atmospheric and underwater propagation of bullfrog vocalizations. *Bioacoustics*, 9, 257–280.

Bouchard, J., Ford, A. T., Eigenbrod, F., & Fahrig, L. (2009). Behavioral responses of northern leopard frogs (*Rana pipiens*) to roads and traffic: Implications for population persistence. *Ecology and Society*, 14, 23. Available at http://www.ecologyandsociety.org/vol14/iss2/art23/.

Brittan-Powell, E. F., Christensen-Dalsgaard, J., Tang, Y., Carr, C., & Dooling, R. J. (2010). The auditory brainstem response in two lizard species. *The Journal of the Acoustical Society of America*, 128, 787–794.

Brumm, H., & Zollinger, S. A. (2011). The evolution of the Lombard effect: 100 years of psychoacoustic research. *Behaviour*, 148, 1173–1198.

Caldwell, M. S., & Bee, M. A. (2014). Spatial hearing in Cope's gray treefrog: Open and closed loop experiments on sound localization in the presence and absence of noise. *Journal of Comparative Physiology A: Neuroethology, Sensory, Neural, and Behavioral Physiology,* 200, 265–284.

Caldwell, M. S, Johnston, G. R., McDaniel, J. G., & Warkentin, K. M. (2010). Vibrational signaling in the agonistic interactions of red-eyed treefrogs. *Current Biology*, 20, 1012–1017.

Capranica, R. R., & Moffat, A. J. M. (1983). Neurobehavioral correlates of sound communication in anurans. In J. P. Ewert, R. R. Capranica, & D. Ingle (Eds.), *Advances in Vertebrate Neuroethology* (pp. 701–730). New York: Plenum Press.

Crawford, A. C., & Fettiplace, R. (1980). The frequency selectivity of auditory nerve fibers and hair cells in the cochlea of the turtle. *The Journal of Physiology*, 306, 79–125.

Cunnington, G. M., & Fahrig, L. (2010). Plasticity in the vocalizations of anurans in response to traffic noise. *Acta Oecologica*, 36, 463–470.

Cunnington, G. M., & Fahrig, L. (2013). Mate attraction by male anurans in the presence of traffic noise. *Animal Conservation*, 16, 275–285.

Dooling, R. J., Lohr, B., & Dent, M. L. (2000). Hearing in birds and reptiles. In R. J. Dooling, A. N. Popper, & R. R. Fay (Eds.), *Comparative Hearing: Birds and Reptiles* (pp. 308–359). New York, Springer-Verlag.

Ehret, G., & Gerhardt, H. C. (1980). Auditory masking and effects of noise on responses of the green treefrog (*Hyla cinerea*) to synthetic mating calls. *Journal of Comparative Physiology A: Neuroethology, Sensory, Neural, and Behavioral Physiology,* 141, 13–18.

Eigenbrod, F., Hecnar, S. J., & Fahrig, L. (2009). Quantifying the road-effect zone: threshold effects of a motorway on anuran populations in Ontario, Canada. *Ecology and Society*, 14, 24. Available at http://www.ecologyandsociety.org/vol14/iss1/art24/.

Ey, E., & Fischer, J. (2009). The "acoustic adaptation hypothesis"—A review of the evidence from birds, anurans and mammals. *Bioacoustics*, 19, 21–48.

Fay, R. R., & Simmons, A. M. (1999). The sense of hearing in fishes and amphibians. In R. R. Fay & A. N. Popper (Eds.), *Comparative Hearing: Fish and Amphibians* (pp. 269–318). New York: Springer-Verlag.

Feng, A. S., Narins, P. M., Xu, C.-H., Lin, W.-Y., Yu, Z.-L., Qiu, Q., Xu, Z.-M., & Shen, J.-X. (2006). Ultrasonic communication in frogs. *Nature*, 440, 333–336.

Ferrara, C. R., Vogt, R. C., Sousa-Lima, R. S., Tardio, B. M. R., & Bernardes, V. C. C. (2014). Sound communication and social behavior in an Amazonian river turtle (*Podocnemis expansa*). *Herpetologica*, 70, 149–156.

Freedman, E. G., Ferragamo, M., & Simmons, A. M. (1988). Masking patterns in the bullfrog (*Rana catesbeiana*). II: Physiological effects. *The Journal of the Acoustical Society of America*, 84, 2081–2091.

Gerhardt, H. C. (1975). Sound pressure levels and radiation patterns of the vocalizations of some North American frogs and toads. *Journal of Comparative Physiology A: Neuroethology, Sensory, Neural, and Behavioral Physiology*, 102, 1–12.

Gerhardt, H. C., & Huber, F. (2002). *Acoustic Communication in Insects and Frogs: Common Problems and Diverse Solutions*. Chicago: University of Chicago Press.

Goense, J. B. M., & Feng, A. S. (2012). Effects of noise bandwidth and amplitude modulation on masking in frog auditory midbrain neurons. *PLoS ONE*, 7(2), e31589. https://doi.org/10.1371/journal.pone.0031589.

Goutte, S., Dubois, A., & Legendre, F. (2013). The importance of ambient sound level to characterize anuran habitat. *PLoS ONE*, 8, e78020. https://doi.org/10.1371/journal.pone.0078020.

Gridi-Papp, M., & Narins, P. M. (2010). Seismic detection and communication in amphibians. In C. E. O'Connell-Rodwell (Ed.), *The Use of Vibrations in Communication: Properties, Mechanisms and Function Across Taxa* (pp. 69–83). Kerala, India: Transworld Research Network.

Halfwerk, W., Jones, P. L., Taylor, R. C., Ryan, M. J., & Page, R. A. (2014). Risky ripples allow bats and frogs to eavesdrop on a multisensory sexual display. *Science*, 343, 413–416.

Halfwerk, W., Lea, A.M., Guerra, M. A., Page, R. A., & Ryan, M. J. (2016). Vocal responses to noise reveal the presence of the Lombard effect in a frog. *Behavioral Ecology*, 27(2), 669–676.

Herrera-Montes, M. I., & Aide, T. M. (2011). Impacts of traffic noise on anuran and bird communities. *Urban Ecosystems*, 14, 415–427. https://doi.org/10.1007/s11252-011-0158-7.

Hof, C., Araujo, M. B., Jetz, W., & Rahbek, W. (2011). Additive threats from pathogens, climate and land-use change for global amphibian diversity. *Nature*, 48, 516–521.

Kaiser, K., & Hammers, J. L. (2009). The effect of anthropogenic noise on male advertisement call rate in the neotropical treefrog, *Dendropsophus triangulum*. *Behaviour*, 146, 1053–1069.

Kaiser, K., Scofield, D. G., Alloush, M., Jones, R. M., Marczak, S., Martineau, K., Oliva, M. A., & Narins, P. M. (2011). When sounds collide: The effect of anthropogenic noise on a breeding assemblage of frogs in Belize, Central America. *Behaviour*, 148, 215–232.

Kaiser, K., Devito, J., Jones, C. G., Marentes, A., Perez, R., Umeh, L., Weickum, R. M., McGovern, K. E., Wilson, E. H., & Saltzman, W. (2015). Effects of anthropogenic noise on endocrine and reproductive function in White's treefrog, *Litoria caerulea*. *Conservation Physiology*, 3(1), cou061. https://doi.org/10.1093/conphys/cou061.

Lengagne, T. (2008). Traffic noise affects communication behaviour in a breeding anuran, *Hyla arborea*. *Biological Conservation*, 141, 2023–2031.

Lewis, E. R., & Narins, P. M. (1985). Do frogs communicate with seismic signals? *Science*, 227, 187–189.

Lopez, P. T., Narins, P. M., Lewis, E. R., & Moore, S. W. (1988). Acoustically induced call modification in the white-lipped frog, *Leptodactylus albilabris*. *Animal Behaviour*, 36, 1295–1308.

Love, E. K., & Bee, M. A. (2010). An experimental test of noise-dependent voice amplitude regulation in Cope's gray treefrog, *Hyla chrysoscelis*. *Animal Behaviour*, 80, 509–515.

Mancera, K. F., Murray, P. J., Lisle, A., Dupont, C., Faucheux, F., & Phillips, C. J. C. (2017). The effects of acute exposure to mining machinery noise on the behavior of eastern blue-tongued lizards (*Tiliqua scincoides*). *Animal Welfare*, 26, 11–24. https://doi.org/10.7120/09627286.26.1.011.

Martin, K. J., Alessi, S. C., Gaspard, J. C., Tucker, A. D., Bauer, G. B., & Mann, D. A. (2012). Underwater hearing in the loggerhead turtle (*Caretta caretta*): A comparison of behavioral and auditory evoked potential audiograms. *Journal of Experimental Biology*, 215, 3001–3009.

Megela-Simmons, A. (1984). Behavioral vocal response thresholds to mating calls in the bullfrog, *Rana catesbeiana*. *The Journal of the Acoustical Society of America*, 76, 676–681.

Moss, C. F., & Simmons, A. M. (1986). Frequency selectivity of hearing in the green treefrog, *Hyla cinerea*. *Journal of Comparative Physiology A: Neuroethology, Sensory, Neural, and Behavioral Physiology*, 159, 257–266.

Narins, P. M. (1982). Effects of masking noise on evoked calling in the Puerto Rican coqui (Anura: Leptodactylidae). *Journal of Comparative Physiology A: Neuroethology, Sensory, Neural, and Behavioral Physiology*, 147, 439–446.

Narins, P. M. (1987). Coding of signals in noise by amphibian auditory nerve fibers. *Hearing Research*, 26, 145–154.

Narins, P. M., & Zelick, R. (1988). The effects of noise on auditory processing and behavior in amphibians. In B. Fritszch, M. J. Ryan, W. Wilczynski, T. Hetherington, & W. Walkowiak (Eds.), *The Evolution of the Amphibian Auditory System* (pp. 511–536). New York: John Wiley.

Narins, P. M., & Wagner, I. (1989). Noise susceptibility and immunity of phase locking in amphibian auditory-nerve fibers. *The Journal of the Acoustical Society of America*, 85, 1255–1265.

Narins, P. M., & Clark, G. A. (2016). Principles of matched filtering with auditory examples from selected vertebrates. In G. von der Emde & E. Warrant (Eds.), *The Ecology of Animal Senses: Matched Filtering for Economical Sensing* (pp. 111–140). Heidelberg: Springer-Verlag.

Narins, P. M., Feng, A. S., Schnitzler, H. U., Denzinger, A., Suthers, R. A., Lin, W., & Xu, C.-H. (2004). Old World frog and bird vocalizations contain prominent ultrasonic harmonics. *The Journal of the Acoustical Society of America*, 115, 910–913.

Narins, P. M., Wilson, M., & Mann, D. A. (2014). Ultrasound detection in fishes and frogs: Discovery and mechanisms. In C. Köppl, G. A. Manley, A. N. Popper, & R. R. Fay (Eds.), *Insights from Comparative Hearing Research* (pp. 133–156). New York: Springer-Verlag.

Narins, P. M., Stoeger-Horwath, A. & O'Connell-Rodwell, C. (2016). Infrasound and seismic communication in the vertebrates with special emphasis on the Afrotheria: An update and future directions. In R. A. Suthers, W. T. Fitch, R. R. Fay, & A. N. Popper (Eds.), *Vertebrate Sound Production and Acoustic Communication* (pp. 191–228). New York: Springer-Verlag.

Nelson, D. V., Klinck, H., Carbaugh-Rutland, A., Mathis, C. L., Morzillo, A. T., & Garcia, T. S. (2017). Calling at the highway: The spatiotemporal constraint of road noise on Pacific chorus frog communication. *Ecology and Evolution*, 7, 429–440. https://doi.org/10.1002/ece3.2622.

Parris, K. M., Velik-Lord, M., & North, J. M. A. (2009). Frogs call at a higher pitch in traffic noise. *Ecology and Society*, 14, 25. Available at http://www.ecologyandsociety.org/vol14/iss1/art25/.

Pechmann, J. H. K., Scott, D. E., Semlitsch, R. D., Caldwell, J. P., Vitt, L. J., & Gibbons, J. W. (1991). Declining amphibian populations: The problem of separating human impacts from natural fluctuations. *Science*, 253, 892–895.

Penna, M., & Zúñiga, D. (2014). Strong responsiveness to noise interference in an anuran from the southern temperate forest. *Behavioral Ecology and Sociobiology*, 68, 85–97.

Penna, M., Pottstock, H., & Velásquez, N. (2005). Effect of natural and synthetic noise on evoked vocal responses in a frog of the temperate austral forest. *Animal Behaviour*, 70, 639–651.

Popper A. N., Hawkins, A. D., Fay, R. R., Mann, D. A., Bartol, S., Carlson, T. J., Coombs, S., Ellison, W. T., Gentry, R. L., Halvorsen, M. B., Lokkeborg, S., Rogers, P., Southall, B. L., Zeddies, D. G., & Tavolga, W. N. (2014). *ASA S3/SC1. 4 TR-2014 Sound Exposure Guidelines for Fishes and Sea Turtles: A Technical Report Prepared by ANSI-Accredited Standards Committee S3/SC1 and Registered with ANSI*. New York: Springer International Publishing.

Schwartz, J. J., & Wells, K. D. (1983a). An experimental study of acoustic interference between two species of neotropical treefrogs. *Animal Behaviour*, 31, 181–190.

Schwartz, J. J., & Wells, K. D. (1983b). The influence of background noise on the behavior of a neotropical treefrog, *Hyla ebraccata*. *Herpetologica*, 39, 121–192.

Schwartz, J. J., & Gerhardt, H. C. (1989). Spatially mediated release from masking in an anuran amphibian. *Journal of Comparative Physiology A: Neuroethology, Sensory, Neural, and Behavioral Physiology*, 166, 37–41.

Schwartz, J. J., & Gerhardt, H. C. (1998). The neuroethology of frequency preferences in the spring peeper. *Animal Behaviour*, 56, 55–69.

Schwartz, J. J., & Bee, M. A. (2013). Anuran acoustic signal production in noisy environments. In H. Brumm (Ed.), *Animal Communication and Noise* (pp. 91–132). Berlin Heidelberg: Springer-Verlag.

Schwartz, J. J., Buchanan, B. W., & Gerhardt, H. C. (2001). Female mate choice in the gray treefrog (*Hyla versicolor*) in three experimental environments. *Behavioral Ecology and Sociobiology*, 49, 443–455.

Shen, J.-X., Xu, Z.-M., Feng, A., and Narins, P. M. (2011). Large odorous frogs (*Odorrana graminea*) produce ultrasonic calls. *Journal of Comparative Physiology A: Neuroethology, Sensory, Neural, and Behavioral Physiology*, 197, 1027–1030. https://doi.org/10.1007/s00359-011-0660-7.

Simmons, A. M. (1988). Masking patterns in the bullfrog (*Rana catesbeiana*). I: Behavioral effects. *The Journal of the Acoustical Society of America*, 83, 1087–1092.

Simmons, A. M., Schwartz, J. J., & Ferragamo, M. (1992). Auditory nerve representation of a complex communication sound in background noise. *The Journal of the Acoustical Society of America*, 91, 2831–2844.

Sun, J. W. C., & Narins, P. M. (2005). Anthropogenic sounds differentially affect amphibian call rate. *Biological Conservation*, 121, 419–427.

Tennessen, J. B., Parks, S. E., & Langkilde, T. (2014). Traffic noise causes physiological stress and impairs breeding migration behavior in frogs. *Conservation Physiology*, 2, cou032. https://doi.org/10.1093/conphys/cou032.

Vargas-Salinas, F., & Amézquita, A. (2013). Traffic noise correlates with calling time but not spatial distribution in the threatened poison frog *Adinobates bombetes*. *Behaviour*, 150, 569–584.

Vargas-Salinas, F., Cunnington, G. M., Amézquita, A., & Fahrig, L. (2014). Does traffic noise alter calling time in frogs and toads? A case study of anurans in eastern Ontario, Canada. *Urban Ecosystems*, 17, 945–953.

Velez, A., Schwartz, J. J., & Bee, M. A. (2013). Anuran acoustic signal perception in noisy environments. In H. Brumm (Ed.), *Animal Communication and Noise* (pp. 133–185). Berlin Heidelberg: Springer-Verlag.

Vergne, A. L., Pritz, M. B., & Mathevon, N. (2009). Acoustic communication in crocodilians: From behaviour to brain. *Biological Reviews*, 84, 391–411.

Wollerman, L., & Wiley, H. (2002). Background noise from a natural chorus alters female discrimination of male calls in a Neotropical frog. *Animal Behaviour*, 63, 15–22.

Zelick, R., & Narins, P. M. (1983). Intensity discrimination and the precision of call timing in two species of neotropical treefrogs. *Journal of Comparative Physiology A: Neuroethology, Sensory, Neural, and Behavioral Physiology*,153, 403–412.

Zelick, R., & Narins, P. M. (1985). Characterization of the advertisement call oscillator in the frog *Eleutherodactylus coqui*. *Journal of Comparative Physiology A: Neuroethology, Sensory, Neural, and Behavioral Physiology*,156, 223–229.

Chapter 8
Impact of Man-Made Sound on Birds and Their Songs

Wouter Halfwerk, Bernard Lohr, and Hans Slabbekoorn

Abstract Vocalizing birds are ubiquitous and often prominent in areas that are reached by noisy human activities. Birds have therefore been studied for the effects of man-made sound on song production and perception, physiological stress, distribution range, breeding density, and reproductive success. There are examples of birds that sing louder, higher, and longer when ambient-noise levels are elevated due to human activities. This may lead to perceptual advantages through masking release, although song modifications may also lead to a functional compromise. Fitness benefits of noise-dependent modifications have not been proven yet. Masking effects are reported for outdoor and indoor studies, but data on physiological consequences are not widespread yet. There are also still only few experimental studies on more long-term consequences of man-made sound on development, maturation, and fitness. Observational data on species distributions and densities show that there are birds that persist at noisy sites but also that artificially elevated noise levels can have detrimental consequences for particular species. Birds in noisy localities may move away or stay and fare less well. Furthermore, the effects of noise pollution can go beyond single species because all species may be more or less negatively affected, but the effect on one species may also positively or negatively affect another. The variety in sensitivity among species and the diversity in impact and counterstrategies have made birds both cases of concern and popular model species for fundamental and applied research.

W. Halfwerk (✉)
Department of Ecological Sciences, Faculty of Earth and Life Sciences, VU University Amsterdam, Amsterdam, The Netherlands
e-mail: W.H.Halfwerk@vu.nl

B. Lohr
University of Maryland, Baltimore County (UMBC), Baltimore, MD, USA
e-mail: blohr@umbc.edu

H. Slabbekoorn
Faculty of Science, Institute of Biology Leiden (IBL), Leiden University, Leiden, The Netherlands
e-mail: H.W.Slabbekoorn@biology.leidenuniv.nl

H. Slabbekoorn et al. (eds.), *Effects of Anthropogenic Noise on Animals*, Springer Handbook of Auditory Research 66,
https://doi.org/10.1007/978-1-4939-8574-6_8

Keywords Avian song · Coping strategy · Experimental exposure · Fitness consequences · Lombard effect · Pitch shift · Signal interference · Signal-to-noise ratio · Vocal plasticity

8.1 The Acoustic Life of Birds

Birds generate a great diversity of sounds (see Fig. 8.1), varying among and within species and across context and arousal level (Kroodsma and Miller 1996; Marler and Slabbekoorn 2004). Avian vocalizations have been well studied in terms of underlying neurobiology, physiology, morphology, function, and evolution. Each species sings a species-specific song that can be relatively simple and repetitive or highly complex and variable. Some species repeat a single-song type, and others have a large repertoire of songs. Song variants can be high or low in frequency, tonal and melodious, or more "noisy" and raw, loud, or faint (Kroodsma and Miller 1996; Catchpole and Slater 2008). Besides song, many species also produce a variety of calls that are typically less loud and less complex than songs, although there are many exceptions to this rule (Marler and Slabbekoorn 2004).

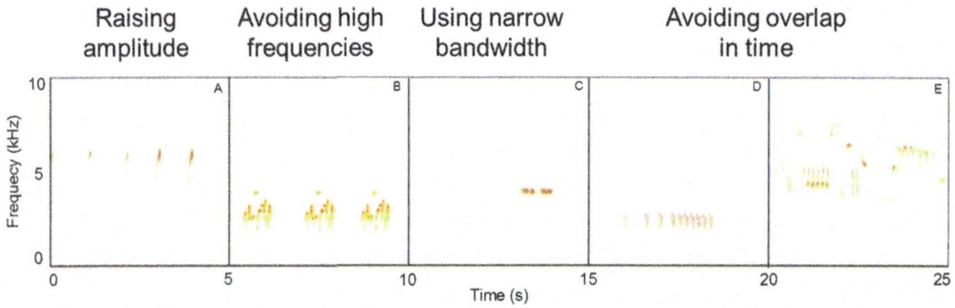

Fig. 8.1 Sonograms of examples of acoustic variety in birdsong for species that were also shown to respond flexibly to natural ambient sounds (dynamic singing strategies; see Sects. 8.2.3 and 8.3.6). (**a**) Song of the blue-throated hummingbird (*Lampornis clemenciae*), which was shown to adjust the amplitude to the sound level of nearby forest streams (Pytte et al. 2003). (**b**) Song of the gray-breasted woodwren (*Henichorina leucophrys hilaris*), which was shown to avoid high frequencies that are in the range of rainforest cicadas (Dingle et al. 2008). (**c**) Song of the green hylia (*Hylia prasina*), which was shown to concentrate energy in a narrow bandwidth, thereby avoiding competition across a wider range and exploiting reverberation effects (Slabbekoorn and Smith 2002b). (**d**) Song of the wrentit (*Chamaea fasciata*), which was shown to sing when nearby, heterospecific Bewick's wrens (*Thryomanes bewickii*) end their songs (Cody and Brown 1969). (**e**) Song of the Eurasian wren (*Troglodytes troglodytes*), which was also shown to avoid temporal overlap with conspecifics by singing after competitor songs have ended, as shown through playback trials (Yang et al. 2014)

8.1.1 How and Why Birds Sing

The source of most sounds produced by birds is the syrinx (Podos et al. 2004; Elemans et al. 2015). Sounds are generated when air is forced through the bronchi into the trachea and membranes between cartilaginous rings start to vibrate. Temporal variation in bird sounds is determined by the onset and offset of the air-flow, and amplitude and frequency depend on air sac pressure, membrane size, and flexibility. Songs and calls develop with age and change over time in all birds (Marler 1970; Kroodsma 2004). Songbirds and a number of nonsongbird taxa, such as hummingbirds and parrots, learn at least some of their vocalizations from conspecific tutors. Songbirds, for example, can memorize acoustic features of singing adults heard during a particular early life stage and many species adjust their developing songs during a later motor stage to match the memorized model. Some species retain this flexibility throughout their life (Kroodsma and Miller 1996; Catchpole and Slater 2008).

Songs typically have dual functions, which are often not easily separated (Kroodsma and Byers 1991; Collins 2004). Nevertheless, songs predominantly serve in relatively long-range communication to attract mates and maintain pair bonds or to deter competitors and defend territorial boundaries. It is often exclusively the male bird that sings in temperate zones, but in more tropical regions, both males and females may join in solo singing activities and duets (Riebel et al. 2005; Hall 2009; Odom et al. 2014). Calls can sometimes serve the same function as songs, but they are more often associated with specific contexts such as agonistic encounters with competitors at short range or communication of alarm in the presence of predators (Templeton and Greene 2007; Suzuki et al. 2016). Calls are also often used in the context of maintaining contact among group members, such as in dense habitat or during nocturnal flight, or when finding food or feeding chicks. There are also examples for both songs and calls that occur in situations of serious distress or under high risk of predation, in which cases the sounds may function to attract other birds of the same or other species or to threaten or confuse the predator directly.

The large diversity in songs and calls among and within bird species has evolved in the context of constraints and opportunities (Slabbekoorn and Smith 2002a; Podos et al. 2004). There is often a direct link between song and fitness through acoustic functions that are critical for survival and reproduction (McGregor et al. 1981; Hasselquist et al. 1996). Being vocal is often advantageous over being silent, and some vocal variants have a selective advantage over others (Cardoso and Atwell 2011; Halfwerk et al. 2011a). Evolution of communicative signals is shaped by the range of sounds that a bird can produce given its size, brain, and syringeal capacities, the relative costs of vocal variation, and the range of sounds that relevant receivers can hear. Consequently, vocalizations are subject to phylogenetic constraints related to both sound production and perception, although within the range of potential variation, learned vocalizations may change relatively quickly via ontogenetic plasticity and cultural evolution (Lachlan et al. 2013; Halfwerk et al. 2016a).

Furthermore, signal efficiency will depend on the acoustic habitat in which birds use their songs (see Larsen and Radford, Chap. 5). Propagation and interference properties of the species-typical habitat will therefore also affect the direction and speed of evolutionary change (Wiley and Richards 1978; Slabbekoorn and Smith 2002a).

8.1.2 The Natural Acoustic Habitat of Birds

Birds live in a wide variety of habitats, from bare deserts and icy flats to dense shrubs and complex rainforests. None of these bird habitats are silent (Klump 1996; Slabbekoorn 2004). Biotic and abiotic sound sources typically fill the air with diverse acoustic patterns, often with stereotypic level fluctuations during diurnal and seasonal cycles. Biotic sources concern all sound-generating species of local animal communities, which may include other bird, amphibian, and insect species and sometimes also mammalian species, especially in tropical environments (Halfwerk et al. 2016b; Stanley et al. 2016). Abiotic sources include, for example, river and coastal water turbulence or wind and rain, which all heavily depend on weather conditions and local habitat features (Dubois and Martens 1984; Halfwerk and Slabbekoorn 2015).

The acoustic structure of natural sound varies considerably. Although many animals produce more or less stereotypic but intermittent sounds, frog or insect assemblages can generate continuous sounds for long periods of the day and night. Frequencies between 2.0 and 6.0 kHz are, for example, especially prominent in tropical habitats (Planque and Slabbekoorn 2008; Stanley et al. 2016). Wind, running water, and rain-related noise spectra typically cover a wide range of frequencies (Dubois and Martens 1984; Penna et al. 2005). Wind and rain may cause rustling leaves and dripping water clicks that may spread relatively equally across the spectrum, whereas high wind conditions, ocean surf, or turbulent water in fast-flowing rivers are typically also wideband sounds but biased toward frequencies below 2.0 kHz.

Consequently, habitats vary considerably in the diversity and levels of sound present throughout the day and across seasons (Slabbekoorn 2004; Bormpoudakis et al. 2013). Birds in temperate zones typically have distinct vocal activity peaks during the dawn chorus and, to a lesser extent, the dusk chorus, and this is usually restricted to the breeding season (Mace 1987; Cuthill and Macdonald 1990). Calling frogs and insects also have their diurnal and seasonal cycles, especially in subtropical and tropical environments, and they may vary in spectrum and effort in response to spatial and temporal fluctuations in temperature (Runkle et al. 1994; Sueur and Sanborn 2003). Habitat-specific sound profiles are also reported to vary consistently with latitude and altitude (Slabbekoorn 2004) and may even yield comparable selection regimens for birds of similar habitat across continents (Cardoso and Price 2010).

8.1.3 The Human-Altered Acoustic Environment

The acoustic environment of birds can be altered by habitat degradation and fragmentation and by increasing amounts of human activities within, adjacent to, or above the bird habitat (Reijnen and Foppen 2006; Fahrig and Rytwinski 2009). Habitat degradation can alter sound transmission properties of the habitat as well as the composition and activity patterns of local, vocally communicating animal communities (Smith et al. 2005). Habitat fragmentation can also result in reduced genetic or acoustic connectivity (Laiolo and Tella 2005). Birds may sing less and sing differently, for example, when in small habitat fragments compared with birds of large, contiguous patches of suitable habitat (Rivera-Gutierrez et al. 2010). Furthermore, degradation and fragmentation can induce and elevate a range of edge effects because the human impacts and presence are typically higher close to the habitat edge (Halfwerk and Slabbekoorn 2015; Swaddle et al. 2015). One prominent edge effect is the elevation of ambient-noise levels (Slabbekoorn and Ripmeester 2008; Francis and Barber 2013).

Industrial-, recreational-, or traffic-related activities produce man-made sounds that can penetrate into the bird habitat (see Fig. 8.2). Highways may introduce diurnal cycles of rising and waning noise related to traffic load and daily human activities (Halfwerk et al. 2011b; Arroyo-Solis et al. 2013). Similarly, train traffic causes noisy events that may be briefer, except for some areas with long and loud cargo trains (Halfwerk et al. 2011b). Aircraft sounds can lead to some of the most severe sound exposure patterns for areas surrounding military or public airports: predictable and intermittent but extremely loud (Gil et al. 2014). A full habitat turnover takes place when the natural habitat is exploited for urban development. Villages and cities typically still have vegetation but usually with a radically different species composition and canopy cover and height. Furthermore, buildings and tarmac alter physical transmission properties of sounds dramatically, and all sorts of traffic and motorized activity make city life very noisy in ways that have similarities and differences to the natural world of pristine moorlands or virgin rainforests (Warren et al. 2006; Slabbekoorn et al. 2007).

The acoustic structure of man-made sound sources varies but has some very dominant features (Slabbekoorn and Ripmeester 2008; Barber et al. 2009). Although there are impulsive sounds that may extend into higher frequency ranges in specific cases, most industrial machinery and vehicle engines generate sounds that are broadband and biased to relatively low frequencies (below 2.0 kHz). Although highways, train tracks, and airports typically have predictable sound exposure patterns (Halfwerk et al. 2011b), sound levels in urban areas can vary in time and space in a highly unpredictable way (Luther and Baptista 2010; Arroyo-Solis et al. 2013). Highways may be acoustically similar in some ways to coastline surf or rivers with strong currents. Train tracks or moderately noisy urban neighborhoods may reflect sound conditions in woodlands that can be affected acoustically by a breeze, drafts of wind, or a sudden rain shower. However, buses can suddenly come around a corner and a busy street may be separated by tall buildings from a very quiet urban

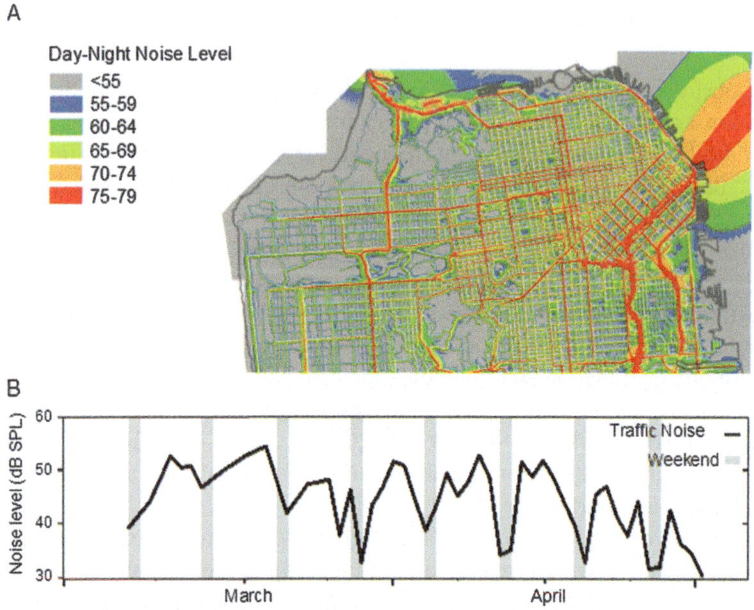

Fig. 8.2 Spatiotemporal variation in the occurrence of man-made sounds. (**a**) Sound map from the San Francisco Department of Public Health, CA, showing spatial variation in noise levels measured over a whole day (in dB day-night equivalent level [Ldn]: average dB sound pressure level [SPL], A-weighted over 24 h; the level of this common unit is assessed by an artificial increase of 10 dB between 10 p.m. and 7 a.m. to account for the nighttime decrease in community background noise). Car traffic over the San Francisco-Oakland Bay Bridge (*top right* of map) visibly spreads sound over a large area, whereas sound levels in the inner city reflect a network of major roads and the delimiting effect of urban canyons (related to Luther and Baptista 2010). (**b**) Traffic sound levels (in dB SPL A-weighted) recorded throughout the breeding season on a major two-lane highway in The Netherlands (A12: Utrecht-Arnhem). Note the reduced levels during the weekend and a gradual decrease over the season that is related to an increase in temperature as well as to a higher density of foliage (based on Halfwerk et al. 2011b)

canyon. Noisy cities are therefore probably the most divergent from any natural habitat type, especially in terms of the heterogeneous nature of temporal and spatial variation.

8.2 Are Birds Affected by Man-Made Sounds?

All bird species rely to some extent on acoustic signals and cues to find their food, avoid predators, or communicate with conspecifics (Kroodsma and Miller 1996). Furthermore, many species show specific acoustic adaptations in either their communication signals or perceptual sensitivity depending on the habitats they live in or the functional niche they occupy. Birds living near fast-flowing rivers, for example, have evolved high-pitched songs of narrow bandwidth to avoid spectral overlap with low-frequency river noise (Dubois and Martens 1984). Several owl species, such as the barn owl (*Tyto alba*), are well-known for their habit and ability to hunt by ear (Payne 1971; Knudsen and Konishi 1979). Given the apparent importance of acoustic signals and cues in the life of birds, there is a wide range of effects of man-made sounds on birds and their songs.

8.2.1 Population-Level Variation with Ambient-Noise Levels

Some of the first circumstantial evidence that man-made sound can affect birds came from early roadside-effect studies (van der Zande et al. 1980; Reijnen and Foppen 1995). These studies reported that breeding densities and species diversity decreased closer to highways. Furthermore, these earlier studies hypothesized that noise levels could be the causal factor driving declines near roads (Reijnen and Foppen 2006). One of the major problems with these correlational studies, however, was that many confounding variables associated with road proximity could explain the same effect (Halfwerk et al. 2011b; McClure et al. 2013). Among other causes for density declines are factors such as light levels, chemical pollution, habitat-edge effects, and mortality through collision with cars (Halfwerk et al. 2011b). All of these factors tend to correlate positively with noise levels, albeit to various degrees, and all of these factors can potentially reduce bird breeding numbers and species diversity.

More recent studies have been able to control for these confounding factors. These studies typically focused on noise pollution from compressor stations associated with drilling for gas and oil. Bayne et al. (2008) looked at breeding densities in relation to two types of stations, one type actively pumping gas and consequently emitting sound levels (root-mean-square [rms]) between 75 and 90 dB(A) close to the source and another type not actively pumping and therefore emitting no sound. Breeding densities of several bird species were reduced 50% on average near noisy stations compared with quiet control stations. Other studies on gas compressor stations (Francis et al. 2009) have confirmed reduced breeding densities and revealed several species-specific effects that could help to understand how birds are affected. Birds singing higher frequency songs, for example, were less affected by the presence of a noisy station (Francis et al. 2012).

Similar patterns have been reported for roadside studies, with species singing higher frequency songs less negatively impacted than species singing lower frequency songs (Rheindt 2003; Goodwin and Shriver 2011), which may also apply to owl species (Senzaki et al. 2016). Song frequency is also strongly related to body size, however, and earlier studies on compressor stations (e.g., Fig. 8.3) could not distinguish between body size and song frequency in explaining noise-dependent effects (Francis et al. 2012). Furthermore, recent comparative studies indicated that singing high is not a guarantee for urban success (Moiron et al. 2015) and that species abundance in noisy areas may be related to song frequency as well as diet,

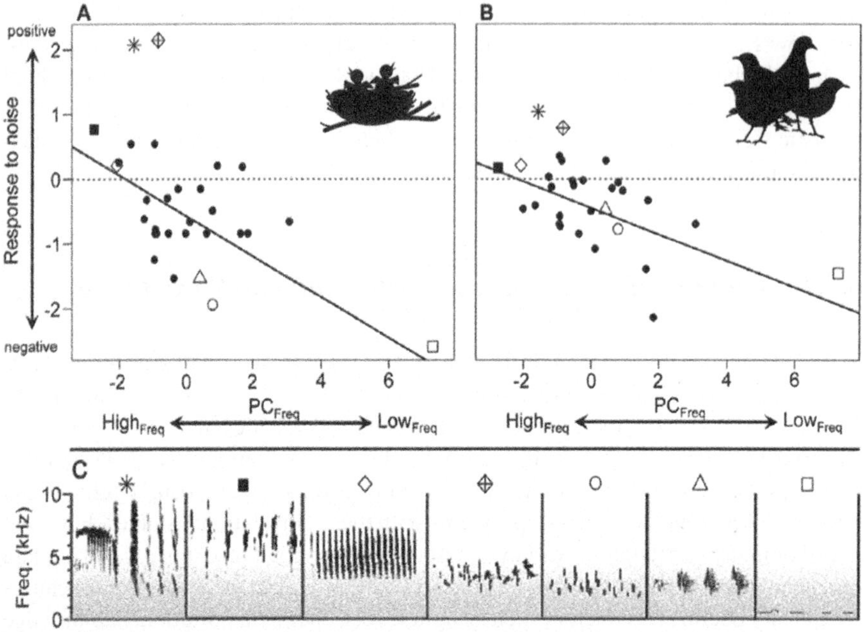

Fig. 8.3 The effect of noise pollution on species-specific densities. Nesting (**a**) and abundance (**b**) data of responses to elevated sound levels (gas-compressor station noise) per species. Species that were negatively affected by the low-frequency noise sang relatively low-frequency songs. These species were also found to be bigger in size. PC$_{Freq}$, first principal component of a multivariate analyses with frequency (Freq) as the highest loading. (**c**) Spectrograms of a subset of species ordered according to frequency. *Asterisks*, black-chinned hummingbird; *solid squares*, bushtit; *open diamonds*, chipping sparrow; *crossed diamonds*, house finch; *open circles*, black-headed grosbeak; *open triangles*, western tanager; *open squares*, mourning dove; *solid circles*, all other species. From Francis et al. (2012)

suggesting that multiple factors may drive the correlations between man-made noise levels and population densities (Francis 2015).

8.2.2 Acoustic Correlates with Man-Made Sound Levels

Man-made noise may not only influence the relative breeding success of birds but may also impact their song structure. Bergman (1982) was one of the first to report on a change over time in the occurrence of birdsong variants and to suggest a link to man-made sounds. He scored the occurrence of two-note and three-note songs in great tits (*Parus major*), and it appeared that they sang songs with fewer notes in areas that had become noisier. Many years later, this idea was tested with more advanced tools and a better replicated sampling design across Europe, but investigators failed to find a similar pattern (Slabbekoorn and den Boer-Visser 2006). Skiba

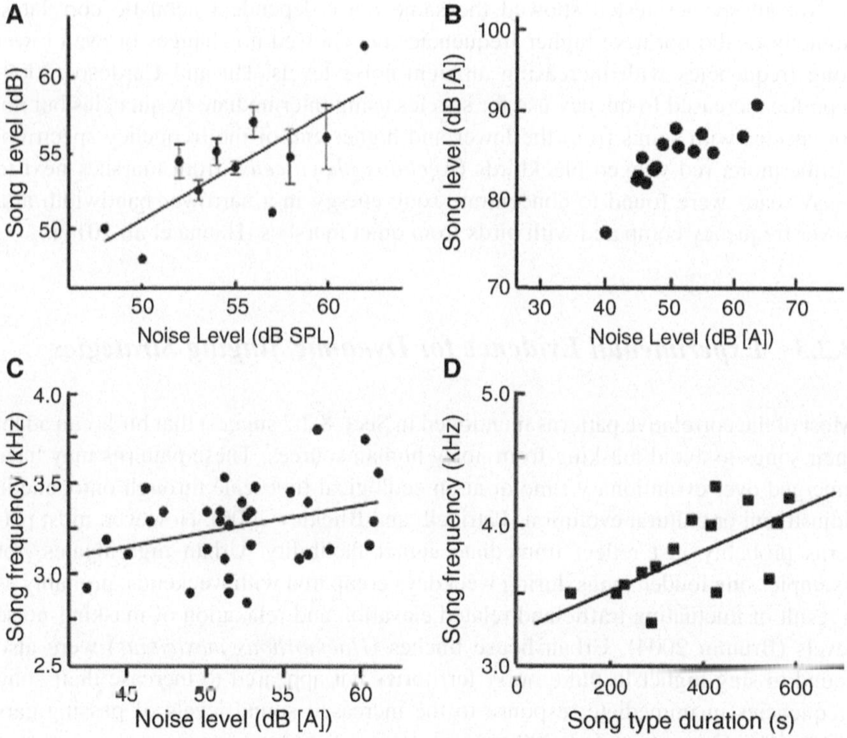

Fig. 8.4 Noise-related song changes. (**a**) Hummingbirds sing louder with increased levels of waterfall noise (after Pytte et al. 2003). (**b**) Nightingales sing louder with increased levels of traffic noise (after Brumm 2004). Great tits sing higher pitched songs with increased levels of traffic noise (**c**) and do this by singing the lower frequency song types for shorter durations during more noisy conditions (**d**). After Slabbekoorn and Peet (2003) and Halfwerk and Slabbekoorn (2009)

(2000) also hypothesized that learned rain calls of the chaffinch (*Fringilla coelebs*) might show acoustic variation related to local ambient-noise levels but a collection of recordings of calls and ambient-noise levels did not yield any significant correlation.

It was at the start of the twenty-first century that several studies started to find significant effects of local noise levels on individual singing behavior (see Fig. 8.4). Brumm (2004) found that nightingales (*Luscinia megarhynchos*) in Berlin were singing louder in noisy territories than in quieter ones (Fig. 8.4b). Slabbekoorn and Peet (2003) did not measure song amplitude but focused on spectral traits. They found that acoustic frequencies of individual great tit songs within the same urban population depended on the relative ambient-noise levels in their territories. Birds in quieter areas were singing with a lower minimum frequency than nearby birds at noisier locations (Fig. 8.4c). This pattern was replicated afterward in great tits in England, Spain, and Japan and also in many other urban bird species across the globe (reviewed in Slabbekoorn 2013).

Not all species tested showed the same noise-dependent acoustic correlates. Some birds did not have higher frequencies but showed no changes or even lower song frequencies with increasing ambient-noise levels. Hu and Cardoso (2009) reported increased frequency use for species using intermediate frequencies but not for species with songs from the lower and higher end of the frequency spectrum. Furthermore, red-winged blackbirds (*Agelaius phoeniceus*) from marshes next to noisy roads were found to concentrate song energy in a narrower bandwidth at a lower frequency compared with birds from quiet marshes (Hanna et al. 2011).

8.2.3 Experimental Evidence for Dynamic Singing Strategies

Most of the correlative patterns mentioned in Sect. 8.2.2 suggest that birds can adapt their songs to avoid masking from noisy human sources. These patterns may have emerged over evolutionary time or at an ecological timescale through ontogenetic adjustment or cultural evolution (Patricelli and Blickley 2006). However, most patterns probably just reflect immediate signal flexibility. Urban nightingales, for example, sing louder songs during weekdays compared with weekends, probably as a result of fluctuating traffic and related elevation and relaxation of masking-noise levels (Brumm 2004). Urban house finches (*Haemorhous mexicanus*) were also found to sing higher in more noisy territories but appeared to increase their song frequencies in immediate response to the increased sound levels of passing cars (Bermúdez-Cuamatzin et al. 2009).

Several studies have now also provided experimental evidence for immediate ambient-noise level-dependent changes in various species (Halfwerk and Slabbekoorn 2009; Bermudez-Cuamatzin et al. 2010; reviewed in Slabbekoorn 2013). Great tits switch relatively quickly to another song type when they sing a relatively low-frequency song and experimental noise exposure involves traffic-like low-frequency sound (see Fig. 8.4d) but continue to sing the same song type if it is a relatively high song. Chiffchaffs (*Phylloscopus colybyta*) shift their songs upward in frequency when highway noise is played back to birds (Verzijden et al. 2010). A recent study on black-capped chickadees (*Poecile atricapillus*) suggests that familiarity with fluctuations in traffic-noise level can also affect the dynamic response pattern (Lazerte et al. 2016). Local ambient-noise levels correlated positively with the song frequency shift in response to experimental exposure: birds in noisier territories switched song frequencies upward, whereas birds in the quieter territories, and therefore perhaps less experienced birds, switched downward.

The most obvious explanation for dynamic singing strategies seems to be one of adaptive masking avoidance (Slabbekoorn and Peet 2003; Brumm 2004; Brumm and Slabbekoorn 2005). Louder and higher songs improve signal-to-noise levels and counteract the detrimental signaling conditions during noisy periods. However, the benefits are often assumed and the data are usually still lacking for associated signal efficiency improvements, let alone fitness benefits (Nemeth and Brumm 2010; Slabbekoorn et al. 2012). Furthermore, it is also important to take into account the costs of singing altered songs, such as potentially increased energy expenditure or potentially reduced attractiveness (Halfwerk et al. 2011a; Read et al. 2014). An alternative to an adaptive explanation could be that elevated sound levels just cause some change or disturbance that affects singing. Birds in noisier areas may, for example, experience less acoustic competition from neighboring rivals and change their singing style in response (Ripmeester et al. 2010; Hamao et al. 2011). High levels of noise could also increase perceived predation risk or induce stress and thereby indirectly affect song production (Quinn et al. 2006; Halfwerk et al. 2012).

Both correlational and experimental approaches have helped to identify two major categories by which man-made sound can affect birds: (1) *perceptual interference of signals and cues* and (2) *physiological and behavioral disturbance.* Sections 8.3 and 8.4 review evidence for each of these categories, followed by discussions on evolutionary and ecological implications.

8.3 Perceptual Interference of Signals and Cues

Ambient noise can interfere with the detection and processing of important signals and cues and thereby affect an individual's ability to survive and reproduce (Brumm and Slabbekoorn 2005; Halfwerk and Slabbekoorn 2015). Ambient noise can, for example, mask the acoustic cues of potential prey, an approaching predator, or the alarm calls of surrounding songbirds that have detected a nearby hawk (Mason et al. 2016; Templeton et al. 2016). Ambient noise may also reduce song detection by neighbors and thereby reduce a male's ability to defend his territory. Acoustic interference may also limit the area over which males can attract females with their songs or even reduce the attractiveness of their sexual signals. The perceptual impact of man-made sound on birds (see Fig. 8.5) depends ultimately on a combination of sound characteristics (e.g., amplitude, frequency, periodicity), species-specific perceptual traits (e.g., hearing sensitivities, critical masking ratios), and the overlap in time and space between the production of man-made sounds and species-specific behavioral activities (e.g., singing or feeding). First, the species-specific auditory abilities and perceptual impacts of ambient noise on birds are reviewed and then the perceptual and behavioral strategies that birds could use to avoid these impacts.

8.3.1 What Birds Hear: The Audiogram

The minimum audible sound pressure across a bird's frequency range of hearing constitutes its audiogram. Such hearing thresholds may be determined through behavioral, physiological, or anatomical methodologies (Dooling and Okanoya 1995; Dooling et al. 2000). Audiograms determined using behavioral thresholds are available for about 50 species of birds and thus far show a remarkable consistency

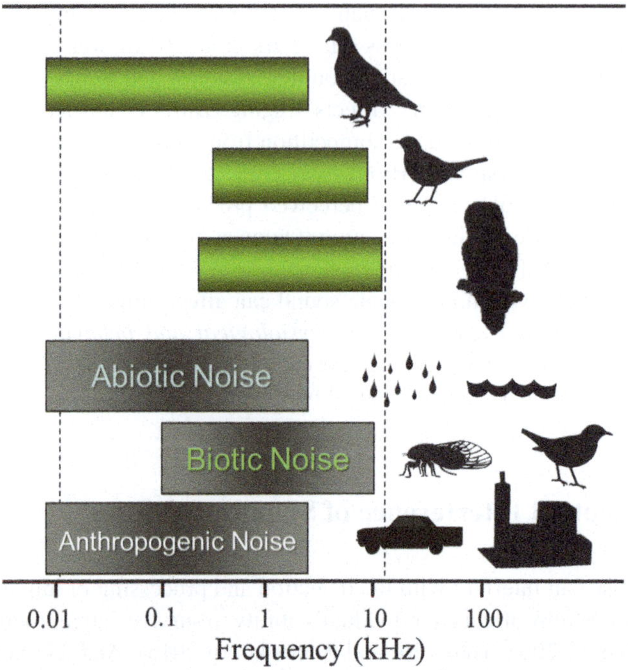

Fig. 8.5 Hearing ranges of different bird species groups (doves, songbirds, and owls) to illustrate typical variety among bird species and their overlap with environmental sounds (of abiotic, biotic, or man-made origin). *Vertical dashed lines*, human hearing range. Note that most birds do not extend beyond or even stay well within the human hearing range, although some bird species like owls may have a considerably lower absolute threshold and thereby hear better than us (reviews in Fay 1988; Dooling et al. 2000)

in overall shape and sensitivity for most of these species (Dooling et al. 2000). Generally, birds are not quite as sensitive as humans when hearing at their frequency of best sensitivity (with best hearing typically around 10 dB sound pressure level [SPL] between 2 and 4 kHz), and their sensitivity falls off rapidly compared with human hearing at frequencies below and especially above this range. The broad similarity across species reflects the relatively low variability in cochlear anatomy among bird taxa studied to date (Gleich and Manley 2000).

Most small birds that have been tested fall in the order Passeriformes (~20 species), which is likely due to their tractability in operant tests. A few other small, seed-eating birds from other groups (e.g., parrots, pigeons, chickens) have been tested for their auditory capacity (about 15 species; Dooling et al. 2000). Songbirds, as a subgroup of the Passeriformes, make up the largest taxon of birds that have been tested for their auditory abilities, and given the importance of vocal signaling and vocal communication in their breeding biology, it is not surprising that there is generally a good match between the frequency of best hearing and the frequency of peak power in their vocal communication signals.

The broadly similar behavioral audiograms of small-bird species contrast with the hearing abilities of owls, however. Owls, which, unlike other raptors, are generally nocturnal predators, are substantially more sensitive than other birds, especially at high frequencies, with best thresholds close to −20 dB SPL (Fay 1988; Dooling et al. 2000). In owls, hearing is primarily adapted for prey detection rather than communication, resulting in both this enhanced sensitivity and best hearing frequencies that are not closely matched to the peak power in their vocalizations. Owl hoots usually have fundamental frequencies that are below their most sensitive range of hearing.

8.3.2 *Masking Issues: Critical Ratios Across Species*

Ambient noise can mask acoustic signals for communication as well as environmental cues exploited for detecting predators or finding prey (Dooling et al. 2000; Dooling and Blumenrath 2013). The most basic measure of auditory masking is the critical ratio, defined as the ratio between the power in a pure tone at threshold and the spectrum level (power per hertz) of the background noise when the tone is just masked (Scharf 1970). The critical ratio function maps this ratio across frequencies. Thus, sounds below the critical ratio in a given level of background noise at a given frequency are inaudible (see Dooling and Leek, Chap. 2). Critical ratio functions have been measured behaviorally for about 15 species of birds, again including songbirds, parrots, and pigeons (Dooling et al. 2000; Dooling and Blumenrath 2013). They are generally measured using operant conditioning techniques (Noirot et al. 2011).

Most bird species show an increase in critical ratio of approximately 3 dB per octave (doubling of frequency) over a broad spectral range. This increase is a consequence of inner ear mechanics and the logarithmic representation of frequency along the basilar membrane in the cochlea (Gleich and Manley 2000). As in mammals, the bird's auditory system functions as a bank of overlapping filters that increase in size (auditory filter bandwidth) with increases in frequency (Dooling et al. 2000). Thus, as auditory filters get larger at a higher frequency, they admit more noise and signal levels must be louder for a higher frequency signal to be detectable. Exceptions to the general rule of a monotonic increase of approximately 3 dB per octave include parrots such as budgerigars (*Melopsittacus undulatus*) and orange-fronted conures (*Aratinga canicularis*; Okanoya and Dooling 1987; Wright et al. 2003) as well as great tits (Langemann et al. 1998).

The parrot species tested thus far show relatively constant critical ratios with increased frequency until about 3–4 kHz, at which point the critical ratios increase rapidly (Okanoya and Dooling 1987; Wright et al. 2003). Thus, these parrots have relatively low critical ratios between about 2–4 kHz compared with other bird species. Wright et al. (2003) speculated that this may have functional consequences in noisy environments for detecting their long-distance contact calls that have rapid amplitude and frequency modulations. The critical ratio of great tits is relatively constant to even higher frequencies (Langemann et al. 1998), which may provide them with an adaptive advantage relative to other species in the detectability of high-pitched alarm calls used in the context of aerial predators. Furthermore, great tits have also been shown to have perceptual plasticity in how they use different signal features in different ambient-noise conditions. Although discrimination among different song types in woodland noise is based on aspects of the whole song, including the high-frequency elements, discrimination in urban noise relies entirely on features of the high-frequency elements (Pohl et al. 2012).

Once again, owls also show a different pattern than other bird species with respect to their critical ratios (Dooling et al. 2000). Barn owls were shown to exhibit a more rapid increase in critical ratio of about 5 dB per octave, and their absolute threshold values are considerably lower than those of other birds (Konishi 1973). This exceptional critical ratio function is likely the consequence of strong selection pressures for nocturnal prey detection in the acoustically challenging natural habitats in which owls forage (i.e., faint high-frequency cues and relatively noisy background).

8.3.3 How Far Audible: The Active Space

The concept of a signal's active space has been defined as the maximum detection distance for biologically relevant acoustic signals (Marten and Marler 1977; Brenowitz 1982). Arguments have been made that this definition should be reconfigured to include discrimination and recognition thresholds because such thresholds are more directly relevant to estimating the distances over which the "message" within an acoustic signal is likely to be perceived (Lohr et al. 2003; Dooling and Blumenrath 2013). Critical ratio functions are directly applicable to deriving a first estimate of the "active space" (see Fig. 8.6) and have been used to estimate the range over which acoustic signals can convey biologically meaningful information in noisy habitats (Pohl et al. 2009; Nemeth and Brumm 2010).

Studies measuring thresholds for natural signals (and synthetic signals designed to mimic natural signals) in ambient noise with a variety of spectral shapes have served to identify some common features for auditory signal detection (Lohr et al. 2003; Pohl et al. 2009). The most salient issue concerns the auditory bandwidth overlapping and surrounding the spectral region of the signal. Although there may be some upward spread of masking from lower frequency channels toward higher frequency channels (Moore et al. 1997), it is the signal-to-noise ratio in the spectral region of the signal that will largely determine the signal-to-noise ratio at threshold (Lohr et al. 2003; Pohl et al. 2009). Thus, if a signal's power is concentrated in a single-frequency band rather than being spread across a range of frequencies (as in a broadband or harmonic signal), that signal will have a lower threshold in a given level of background noise (cf. Hanna et al. 2011). The region of auditory filtering is especially important because the spectral shape of ambient noise varies considerably among different natural and man-made habitats. Traffic and urban noise (Slabbekoorn and Peet 2003; Pohl et al. 2009), for example, are broadband, and the sound spectra are similar to those of natural streams or vegetation moved by wind (see, e.g., Feng et al. 2006). Man-made sounds, however, are typically biased more toward lower frequencies (<2000 Hz) than most natural sounds.

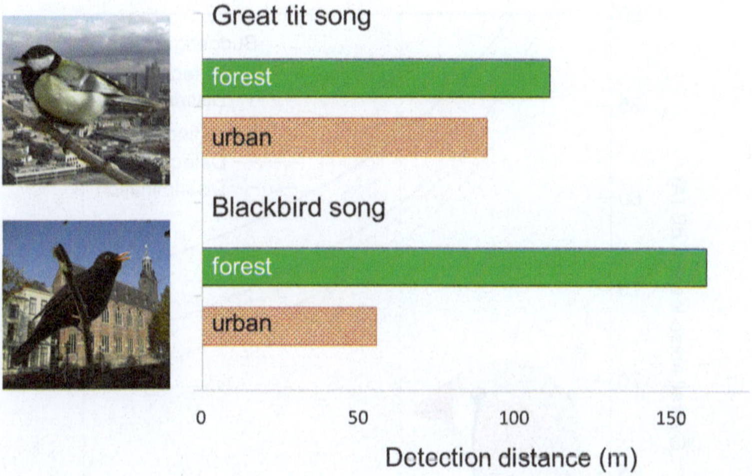

Fig. 8.6 The effect of song frequency and background noise on maximum detection distances of two bird species that occupy different song frequency ranges. Great tits have higher pitched songs than blackbirds and have therefore lower maximum detection ranges in the forest habitat due to attenuation. The pattern is reversed in the urban habitat where the low-frequency song of blackbirds is more affected by the masking of background noise. Based on Nemeth and Brumm (2010), (Bird pictures by Herman Berkhoudt)

The temporal resolving power of the avian auditory system must also be considered when estimating masking thresholds and the active space of signals. Lohr et al. (2003) showed that measuring the instantaneous peak of the amplitude envelope of a natural sound signal provided a better estimate of the signal-to-noise ratio at threshold than rms measures that integrate signal power over a time constant. Although small birds are similar to humans in a number of temporally related auditory tasks, including thresholds for temporal integration, duration discrimination, gap detection, and detection of amplitude modulation (Lohr and Dooling 1998; Dooling et al. 2000), birds are superior to humans in their ability to discriminate rapid changes in the temporal fine structure of sounds (Lohr and Dooling 1998). As a consequence, birds may be able to integrate over very short time periods, making peak sound pressure level the most accurate measurement for estimating signal-to-noise ratios at threshold. Consequently, sound signals with rapidly modulated amplitude envelopes will be more detectable than those with smooth envelopes for a given rms SPL (Lohr et al. 2003).

8.3.4 Masking of Higher Level Cognitive Processes

Birds not only need to detect sounds but also to extract any relevant information. Thus, when a sound is detected, the next steps in acoustic processing involve sound localization, recognition of the sound source (whether made by certain species or individuals), and classification of additional information (e.g., size of the signaler or

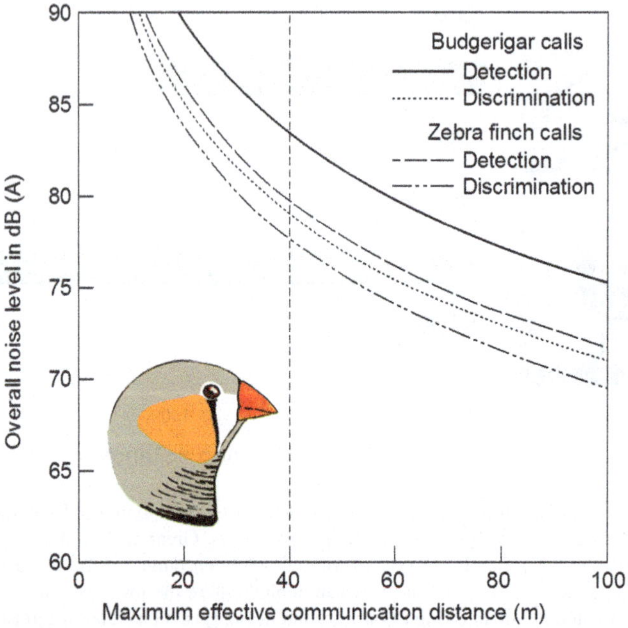

Fig. 8.7 Theoretical maximum communication distances based on detection and discrimination thresholds for budgerigar and zebra finch calls in traffic-spectrum noise. *Curves* illustrate distances based on detection thresholds and discrimination thresholds and assume an excess attenuation of 5 dB/100 m and a source intensity of 95 dB SPL noise. *Vertical dashed line* represents a distance of 40 m. In most cases, discrimination threshold distances (*dotted lines*) probably represent the maximum limit for communication distances (or "active space") for a given call type. Based on Lohr et al. (2003)

whether a sound was a sexual or warning signal). At its most basic level, such processing might involve an increasingly challenging set of perceptual tasks proceeding from simple detection through discrimination of sounds to recognition and identification of specific acoustic signals with particular messages. At each level, as the auditory task becomes increasingly difficult, the masking effects of noise are likely to increase, with higher signal-to-noise ratios required at each step (Klump 1996; Dooling and Blumenrath 2013; see Fig. 8.7). Zebra finches (*Taeniopygia guttata*) and budgerigars, for example, require an approximately 2- to 5-dB increase in signal-to-noise ratio to discriminate among a set of species-typical calls compared with simply being able to detect those calls in background noise (Lohr et al. 2003). A similar phenomenon occurs in the perception of human speech signals. The detection of speech sounds in a noisy background typically involves the lowest signal-to-noise ratios at threshold, whereas discrimination of different speech sounds requires higher signal-to-noise ratios than detection and recognition of specific speech sounds requires even higher signal-to-noise ratios than discrimination (Miller 1951).

The challenge of extracting higher level information from masked acoustic signals is perhaps best illustrated by the "cocktail party problem" (Bee and Micheyl 2008; see Dent and Bee, Chap. 3). This long-standing problem derives from questions about how human listeners are able to "hear out" a particular voice in a background of other human speech sounds that can produce both "energetic masking" (direct acoustic interference) and "informational masking" (cognitive interference). A number of other vertebrates, including birds, face an analogous problem if they aggregate in large, vocal groups during at least part of the year and the ability to identify the sounds of specific individuals is at a premium (Jouventin et al. 1999; Leonard and Horn 2005). European starlings (*Sturnus vulgaris*) serve as a good example and have been shown to be capable of discriminating the songs of specific individual starlings when presented against a background of other individuals' songs (Hulse et al. 1997; Wisniewski and Hulse 1997).

Human psychophysical studies in the context of auditory scene analysis have provided an important conceptual and experimental framework for understanding the cocktail party problem (Bregman and Campbell 1971). In this context, relevant sounds are both segregated from similar background sounds and integrated when arising from the same source at different times, through a variety of cues including similar frequency ranges, spatial locations, onset and offset times, and common amplitude modulations. This process of auditory segregation and integration results in the formation of auditory "streams" or "objects" that then provide a higher order cognitive basis for the recognition of a particular individual's vocal signals in a noisy background. The increasing complexity of such auditory processes can be viewed on a functional continuum from simple detection at one end to auditory stream segregation, auditory object formation, and, finally, recognition at the other end (Knudsen and Gentner 2010).

8.3.5 Perceptual Strategies for Noisy Conditions

There are several ways in which the signal-to-noise ratio can be improved using perceptual strategies in time and space in noisy habitats. Most sounds, including those from human activities, are not temporally continuous and birds may employ "dip listening," in which individuals take advantage of transient decreases in the background noise to catch "acoustic glimpses" of target signals. Few studies have demonstrated dip listening in vertebrates other than humans using natural communication signals and natural background noise (Ronacher and Hoffmann 2003; Vélez et al. 2013). However, fluctuations in background noise level have been shown to enhance the detectability of tones against narrowband ambient noise in a number of birds under laboratory conditions (Klump 1996; Langemann and Klump 2007).

Similarly, signal and noise sources are often separated in space in natural habitats, and such spatial separation results in enhanced signal perception, a phenomenon known as "spatial release from masking" (Klump 1996). The release can be substantial and is due to directional sensitivity of the ears as well as to differences in the phase relationship of the signal and masking noise at the two ears (the binaural masking-level difference). In budgerigars, for example, separation of signal and noise sources by 90° results in up to 10 dB of masking release for tones (Dent et al. 1997) and 20–30 dB of masking release for different types of natural vocal signals (Dent et al. 2009).

Another perceptual strategy exploits the fact that broadband ambient noise simultaneously enters several auditory channels, while many bird sounds excite only a single acoustic channel at a given time (because many bird vocalizations are pure tones or narrowband signals). If signals and broadband ambient noise are subject to different temporal fluctuations in level, birds may be able to achieve an increase in the signal-to-noise ratio through comodulation masking release (Hall et al. 1984). If temporal fluctuations in ambient noise are coherently amplitude modulated across auditory channels, even spectral regions distant from those of the signal can result in a substantial release from masking (Dooling et al. 2000). Several studies have now shown comodulation masking release in birds, including European starlings (Klump and Langemann 1995) and hooded crows (*Corvus corone cornix*; Jensen 2007). Man-made sounds that cover wide frequency ranges offer the most potential for hearing benefits because the wider the bandwidth, the larger the masking release.

Finally, many acoustic signals are produced simultaneously with signals in other modalities. Birds often signal visually as well as acoustically, for example, and such complex, multimodal signals can improve the transmission efficacy of signals under noisy conditions (Brumm and Slabbekoorn 2005). Partan and Marler (2005) provided a framework for understanding how information may change in a multimodal signaling context depending on whether signals in different modalities carry the same message (are redundant) or carry different messages (are nonredundant). In the case of redundant signals in different modalities, a major functional consequence of that redundancy is thought to be the potential for enhancing that signal's detectability (Partan and Marler 2005). Halfwerk and Slabbekoorn (2015) extended the framework and included the notion that pollution is often also multimodal.

8.3.6 Behavioral Strategies for Noisy Conditions

Dynamic singing strategies (see Sect. 8.2.3; Fig. 8.1) are probably the most straightforward behavioral strategies that singing birds can use to make sure that their songs are detected by the intended receivers (Brumm and Todt 2002; Brumm and Slabbekoorn 2005). Singing louder and higher can improve song detection and recognition. Nightingales increase song amplitude when exposed to broadband noise, a phenomenon known as the Lombard effect that is thought to have evolved in terrestrial vertebrates to improve signal-to-noise ratios (Brumm and Todt 2002). Great tits, on the other hand, make use of noise-dependent frequency shifts (Halfwerk and Slabbekoorn 2009), which have also been shown to improve signal detection both in the laboratory and in the field (Halfwerk et al. 2011a; Pohl et al. 2012). The benefits of frequency shifting depend on the associated change in signal-to-noise ratio and the species-specific critical ratio function (see Sect. 8.3.2). As mentioned in Sect. 8.3.2, great tits possess a remarkably flat critical ratio function, and an increase in frequency will consequently not come at the widespread cost of increased detection thresholds (Langemann et al. 1998).

Other behavioral strategies for noisy signaling conditions involve singing closer to the intended receiver, singing from a larger distance to the sound source, singing at higher song posts, or waiting for a period with relatively low levels of ambient noise. Great tit males that sing toward a mate sitting in her nest box have been found to sing closer to these nest boxes when traffic sound was broadcast inside (Halfwerk et al. 2012). This change in song post was associated with an increase in song amplitude levels recorded inside the nest box and thus improved signal-to-noise ratios. Interestingly, males themselves were not exposed to additional sound and thus had to rely on female feedback to move closer. In another study on European robins (*Erithacus rubecula)*, birds were found to move away to song posts at a larger distance from experimental exposure to traffic noise (McLaughlin and Kunc 2013). Birds can also sing from higher song posts to improve the active space of their songs because song post height is negatively correlated with attenuation and degradation, in particular in vegetated habitats (Mathevon et al. 1996). It remains to be tested whether this behavior improves both signal transmission of a bird's own song and signal perception of the song of neighbors in noisy areas.

Singing birds may also employ temporal strategies to avoid overlap with high levels of noise (Fuller et al. 2007; Planque and Slabbekoorn 2008). Fuller et al. (2007) were the first to suggest that birds in noisy urban areas could shift the timing of their song to periods with relative low noise levels to avoid masking. They found nocturnal singing of European robins to be correlated with daytime noise levels in their territories. Noise was even a better predictor of nocturnal singing than artificial light levels. A related finding was that bird communities around airports advance the start of their dawn chorus (Gil et al. 2014). Arroyo-Solis et al. (2013) used an experimental approach within a city and also reported that several urban species started singing earlier in response to elevated sound levels. Whether singing at night or singing earlier in the morning actually improves signal-to-noise ratios in areas with fluctuating noise levels remains to be tested. Additional data are required on whether singing earlier results in a shift or an increase in the duration of singing and whether the costs outweigh the benefits of the altered communication conditions.

Temporal avoidance of masking at very short timescales was investigated for Eurasian wrens (*Troglodytes troglodytes*), who were able to avoid overlap with conspecific songs (Yang et al. 2014; see Fig. 8.1) but did not show any evidence for using this capacity to avoid traffic noise or intermittent playback of wideband noise (Yang and Slabbekoorn 2014).

8.3.7 Nonmasking Perceptual Impact of Man-Made Sound

Ambient noise can also influence the detection of signals and cues through nonmasking effects, even at very low levels. In human terms, such effects due to perceptible but low-level background noise fall under "annoyance" and form the basis for noise pollution ordinances as well as decreased property values near loud sound sources such as airports and military bases (Basner et al. 2014). Similarly, man-made sounds may "annoy" birds or cause distraction and thereby interfere with perceptual and cognitive processing (Chan et al. 2010; Chan and Blumstein 2011). So far, the evidence for a distraction effect on cognitive processes has only been found for fish (Simpson et al. 2014) and crabs (Chan et al. 2010) but could potentially have a large effect on the performances of birds, in particular, in a foraging context.

Many bird species have good spatial memory that helps them to remember where they recently depleted food patches or stored food items such as acorns or seeds (Hilton and Krebs 1990). Noise distraction can interfere with the formation or retrieval of these spatial memories and thereby impact avian foraging efficiency. A distraction effect of noise can also influence the detection and processing of other sensory cues (Chan et al. 2010; Simpson et al. 2016). Quinn et al. (2006), for example, exposed groups of feeding chaffinches to high levels of white noise and found birds to increase their vigilance behavior. Similar effects of noise exposure on vigilance behavior have been seen in white-crowned sparrows (*Zonotrichia leucophrys*; Ware et al. 2015) and suggest that noise may also influence visual processing. Such "cross-modal" effects of noise can have severe fitness consequences for birds, especially in the context of predation (Halfwerk and Slabbekoorn 2015).

Two other studies have addressed the effect of noise pollution in the context of predation risk and antipredator responses more specifically. Meillère et al. (2015) studied a rural population of house sparrows (*Passer domesticus*) and showed that noise-exposed females flushed earlier when approached by an experimenter. Owens et al. (2012) demonstrated that two tit species, Carolina chickadees (*Poecile carolinensis*) and tufted titmice (*Baeolophus bicolor*), were less likely to approach a novel object during a prolonged sound exposure treatment, suggesting an impact on risk-taking behavior. These latter studies again suggest that man-made sounds can influence visual processing (assuming birds relied on visual cues). However, whether the ambient noise-induced behavioral changes are caused by general disturbance or reflect an adaptive perceptual strategy remains to be tested (Halfwerk and Slabbekoorn 2015).

8.4 Physiological and Behavioral Disturbances

Birds may be affected more broadly by man-made sounds than simply by the perceptual impact on the detection and processing of auditory signals and cues. Sounds in the surrounding habitat may reduce the singing or foraging efficiency of birds or make them more vulnerable to predation, which may obviously all lead to negative consequences in terms of survival and reproduction (e.g., Kight and Swaddle 2011; Francis and Barber 2013). However, a broader impact with the potential for negative effects on fitness can occur through physiological and behavioral disturbance. Physiological effects often include involuntary changes in stress hormone levels potentially associated with a number of externally visible or measurable consequences, such as changes in expiration rate, heartbeat, or activity patterns. Behavioral effects often include a slowdown or interruption of activities or spatial deterrence, which may lead to both a less optimal performance of normal activities or missed opportunities.

8.4.1 Physiological Disturbance by Man-Made Sound

Man-made sound can cause a change in an animal's physiology at any audible level. At its loudest levels, acoustic exposure can cause damage to anatomical structures in the inner ear, such as the sensory hair cells, a phenomenon well-known in humans (Rabinowitz 2000). Such damage may be permanent, as in mammals (Rivolta and Holley 2008), or transitory and lead to temporary threshold shifts, as in birds (Ryals et al. 1999; see Saunders and Dooling, Chap. 4). Typically, birds do not encounter sounds sufficiently loud to cause auditory damage in natural habitats, although they may be sufficiently close to continuous or impulsive man-made sounds for at least temporary threshold shifts to occur under some circumstances (Dooling and Blumenrath 2013).

At intermediate and even very moderate levels, sounds from human activities can influence the physiology of birds by causing changes in stress-related hormonal activity (Sapolsky et al. 2000). Acute changes are considered the primary stress response, which usually fall within an adaptive dynamic range as part of natural response patterns to external stimuli. In contrast, chronic changes are considered the secondary stress response that affects energy regulation in the body and may undermine digestive and reproductive activity, immunocompetence, general homeostasis and thereby growth, survival, and reproductive success.

Few studies have related noisy human activities to physiological stress levels in birds (Blickley et al. 2012a; Crino et al. 2013). These studies typically focused on chronic changes in corticosterone (CORT; the avian equivalent of human cortisol) levels. CORT can be obtained from fecal, feather, or blood tissue, although the different sampling techniques are known to provide different information on an animal's stress physiology as well as the duration during which animals might show elevated stress levels (Sheriff et al. 2011). Blickley et al. (2012a) exposed displaying males of the greater sage-grouse (*Centrocercus urophasianus*) to noise as well as control conditions and found males displaying at noisy sites had higher levels of fecal CORT (Fig. 8.8). Strasser and Heath (2013), on the other hand, focused on blood CORT levels obtained within 5 min after capture of adult American kestrels (*Falco sparverius*). They found that females breeding in closer proximity to roads as well as to human settlements had higher levels of baseline CORT.

The effects of noise on stress-related avian physiology are, however, far from clear. Tempel and Gutiérrez (2003), for example, found no changes in fecal CORT of owls who were experimentally exposed to sounds of a chain saw nor did owls living near roads show elevated CORT levels (Tempel and Gutiérrez 2004). Noise-exposed nestling house sparrows also showed no elevated CORT levels compared with control birds (Angelier et al. 2015). Finally, Crino et al. (2013) exposed white-crowned sparrow nestlings for five consecutive days to noise and found reduced

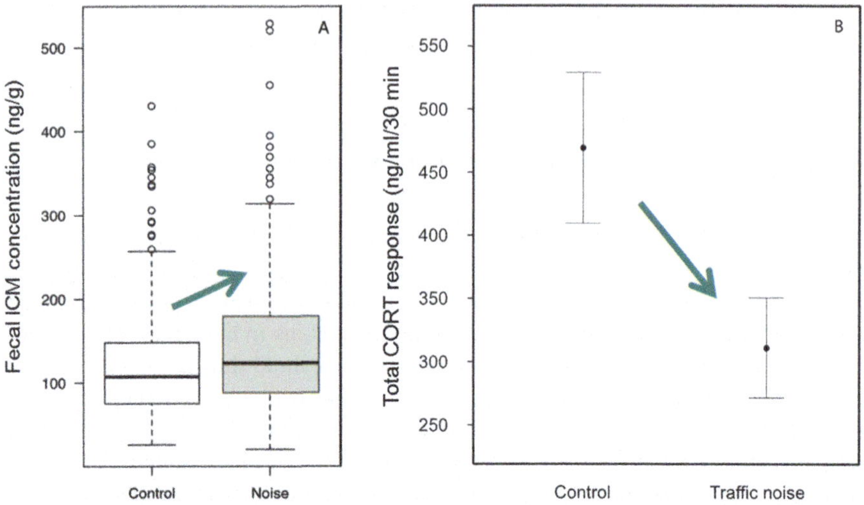

Fig. 8.8 Opposing effect of man-made noise on physiological stress response. Sage-grouse show an increase in corticosterone (CORT; measured from immunoreactive metabolites [ICM]) in response to noise playback (**a**), whereas white-crowned sparrows show a decrease in cortiscosterone with elevated noise levels (**b**). (**a**) Boxplots with outliers (from Blickley et al. 2012a); (**b**) mean values with standard errors (from Crino et al. 2013)

levels of plasma CORT compared with a control group of nestlings. Obviously, more studies are required to gain understanding of acoustic causes and fitness consequences of chronic stress as reflected by CORT levels.

8.4.2 Behavioral Disturbance by Man-Made Sound

Several studies on bird breeding densities revealed that birds appear to avoid man-made sounds (see Sect. 8.2.1), which can be seen as spatial deterrence and indirect evidence of disturbance (e.g., Francis et al. 2009; Goodwin and Shriver 2011). Experimental exposure studies have confirmed this at various scales for birds involved in singing activity at their territories (McLaughlin and Kunc 2013) and while foraging at a migratory stopover site (McClure et al. 2013). Furthermore, greater sage-grouse were shown to prefer to display at quiet control sites compared with those in experimentally elevated sound levels (Blickley et al. 2012b), whereas great tits were shown to prefer to breed in quiet compared with noisy nest boxes (Halfwerk et al. 2016c). All of these studies suggest that the acoustic environment plays a critical role in habitat preferences and that man-made sounds reduce the amount of suitable habitat for critical bird behaviors such as breeding and foraging.

If man-made sounds do not deter birds from a particular area, behavioral changes can still have negative consequences for birds that persist in noisy areas. Birds have, for example, been shown to feed less, to stop singing, or to fly off when exposed to experimental sound exposure (Quinn et al. 2006; Halfwerk and Slabbekoorn 2009). In another type of study, house sparrows and great and blue tits (*Cyanistes caeruleus*) as well as tree swallows (*Tachycineta bicolor*) were shown to provide less food to their offspring with elevated noise levels (Schroeder et al. 2012; Naguib et al. 2013). It is currently still unclear whether these changes in parental feeding rates are caused by acoustic avoidance by the parents or driven by changes in parent-offspring communication (Leonard and Horn 2005, 2012; Lucass et al. 2016). Although there are still few studies showing straightforward behavioral changes that would provide clear insight into the causal relationship between man-made sound and detrimental consequences for survival or reproduction, current studies do indicate that the deterrence and interruption effects of such sounds are likely to be widespread and potentially harmful.

8.5 Evolutionary and Ecological Consequences

The omnipresence of noise pollution and the variety and spread of effects from man-made sounds on birds make it likely that there will be evolutionary and ecological consequences that impact many species. Noisy conditions in an urban habitat or along highways may not only drive cultural evolution in species that learn their vocal signals (Luther and Baptista 2010; Slabbekoorn 2013) but may also drive genetic changes (Ghalambor et al. 2007). Providing evidence for contemporary evolutionary changes caused by man-made sounds remains a challenge for the future, but a first step would be to show the effects on individual fitness. Therefore, Sect. 8.5.1 examines studies on reproductive success and survival. Selection on communication signals depends on many factors and is strongly influenced by the ecological setting and the composition of the community of predators, prey, and competitors.

8.5.1 Sound Impact on Individual Fitness

It is important to study birds in different life stages and engaging in different activities to understand whether and how exposure to man-made sounds affects fitness. Habib et al. (2007) found lower pairing success for birds breeding near noisy gas-compressor stations. Reijnen and Foppen (1991) also found lower pairing success near noisy highways, suggesting that noise impairs mate attraction. A study on female domestic canaries (*Serinus canaria*) confirmed such an effect in the laboratory because they usually have a preference for low- over high-frequency songs with so-called "sexy syllables," but this preference disappeared when exposed to experimental traffic noise (des Aunay et al. 2014). Similarly, the susceptibility to cuckoldry was lower for great tit males that sang relatively low-frequency songs, but a signaling advantage of low-frequency relative to high-frequency songs was shown to fade under noisy conditions (Halfwerk et al. 2011a). In another study on male-male interactions in northern cardinals (*Cardinalis cardinalis*), Luther and Magnotti

(2014) also found an impact of ambient noise. Regular low-frequency songs lost the advantage they have over upward shifted songs under quiet conditions when birds were exposed to noisy conditions.

The impact of man-made sound on life-history stages related to egg laying, incubation, and providing offspring care is unfortunately far less clear (also see Sect. 8.4.2). Potvin and MacDougall-Shackleton (2015) exposed breeding zebra finches to experimental noise and found higher rates of embryo mortality as well as lower body weight of hatchlings, presumably caused by changes in female incubation rates. Differences in the number of successfully fledged young between treatment and control groups, however, were not significant. Halfwerk et al. (2011b) found that great tits laid smaller clutches and fledged fewer young in noisier territories, but an experimental follow-up study failed thus far to prove a causal link between ambient-noise levels and reproductive success (Halfwerk et al. 2016c). Similarly, a case study on house sparrows reported lower fledging success for birds breeding close to a noisy generator (Schroeder et al. 2012), but experimental exposure again failed to find a causal effect (Angelier et al. 2015).

Birds also need to stay alive to successfully reproduce, and man-made sounds have been shown to affect various life-history traits that are crucial for survival. Noisy human activities can, for example, reduce the use of signals and cues involved in scanning for predators (Quinn et al. 2006). Noise can also affect foraging efficiency, either directly by masking cues of potential prey or indirectly by forcing birds to spend more time on vigilance behavior that cannot be spent on foraging (Quinn et al. 2006). Finally, man-made sound has been shown to affect birds during their migration to the wintering grounds, with some species avoiding noisy stopover sites and other species showing slower weight gain rates under noisy conditions (Ware et al. 2015).

8.5.2 Sound Impact at Community and Ecosystem Levels

The first studies that clearly showed the potential impact of man-made sounds at the community level came from often remote drilling stations in the oil and gas industry (Bayne et al. 2008). Francis et al. (2009) showed, for example, that some bird species had lower breeding densities in noisy areas compared with quiet areas, whereas other species showed an opposite pattern. They argued that this latter, somewhat counterintuitive positive effect of ambient noise could be explained by reduced egg predation. Outfitting dummy nests with camera traps confirmed that western scrub jays (*Aphelocoma californica*) were less likely to prey on eggs in noisy areas, revealing how some birds can benefit from noise. These results clearly showed how noise pollution could affect species interactions and cause changes at the level of the community.

Species obviously do not operate in isolation and their success depends on a whole suite of other species (see Fig. 8.9), including their predators and prey, as well as their ecological competitors (Slabbekoorn and Halfwerk 2009; Francis and

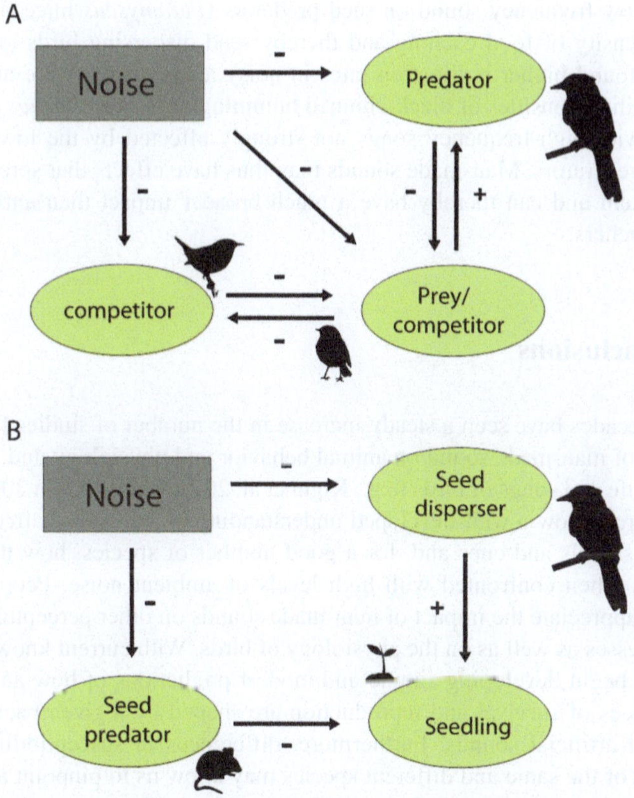

Fig. 8.9 Man-made noise affects community ecology. (**a**) Man-made sound can have direct and indirect effects on species within bird assemblages. Man-made sound can, for example, reduce predator densities and thereby cause an indirect positive effect on prey species. (**b**) Man-made sound may also affect the whole ecosystem through wide-ranging taxa, for example, through a direct, negative impact on seed dispersers (reducing a positive effect on seedlings) or seed predators (reducing a negative effect on seedlings). Based on Slabbekoorn and Halfwerk (2009) and Francis et al. (2009, 2011)

Barber 2013). If, for example, a predator is more negatively impacted by noise than is its prey, the latter species will receive indirect benefits from breeding in noise. Likewise, if two species that compete over food or other important resources such as breeding sites or shelters differ in their susceptibility to noise, the least affected species will gain a net fitness benefit (Halfwerk et al. 2016c). Such indirect benefits could even outweigh the direct costs of breeding in noisy conditions, depending on the nature of the interaction as well as the relative differences of species in their noise susceptibility (Francis et al. 2009; Slabbekoorn and Halfwerk 2009).

The effects of man-made sounds may even go beyond the bird community through an impact on keystone species that provide important ecosystem services (Fig. 8.9). Francis et al. (2012), for example, found altered vegetation stands through lower rates of seed dispersal in noisy areas. The pattern was explained by a lack of impact of low-frequency sound on seed predators (*Peromyscus* mice) but a lower breeding density of food caching and thereby seed dispersing birds (scrub jays). They also found higher pollination rates in noisy areas, which were attributed to higher breeding densities of black-chinned hummingbirds (*Archilochus alexandri*), a species with high-frequency songs not strongly affected by the low-frequency sounds of generators. Man-made sounds may thus have effects that spread through the ecosystem and can thereby have a much broader impact than anticipated by early researchers.

8.6 Conclusions

The past decades have seen a steady increase in the number of studies focusing on the effects of man-made sound on animal behavior and physiology and, in particular, on the life and songs of birds (e.g., Kight et al. 2012; Slabbekoorn 2013; Francis 2015). There is now a well-developed understanding of how noise affects the perception of signals and cues and, for a good number of species, how they change their songs when confronted with high levels of ambient noise. People are also starting to appreciate the impact of man-made sounds on other perceptual and cognitive processes as well as on the physiology of birds. With current knowledge, it is possible to begin developing simple and modest predictions of how an individual bird's chances of survival and reproduction are shaped by a given background of natural and artificial sounds. Furthermore, differences in susceptibility between individuals of the same and different species may allow us to pinpoint a set of key perceptual, physiological, and behavioral traits that determine whether birds can adapt to man-made sounds (see, e.g., Francis 2015 for such an approach).

Many challenges remain, especially when it comes to assessing costs and benefits of song variation and plasticity in fluctuating conditions of ambient noise (Read et al. 2014; Slabbekoorn 2013). Furthermore, the notion that all signaling interactions among animals are inherently multimodal and that pollution can be multimodal as well guarantees a high degree of complexity for future work but also a high potential for future studies to engage in an integrative way across disciplines (Halfwerk and Slabbekoorn 2015). The widespread popularity of birds and increasing concerns about noise pollution, together with the need to continue expanding our understanding of the impact of man-made sounds on birds from behavioral and physiological effects to ecological and evolutionary consequences, suggest a growing and indispensable future for this area of study.

Compliance with Ethics Requirements
Wouter Halfwerk declares that he has no conflict of interest.
Bernard Lohr declares that he has no conflict of interest.
Hans Slabbekoorn declares that he has no conflict of interest.

References

Angelier, F., Meillère, A., Grace, J. K., Trouvé, C., & Brischoux, F. (2015). No evidence for an effect of traffic noise on the development of the corticosterone stress response in an urban exploiter. *General and Comparative Endocrinology, 232*, 43–50.

Arroyo-Solis, A., Castillo, J. M., Figueroa, E., Lopez-Sanchez, J. L., & Slabbekoorn, H. (2013). Experimental evidence for an impact of anthropogenic noise on dawn chorus timing in urban birds. *Journal of Avian Biology, 44*, 288–296.

Barber, J. R., Crooks, K. R., & Fristrup, K. M. (2009). The costs of chronic noise exposure for terrestrial organisms. *Trends in Ecology & Evolution, 25*, 180–189.

Basner, M., Babisch, W., Davis, A., Brink, M., Clark, C., Janssen, S., & Stansfeld, S. (2014). Auditory and non-auditory effects of noise on health. *Lancet, 383*, 1325–1332.

Bayne, E. M., Habib, L., & Boutin, S. (2008). Impacts of chronic anthropogenic noise from energy-sector activity on abundance of songbirds in the boreal forest. *Conservation Biology, 22*, 1186–1193.

Bee, M. A., & Micheyl, C. (2008). The cocktail party problem: What is it? How can it be solved? And why should animal behaviorists study it? *Journal of Comparative Psychology, 122*, 235–251.

Bergman, G. (1982). Die Veränderung der Gesangmelodie der Kohlmeise *Parus major* in Finnland und Schweden (The change of song pattern of the great tit *Parus major* in Finland and Sweden). *Ornis Fennica, 57*, 97–111.

Bermúdez-Cuamatzin, E., Ríos-Chelén, A. A., Gil, D., & Garcia, C. M. (2009). Strategies of song adaptation to urban noise in the house finch: Syllable pitch plasticity or differential syllable use? *Behaviour, 146*, 1269–1286.

Bermudez-Cuamatzin, E., Rios-Chelen, A. A., Gil, D., & Garcia, C. M. (2010). Experimental evidence for real-time song frequency shift in response to urban noise in a passerine bird. *Biology Letters, 7*, 36–38.

Blickley, J. L., Word, K. R., Krakauer, A. H., Phillips, J. L., Sells, S. N., Taff, C. C., Wingfield, J. C., & Patricelli, G. L. (2012a). Experimental chronic noise is related to elevated fecal corticosteroid metabolites in lekking male greater sage-grouse (*Centrocercus urophasianus*). *PLoS ONE*, 7(11), e50462. https://doi.org/10.1371/journal.pone.0050462.

Blickley, J. L., Blackwood, D., & Patricelli, G. L. (2012b). Experimental evidence for the effects of chronic anthropogenic noise on abundance of greater sage-grouse at leks. *Conservation Biology, 26*, 461–471.

Bormpoudakis, D., Sueur, J., & Pantis, J. D. (2013). Spatial heterogeneity of ambient sound at the habitat type level: Ecological implications and applications. *Landscape Ecology, 28*, 495–506.

Bregman, A. S., & Campbell, J. (1971). Primary auditory stream segregation and perception of order in rapid sequences of tones. *Journal of Experimental Psychology, 89*, 244–249.

Brenowitz, E. A. (1982). The active space of red-winged blackbird song. *Journal of Comparative Physiology A: Neuroethology, Sensory, Neural, and Behavioral Physiology, 147*, 511–522.

Brumm, H. (2004). The impact of environmental noise on song amplitude in a territorial bird. *Journal of Animal Ecology, 73*, 434–440.

Brumm, H., & Todt, D. (2002). Noise-dependent song amplitude regulation in a territorial songbird. *Animal Behaviour, 63*, 891–897.

Brumm, H., & Slabbekoorn, H. (2005). Acoustic communication in noise. In P. J. B. Slater, C. T. Snowdon, T. J. Roper, H. J. Brockmann, & M. Naguib (Eds.), *Advances in the Study of Behavior* (pp. 151–209). San Diego, CA: Academic Press.

Cardoso, G. C., & Price, T. D. (2010). Community convergence in bird song. *Evolutionary Ecology, 24*, 447–461.

Cardoso, G. C., & Atwell, J. W. (2011). Directional cultural change by modification and replacement of memes. *Evolution, 65*, 295–300.

Catchpole, C. K., & Slater, P. J. B. (2008). *Bird Song: Biological Themes and Variations*. Cambridge, UK: Cambridge University Press.

Chan, A. A. Y.-H., & Blumstein, D. T. (2011). Attention, noise, and implications for wildlife conservation and management. *Applied Animal Behaviour Science, 131*, 1–7.

Chan, A. A. Y.-H., Giraldo-Perez, P., Smith, S., & Blumstein, D. T. (2010). Anthropogenic noise affects risk assessment and attention: The distracted prey hypothesis. *Biology Letters, 6*, 458–461.

Cody, M. L., & Brown, J. H. (1969). Song asynchrony in neighbouring bird species. *Nature, 222*(5195), 778–780.

Collins, S. (2004). Vocal fighting and flirting: The functions of birdsong. In P. Marler & H. Slabbekoorn (Eds.), *Nature's Music: The Science of Birdsong* (pp. 39–79). San Diego, CA: Elsevier Academic Press.

Crino, O. L., Johnson, E. E., Blickley, J. L., Patricelli, G. L., & Breuner, C. W. (2013). Effects of experimentally elevated traffic noise on nestling white-crowned sparrow stress physiology, immune function and life history. *Journal of Experimental Biology, 216*, 2055–2062.

Cuthill, I. C., & Macdonald, W. A. (1990). Experimental manipulation of the dawn and dusk chorus in the blackbird *Turdus merula*. *Behavioral Ecology and Sociobiology, 26*(3), 209–216.

des Aunay, G. H., Slabbekoorn, H., Nagle, L., Passas, F., Nicolas, P., & Draganoiu, T. I. (2014). Urban noise undermines female sexual preferences for low-frequency songs in domestic canaries. *Animal Behaviour, 87*, 67–75.

Dent, M. L., Larsen, O. N., & Dooling, R. J. (1997). Free-field binaural unmasking in budgerigars (*Melopsittacus undulatus*). *Behavioral Neuroscience, 111*, 590–598.

Dent, M. L., McClaine, E. M., Best, V., Ozmeral, E., Narayan, R., Gallun, F. J., Sen, K., & Shinn-Cunningham, B. G. (2009). Spatial unmasking of birdsong in zebra finches (*Taeniopygia guttata*) and budgerigars (*Melopsittacus undulatus*). *Journal of Comparative Psychology, 123*, 357–367.

Dingle, C., Halfwerk, W., & Slabbekoorn, H. (2008). Habitat-dependent song divergence at subspecies level in the grey-breasted wood-wren. *Journal of Evolutionary Biology, 21*(4), 1079–1089.

Dooling, R. J., & Okanoya, K. (1995). The method of constant stimuli in testing auditory sensitivity in small birds. In G. M. Klump, R. J. Dooling, R. R. Fay, & W. C. Stebbins (Eds.), *Methods in Comparative Psychoacoustics* (pp. 161–169). Basel: Birkhäuser.

Dooling, R. J., & Blumenrath, S. H. (2013). Avian sound perception in noise. In H. Brumm (Ed.), *Animal Communication and Noise* (pp. 229–250). Berlin: Springer-Verlag.

Dooling, R. J., Lohr, B., & Dent, M. L. (2000). Hearing in birds and reptiles. In R. J. Dooling, R. R. Fay, & A. N. Popper (Eds.), *Comparative Hearing in Birds and Reptiles* (pp. 308–359). New York: Springer-Verlag.

Dubois, A., & Martens, J. (1984). A case of possible vocal convergence between frogs and a bird in Himalayan torrents. *Journal für Ornithologie, 125*, 455–463.

Elemans, C., Rasmussen, J. H., Herbst, C. T., Düring, D. N., Zollinger, S. A., Brumm, H., Srivastava, K., Svane, N., Ding, M., & Larsen, O. N. (2015). Universal mechanisms of sound production and control in birds and mammals. *Nature Communications, 6*, 8978.

Fahrig, L., & Rytwinski, T. (2009). Effects of roads on animal abundance: An empirical review and synthesis. *Ecology and Society, 14*(1), 21.

Fay, R. R. (1988). Comparative psychoacoustics. *Hearing Research, 34*(3), 295–305.

Feng, A. S., Narins, P. M., Xu, C.-H., Lin, W.-Y., Yu, Z.-L., Qiu, Q., Xu, Z.-M., & Shen, J.-X. (2006). Ultrasonic communication in frogs. *Nature, 440*, 333–336.

Francis, C. D. (2015). Vocal traits and diet explain avian sensitivities to anthropogenic noise. *Global Change Biology, 21*, 1809–1820.

Francis, C. D., & Barber, J. R. (2013). A framework for understanding noise impacts on wildlife: An urgent conservation priority. *Frontiers in Ecology and the Environment, 11*, 305–313.

Francis, C. D., Ortega, C. P., & Cruz, A. (2009). Cumulative consequences of noise pollution: Noise changes avian communities and species interactions. *Current Biology, 19*, 1415–1419.

Francis, C. D., Ortega, C. P., & Cruz, A. (2011). Noise pollution filters bird communities based on vocal frequency. *PLoS ONE, 6*, e27052.

Francis, C. D., Kleist, N. J., Ortega, C. P., & Cruz, A. (2012). Noise pollution alters ecological services: Enhanced pollination and disrupted seed dispersal. *Proceedings of the Royal Society B: Biological Sciences*, 279, 2727–2735.

Fuller, R. A., Warren, P. H., & Gaston, K. J. (2007). Daytime noise predicts nocturnal singing in urban robins. *Biology Letters*, 3, 368–370.

Ghalambor, C. K., McKay, J. K., Carroll, S. P., & Reznick, D. N. (2007). Adaptive versus non-adaptive phenotypic plasticity and the potential for contemporary adaptation in new environments. *Functional Ecology*, 21, 394–407.

Gil, D., Honarmand, M., Pascual, J., Pérez-Mena, E., & Garcia, C. M. (2014). Birds living near airports advance their dawn chorus and reduce overlap with aircraft noise. *Behavioral Ecology*, 26, 435–443.

Gleich, O., & Manley, G. A. (2000). The hearing organ of birds and crocodilian. In R. J. Dooling, R. R. Fay, & A. N. Popper (Eds.), *Comparative Hearing in Birds and Reptiles* (pp. 70–138). New York: Springer-Verlag.

Goodwin, S. E., & Shriver, W. G. (2011). Effects of traffic noise on occupancy patterns of forest birds. *Conservation Biology*, 25, 406–411.

Habib, L., Bayne, E. M., & Boutin, S. (2007). Chronic industrial noise affects pairing success and age structure of ovenbirds *Seiurus aurocapilla*. *Journal of Applied Ecology*, 44, 176–184.

Halfwerk, W., & Slabbekoorn, H. (2009). A behavioural mechanism explaining noise-dependent frequency use in urban birdsong. *Animal Behaviour*, 78, 1301–1307.

Halfwerk, W., & Slabbekoorn, H. (2015). Pollution going multimodal: The complex impact of the human-altered sensory environment on animal perception and performance. *Biology Letters*, 11, e20141051.

Halfwerk, W., Bot, S., Buikx, J., van der Velde, M., Komdeur, J., ten Cate, C., & Slabbekoorn, H. (2011a). Low songs lose potency in urban noise conditions. *Proceedings of the National Academy of Sciences of the United States of America*, 108, 14549–14554.

Halfwerk, W., Holleman, L. J. M., Lessells, C. M., & Slabbekoorn, H. (2011b). Negative impact of traffic noise on avian reproductive success. *Journal of Applied Ecology*, 48, 210–219.

Halfwerk, W., Bot, S., & Slabbekoorn, H. (2012). Male great tit song perch selection in response to noise-dependent female feedback. *Functional Ecology*, 26, 1339–1347.

Halfwerk, W., Dingle, C., Brinkhuizen, D. M., Poelstra, J. W., Komdeur, J., & Slabbekoorn, H. (2016a). Sharp acoustic boundaries across an altitudinal avian hybrid zone despite asymmetric introgression. *Journal of Evolutionary Biology*, 29, 1356–1367.

Halfwerk, W., Lea, A. M., Guerra, M., Page, R. A., & Ryan, M. J. (2016b). Vocal responses to noise reveal the presence of the Lombard effect in a frog. *Behavioral Ecology*, 27, 669–676.

Halfwerk, W., Both, C., & Slabbekoorn, H. (2016c). Long-term nestbox noise experiments reveal an impact on nest-site selection but not on reproduction. *Behavioral Ecology*, 27, 1592–1600.

Hall, J. W., Haggard, M. P., & Fernandes, M. A. (1984). Detection in noise by spectro-temporal pattern analysis. *The Journal of the Acoustical Society of America*, 76, 50–56.

Hall, M. L. (2009). A review of vocal duetting in birds. *Advances in the Study of Behavior*, 40, 67–121.

Hamao, S., Watanabe, M., & Mori, Y. (2011). Urban noise and male density affect songs in the great tit *Parus major*. *Ethology Ecology & Evolution*, 23, 111–119.

Hanna, D., Blouin-Demers, G., Wilson, D. R., & Mennill, D. J. (2011). Anthropogenic noise affects song structure in red-winged blackbirds (*Agelaius phoeniceus*). *Journal of Experimental Biology*, 214, 3549–3556.

Hasselquist, D., Bensch, S., & von Schantz, T. (1996). Correlation between male song repertoire, extra-pair paternity and offspring survival in the great reed warbler. *Nature*, 381, 229–232.

Hilton, S. C., & Krebs, J. K. (1990). Spatial memory of four species of *Parus*: Performance in an open-field analogue of a radial maze. *The Quarterly Journal of Experimental Psychology*, 42, 345–368.

Hu, Y., & Cardoso, G. C. (2009). Which birds adjust the frequency of vocalizations in urban noise? *Animal Behaviour*, 79, 863–867.

Hulse, S. H., MacDougall-Shackleton, S. A., & Wisniewski, A. B. (1997). Auditory scene analysis by songbirds: Stream segregation of birdsong by European starlings (*Sturnus vulgaris*). *Journal of Comparative Psychology,* 111, 3–13.

Jensen, K. K. (2007). Comodulation detection differences in the hooded crow (*Corvus corone cornix*), with direct comparison to human subjects. *The Journal of the Acoustical Society of America,* 121, 1783–1789.

Jouventin, P., Aubin, T., & Lengagne, T. (1999). Finding a parent in a king penguin colony: The acoustic system of individual recognition. *Animal Behaviour,* 57, 1175–1183.

Kight, C. R., & Swaddle, J. P. (2011). How and why environmental noise impacts animals: An integrative, mechanistic review. *Ecology Letters,* 14(10), 1052–1061.

Kight, C. R., Saha, M. S., & Swaddle, J. P. (2012). Anthropogenic noise is associated with reductions in the productivity of breeding Eastern Bluebirds (*Sialia sialis*). *Ecological Applications,* 22(7), 1989–1996.

Klump, G. M. (1996). Bird communication in the noisy world. In D. E. Kroodsma & E. H. Miller (Eds.), *Ecology and Evolution of Acoustic Communication in Birds* (pp. 321–338). Ithaca, NY: Cornell University Press.

Klump, G. M., & Langemann, U. (1995). Comodulation masking release in a songbird. *Hearing Research,* 87, 157–164.

Knudsen, D. P., & Gentner, T. Q. (2010). Mechanisms of song perception in oscine birds. *Brain and Language,* 115, 59–68.

Knudsen, E. I., & Konishi, M. (1979). Mechanisms of sound localization in the barn owl (*Tyto alba*). *Journal of Comparative Physiology A: Neuroethology, Sensory, Neural, and Behavioral Physiology,* 133, 13–21.

Konishi, M. (1973). How the owl tracks its prey: Experiments with trained barn owls reveal how their acute sense of hearing enables them to catch prey in the dark. *American Scientist,* 61(4), 414–424.

Kroodsma, D. E. (2004). Diversity and plasticity of bird song. In P. Marler & H. Slabbekoorn (Eds.), *Nature's Music: The Science of Birdsong* (pp. 108–130). San Diego, CA: Elsevier Academic Press.

Kroodsma, D. E., & Byers, B. E. (1991). The function(s) of bird song. *American Zoologist,* 31, 318–328.

Kroodsma, D. E., & Miller, E. H. (1996). *Ecology and Evolution of Acoustic Communication in Birds*. Ithaca, NY: Cornell University Press.

Lachlan, R. F., Verzijden, M. N., Bernard, C. S., Jonker, P.-P., Koese, B., Jaarsma, S., Spoor, W., Slater, P. J., & ten Cate, C. (2013). The progressive loss of syntactical structure in bird song along an island colonization chain. *Current Biology,* 23, 1896–1901.

Laiolo, P., & Tella, J. L. (2005). Habitat fragmentation affects culture transmission: Patterns of song matching in Dupont's lark. *Journal of Applied Ecology,* 42, 1183–1193.

Langemann, U., & Klump, G. M. (2007). Detecting modulated signals in modulated noise: (1) Behavioural auditory thresholds in a songbird. *European Journal of Neuroscience,* 26, 1969–1978.

Langemann, U., Gauger, B., & Klump, G. M. (1998). Auditory sensitivity in the great tit: Perception of signals in the presence and absence of noise. *Animal Behaviour,* 56, 763–769.

Lazerte, S. E., Slabbekoorn, H., & Otter, K. A. (2016). Learning to cope: Vocal adjustment to urban noise is correlated with prior experience in black-capped chickadees. *Proceedings of the Royal Society B: Biological Sciences,* 283(1833), 20161058.

Leonard, M. L., & Horn, A. G. (2005). Ambient noise and the design of begging signals. *Proceedings of the Royal Society B: Biological Sciences,* 272, 651–656.

Leonard, M. L., & Horn, A. G. (2012). Ambient noise increases missed detections in nestling birds. *Biology Letters,* 8, 530–532.

Lohr, B., & Dooling, R. J. (1998). Detection of changes in timbre and harmonicity in complex sounds by zebra finches (*Taeniopygia guttata*) and budgerigars (*Melopsittacus undulatus*). *Journal of Comparative Psychology,* 112(1), 36–47.

Lohr, B., Wright, T. F., & Dooling, R. J. (2003). Detection and discrimination of natural calls in masking noise by birds: Estimating the active space of a signal. *Animal Behaviour, 65*, 763–777.

Lucass, C., Eens, M., & Müller, W. (2016). When ambient noise impairs parent-offspring communication. *Environmental Pollution, 212*, 592–597.

Luther, D., & Baptista, L. (2010). Urban noise and the cultural evolution of bird songs. *Proceedings of the Royal Society B: Biological Sciences, 277*, 469–473.

Luther, D., & Magnotti, J. (2014). Can animals detect differences in vocalizations adjusted for anthropogenic noise? *Animal Behaviour, 92*, 111–116.

Mace, R. (1987). The dawn chorus in the great tit *Parus major* is directly related to female fertility. *Nature, 330*, 745–746.

Marler, P. (1970). Birdsong and speech development: Could there be parallels? There may be basic rules governing vocal learning to which many species conform, including man. *American Scientist, 58*, 669–673.

Marler, P., & Slabbekoorn, H. (2004). *Nature's Music: The Science of Birdsong*. San Diego, CA: Elsevier Academic Press.

Marten, K., & Marler, P. (1977). Sound transmission and its significance for animal vocalization. *Behavioral Ecology and Sociobiology, 2*, 271–290.

Mason, J. T., McClure, C. J. W., & Barber, J. R. (2016). Anthropogenic noise impairs owl hunting behavior. *Biological Conservation, 199*, 29–32.

Mathevon, N., Aubin, T., & Dabelsteen, T. (1996). Song degradation during propagation: Importance of song post for the wren *Troglodytes troglodytes*. *Ethology, 102*, 397–412.

McClure, C. J., Ware, H. E., Carlisle, J., Kaltenecker, G., & Barber, J. R. (2013). An experimental investigation into the effects of traffic noise on distributions of birds: Avoiding the phantom road. *Proceedings of the Royal Society B:Biological Sciences, 280*, 20132290.

McGregor, P. K., Krebs, J. R., & Perrins, C. M. (1981). Song repertoires and lifetime reproductive success in the great tit (*Parus major*). *American Naturalist, 118*, 149–159.

McLaughlin, K. E., & Kunc, H. P. (2013). Experimentally increased noise levels change spatial and singing behaviour. *Biology Letters, 9*, 20120771. https://doi.org/10.1098/rsbl.2012.0771.

Meillère, A., Brischoux, F., & Angelier, F. (2015). Impact of chronic noise exposure on antipredator behavior: An experiment in breeding house sparrows. *Behavioral Ecology, 26*, 569–577.

Miller, G. A. (1951). *Language and Communication*. New York: McGraw-Hill Book Company.

Moiron, M., González-Lagos, C., Slabbekoorn, H., & Sol, D. (2015). Singing in the city: High song frequencies are no guarantee for urban success in birds. *Behavioral Ecology, 26*, 843–850.

Moore, B. C., Glasberg, B. R., & Baer, T. (1997). A model for the prediction of thresholds, loudness, and partial loudness. *Journal of the Audio Engineering Society, 45*, 224–240.

Naguib, M., van Oers, K., Braakhuis, A., Griffioen, M., de Goede, P., & Waas, J. R. (2013). Noise annoys: Effects of noise on breeding great tits depend on personality but not on noise characteristics. *Animal Behaviour, 85*, 949–956.

Nemeth, E., & Brumm, H. (2010). Birds and anthropogenic noise: Are urban songs adaptive? *American Naturalist, 176*, 465–475.

Noirot, I. C., Brittan-Powell, E. F., & Dooling, R. J. (2011). Masked auditory thresholds in three species of birds, as measured by the auditory brainstem response (L). *The Journal of the Acoustical Society of America, 129*, 3445–3448.

Odom, K. J., Hall, M. L., Riebel, K., Omland, K. E., & Langmore, N. E. (2014). Female song is widespread and ancestral in songbirds. *Nature Communications, 5*, e3379.

Okanoya, K., & Dooling, R. J. (1987). Hearing in passerine and psittacine birds: A comparative study of absolute and masked auditory thresholds. *Journal of Comparative Psychology, 101*, 7–15.

Owens, J. L., Stec, C. L., & O'Hatnick, A. (2012). The effects of extended exposure to traffic noise on parid social and risk-taking behavior. *Behavioural Processes, 91*, 61–69.

Partan, S. R., & Marler, P. (2005). Issues in the classification of multimodal communication signals. *American Naturalist, 166*, 231–245.

Patricelli, G. L., & Blickley, J. L. (2006). Avian communication in urban noise: Causes and consequences of vocal adjustment. *Auk*, 123, 639–649.
Payne, R. S. (1971). Acoustic location of prey by barn owls (*Tyto alba*). *Journal of Experimental Biology*, 54, 535–573.
Penna, M., Pottstock, H., & Velasquez, N. (2005). Effect of natural and synthetic noise on evoked vocal responses in a frog of the temperate austral forest. *Animal Behaviour*, 70, 639–651.
Planque, R., & Slabbekoorn, H. (2008). Spectral overlap in songs and temporal avoidance in a Peruvian bird assemblage. *Ethology*, 114, 262–271.
Podos, J., Huber, S. K., & Taft, B. (2004). Bird song: The interface of evolution and mechanism. *Annual Review of Ecology, Evolution, and Systematics*, 35, 55–87.
Pohl, N. U., Slabbekoorn, H., Klump, G. M., & Langemann, U. (2009). Effects of signal features and environmental noise on signal detection in the great tit, *Parus major*. *Animal Behaviour*, 78, 1293–1300.
Pohl, N. U., Leadbeater, E., Slabbekoorn, H., Klump, G. M., & Langemann, U. (2012). Great tits in urban noise benefit from high frequencies in song detection and discrimination. *Animal Behaviour*, 83, 711–721.
Potvin, D. A., & MacDougall-Shackleton, S. A. (2015). Traffic noise affects embryo mortality and nestling growth rates in captive zebra finches. *Journal of Experimental Zoology Part A: Ecological Genetics and Physiology*, 323, 722–730.
Pytte, C. L., Rusch, K. M., & Ficken, M. S. (2003). Regulation of vocal amplitude by the blue-throated hummingbird, *Lampornis clemenciae*. *Animal Behaviour*, 66(4), 703–710.
Quinn, J. L., Whittingham, M. J., Butler, S. J., & Cresswell, W. (2006). Noise, predation risk compensation and vigilance in the chaffinch *Fringilla coelebs*. *Journal of Avian Biology*, 37, 601–608.
Rabinowitz, P. M. (2000). Noise-induced hearing loss. *American Family Physician*, 61, 2759–2760.
Read, J., Jones, G., & Radford, A. N. (2014). Fitness costs as well as benefits are important when considering responses to anthropogenic noise. *Behavioral Ecology*, 25, 4–7. https://doi.org/10.1093/beheco/art102.
Reijnen, R., & Foppen, R. (1991). Effect of road traffic on the breeding site tenacity of male willow warblers (*Phylloscopus trochilus*). *Journal für Ornithologie*, 132, 291–295.
Reijnen, R., & Foppen, R. (1995). The effects of car traffic on breeding bird populations in woodland. IV. Influence of population size on the reduction of density close to a highway. *Journal of Applied Ecology*, 32, 481–491.
Reijnen, R., & Foppen, R. (2006). Impact of road traffic on breeding bird populations. In J. Davenport & J. L. Davenport (Eds.), *The Ecology of Transportation: Managing Mobility for the Environment* (pp. 255–274). Heidelberg: Springer-Verlag.
Rheindt, F. E. (2003). The impact of roads on birds: Does song frequency play a role in determining susceptibility to noise pollution? *Journal für Ornithologie*, 144, 295–306.
Riebel, K., Hall, M. L., & Langmore, N. E. (2005). Female songbirds still struggling to be heard. *Trends in Ecology & Evolution*, 20, 419–420.
Ripmeester, E. A. P., Kok, J. S., van Rijssel, J. C., & Slabbekoorn, H. (2010). Habitat-related birdsong divergence: A multi-level study on the influence of territory density and ambient noise in European blackbirds. *Behavioral Ecology and Sociobiology*, 64, 409–418.
Rivera-Gutierrez, H. F., Matthysen, E., Adriaensen, F., & Slabbekoorn, H. (2010). Repertoire sharing and song similarity between great tit males decline with distance between forest fragments. *Ethology*, 116, 951–960.
Rivolta, M. N., & Holley, M. C. (2008). Gene arrays, cell lines, stem cells, and sensory regeneration in mammalian ears. In R. J. Salvi, A. N. Popper, & R. R. Fay (Eds.), *Hair Cell Regeneration, Repair, and Protection* (pp. 257–307). New York: Springer-Verlag.
Ronacher, B., & Hoffmann, C. (2003). Influence of amplitude modulated noise on the recognition of communication signals in the grasshopper *Chorthippus biguttulus*. *Journal of Comparative Physiology A: Neuroethology, Sensory, Neural, and Behavioral Physiology*, 189, 419–425.

Runkle, L. S., Wells, K. D., Robb, C. C., & Lance, S. L. (1994). Individual, nightly, and seasonal variation in calling behavior of the gray tree frog, *Hyla versicolor*: Implications for energy expenditure. *Behavioral Ecology,* 5, 318–325.

Ryals, B. M., Dooling, R. J., Westbrook, E., Dent, M. L., MacKenzie, A., & Larsen, O. N. (1999). Avian species differences in susceptibility to noise exposure. *Hearing Research,* 131, 71–88.

Sapolsky, R. M., Romero, L. M., & Munck, A. U. (2000). How do glucocorticoids influence stress responses? Integrating permissive, suppressive, stimulatory, and preparative actions. *Endocrine Reviews,* 21, 55–89.

Scharf, B. (1970). Critical bands. *Foundations of Modern Auditory Theory,* 1, 157–202.

Schroeder, J., Nakagawa, S., Cleasby, I. R., & Burke, T. (2012). Passerine birds breeding under chronic noise experience reduced fitness. *PLoS ONE,* 7, e39200. https://doi.org/10.1371/journal.pone.0039200.

Senzaki, M., Yamaura, Y, Francis, C. D., & Nakamura, F. (2016). Traffic noise reduces foraging efficiency in wild owls. *Scientific Reports,* 6, 30602.

Sheriff, M. J., Dantzer, B., Delehanty, B., Palme, R., & Boonstra, R. (2011). Measuring stress in wildlife: Techniques for quantifying glucocorticoids. *Oecologia,* 166, 869–887.

Simpson, S. D., Purser, J., & Radford, A. N. (2014). Anthropogenic noise compromises antipredator behaviour in European eels. *Global Change Biology,* 21, 586–593.

Simpson, S. D., Radford, A. N., Nedelec, S. L., Ferrari, M. C., Chivers, D. P., McCormick, M. I., & Meekan, M. G. (2016). Anthropogenic noise increases fish mortality by predation. *Nature Communications,* 7, e10544.

Skiba, R. (2000). Possible rain call selection in the chaffinch (*Fringilla coelebs*) by noise intensity—An investigation of a hypothesis. *Journal für Ornithologie,* 141, 160–167.

Slabbekoorn, H. (2004). Habitat-dependent ambient noise: Consistent spectral profiles in two African forest types. *The Journal of the Acoustical Society of America,* 116, 3727–3733.

Slabbekoorn, H. (2013). Songs of the city: Noise-dependent spectral plasticity in the acoustic phenotype of urban birds. *Animal Behaviour,* 85, 1089–1099.

Slabbekoorn, H. W., & Smith, T. B. (2002a). Bird song, ecology and speciation. *Philosophical Transactions of the Royal Society B: Biological Sciences,* 357, 493–503.

Slabbekoorn, H., & Smith, T. B. (2002b). Habitat-dependent song divergence in the little greenbul: an analysis of environmental selection pressures on acoustic signals. *Evolution, 56*(9), 1849–1858.

Slabbekoorn, H., & Peet, M. (2003). Ecology: Birds sing at a higher pitch in urban noise. *Nature,* 424, 267.

Slabbekoorn, H., & den Boer-Visser, A. (2006). Cities change the songs of birds. *Current Biology,* 16, 2326–2331.

Slabbekoorn, H., & Ripmeester, E. A. P. (2008). Birdsong and anthropogenic noise: Implications and applications for conservation. *Molecular Ecology,* 17, 72–83.

Slabbekoorn, H., & Halfwerk, W. (2009). Behavioural ecology: Noise annoys at community level. *Current Biology,* 19, R693-R695.

Slabbekoorn, H., Yeh, P., & Hunt, K. (2007). Sound transmission and song divergence: A comparison of urban and forest acoustics. *Condor,* 109, 67–78.

Slabbekoorn, H., Yang, X. J., & Halfwerk, W. (2012). Birds and anthropogenic noise: Singing higher may matter (A comment on Nemeth & Brumm, "Birds and anthropogenic noise: Are urban songs adaptive?"). *American Naturalist,* 180, 142–145.

Smith, T. B., Saatchi, S., Graham, C., Slabbekoorn, H., & Spicer, G. (2005). Putting process on the map: Why ecotones are important for preserving biodiversity. In A. Purvis, J. L. Gittleman, & T. Brooks (Eds.), *Phylogeny and Conservation* (pp. 166–197). Cambridge, UK: Cambridge University Press.

Stanley, C. Q., Walter, M. H., Venkatraman, M. X., & Wilkinson, G. S. (2016). Insect noise avoidance in the dawn chorus of Neotropical birds. *Animal Behaviour,* 112, 255–265.

Strasser, E. H., & Heath, J. A. (2013). Reproductive failure of a human-tolerant species, the American kestrel, is associated with stress and human disturbance. *Journal of Applied Ecology, 50*, 912–919.

Sueur, J., & Sanborn, A. F. (2003). Ambient temperature and sound power of cicada calling songs (Hemiptera: Cicadidae: Tibicina). *Physiological Entomology, 28*, 340–343.

Suzuki, T. N., Wheatcroft, D., & Griesser, M. (2016). Experimental evidence for compositional syntax in bird calls. *Nature Communications, 7*, e10986.

Swaddle, J. P., Francis, C. D., Barber, J. R., Cooper, C. B., Kyba, C. C., Dominoni, D. M., Shannon, G., Aschehoug, E., Goodwin, S. E., & Kawahara, A. Y. (2015). A framework to assess evolutionary responses to anthropogenic light and sound. *Trends in Ecology & Evolution, 30*, 550–560.

Tempel, D. J., & Gutiérrez, R. (2003). Fecal corticosterone levels in California spotted owls exposed to low-intensity chainsaw sound. *Wildlife Society Bulletin*, 31, 698–702.

Tempel, D. J., & Gutiérrez, R. (2004). Factors related to fecal corticosterone levels in California spotted owls: Implications for assessing chronic stress. *Conservation Biology, 18*, 538–547.

Templeton, C. N., & Greene, E. (2007). Nuthatches eavesdrop on variations in heterospecific chickadee mobbing alarm calls. *Proceedings of the National Academy of Sciences of the United States of America, 104*, 5479–5482.

Templeton, C. N., Zollinger, S. A., & Brumm, H. (2016). Traffic noise drowns out great tit alarm calls. *Current Biology, 26*, R1167-R1176.

van der Zande, A. N., ter Keurs, W. J., & van der Weijden, W. J. (1980). The impact of roads on the densities of four bird species in an open field habitat—Evidence of a long-distance effect. *Biological Conservation, 18*, 299–321.

Vélez, A., Schwartz, J. J., & Bee, M. A. (2013). Anuran acoustic signal perception in noisy environments. In H. Brumm (Ed.), *Animal Communication and Noise* (pp. 133–185). Berlin Heidelberg: Springer-Verlag.

Verzijden, M. N., Ripmeester, E. A. P., Ohms, V. R., Snelderwaard, P., & Slabbekoorn, H. (2010). Immediate spectral flexibility in singing chiffchaffs during experimental exposure to highway noise. *Journal of Experimental Biology, 213*, 2575–2581.

Ware, H. E., McClure, C. J., Carlisle, J. D., & Barber, J. R. (2015). A phantom road experiment reveals traffic noise is an invisible source of habitat degradation. *Proceedings of the National Academy of Sciences of the United States of America*, 112(39), 12105–12109.

Warren, P. S., Katti, M., Ermann, M., & Brazel, A. (2006). Urban bioacoustics: It's not just noise. *Animal Behaviour, 71*, 491–502.

Wiley, R. H., & Richards, D. G. (1978). Physical constraints on acoustic communication in atmosphere: Implications for evolution of animal vocalizations. *Behavioral Ecology and Sociobiology, 3*, 69–94.

Wisniewski, A. B., & Hulse, S. H. (1997). Auditory scene analysis in European Starlings (*Sturnus vulgaris*): Discrimination of song segments, their segregation from multiple and reversed conspecific songs, and evidence for conspecific song categorization. *Journal of Comparative Psychology, 111*, 337–350.

Wright, T. F., Cortopassi, K. A., Bradbury, J. W., & Dooling, R. J. (2003). Hearing and vocalizations in the orange-fronted conure (*Aratinga canicularis*). *Journal of Comparative Psychology, 117*, 87–95.

Yang, X.-J., & Slabbekoorn, H. (2014). Timing vocal behavior: Lack of temporal overlap avoidance to fluctuating noise levels in singing Eurasian wrens. *Behavioural Processes, 108*, 131–137.

Yang, X.-J., Ma, X.-R., & Slabbekoorn, H. (2014). Timing vocal behaviour: Experimental evidence for song overlap avoidance in Eurasian wrens. *Behavioural Processes, 103*, 84–90.

Chapter 9
Effects of Man-Made Sound on Terrestrial Mammals

Hans Slabbekoorn, JoAnn McGee, and Edward J. Walsh

Abstract Terrestrial mammals are found in all types of natural habitat, and they are also maintained in large numbers in captivity. Much of what is known about the anatomy and physiology of the peripheral auditory system has been learned by studying a variety of laboratory mammals and a smaller collection of exotic and domesticated species. The influence of noise exposure ranges from overt trauma to cochlear structures to nonauditory physiological effects, including outcomes associated with development and behavior. Although most man-made sounds are insufficiently intense or persistent to cause overt trauma to free-ranging terrestrial mammals, recent studies have shown that noise exposures producing reversible hearing loss can still permanently damage synapses between auditory sensory cells and primary auditory nerve fibers and thereby affect hearing function. Harmful effects of noise exposure on nonauditory functions have also been reported, and work on domesticated animals adds further evidence that exposure to noise can induce stress with effects on physiology and behavior. Studies on free-ranging animals have shown that animals are often deterred from busy roads, industrial areas, or noisy recreational activities and that foraging efficiency declines for at least some herbivore species. The wide-ranging diversity of auditory thresholds and spectral ranges of sound detected by terrestrial mammals adds a dimension of complexity in the effort to understand the impact of man-made noise on animals.

Keywords Ear anatomy · Farm and zoo animals · Inner ear trauma · Mammalian hearing · Masking and counterstrategies · Nonauditory effects · Road disturbance · Vocal plasticity

---

H. Slabbekoorn (✉)
Faculty of Science, Institute of Biology Leiden (IBL), Leiden University, Leiden, The Netherlands
e-mail: H.W.Slabbekoorn@biology.leidenuniv.nl

J. McGee · E. J. Walsh
Developmental Auditory Physiology Laboratory, Boys Town National Research Hospital, Omaha, NE, USA
e-mail: JoAnn.McGee@boystown.org; Edward.Walsh@boystown.org

H. Slabbekoorn et al. (eds.), *Effects of Anthropogenic Noise on Animals*, Springer Handbook of Auditory Research 66,
https://doi.org/10.1007/978-1-4939-8574-6_9

9.1 Introduction

Mammals include roughly 5,500 species (Wilson and Reeder 2005) that successfully navigate every type of terrestrial habitat on the planet. They are predominantly nocturnal or crepuscular, with about 70% of all species having a bias toward nighttime activity or at least a strong preference for the dim light conditions of dawn and dusk. Night vision, smell, and hearing are of key importance for survival and reproduction in many terrestrial mammals. There are also mammal taxa that returned to a more or less aquatic life, and their acoustic world is treated elsewhere (see Erbe, Dunlop, and Dolman, Chap. 10). One significant consequence of the ecological spread of terrestrial mammals is that they are exposed to widely varying natural soundscapes. Mammal habitats range from humid and warm tropical rain forests, exhibiting a complex mix of sounds with great spectral diversity, to the cold and barren, wind-swept and relatively impoverished but often noisy acoustic environment of the arctic tundra (Weir et al. 2012; Singh et al. 2015). The acoustic challenges created by noisy soundscapes can carry significant implications for survival and reproductive success in all habitats.

9.1.1 Ambient Noise in Mammalian Habitats

Sound sources can be classified into abiotic, biotic, and man-made (Brumm and Slabbekoorn 2005), also referred to as the geophony, biophony, and anthrophony, respectively (Pijanowski et al. 2011). All of these environmental sounds can influence the detection, discrimination, and identification of acoustic signals that convey critical information related to intra- and interspecies communication in the context of, for example, mate attraction, territoriality, and predator-prey or progeny-parental interactions. Common examples of natural sound sources experienced by mammals include intermittent, penetrating sounds like occasional high-level, low-frequency thunder; more or less constant noises produced by wind, rushing river rapids and cascading waterfalls; and the highly variable but frequently long-lasting sound of rainfall. These and many other sounds are natural and exhibit highly variable acoustic characteristics that are determined by the geophysical, climatological, and floral composition of the local environment.

Complementing the abiotic sound layer are the varied sounds produced by animals, the biotic sound layer. Even the most pristine environments are typically far from silent. This set of acoustic signals is, in aggregate, spectrally complex and spans a wide range of amplitudes and a wide variety of temporal patterns dependent on taxon and habitat. Examples range from the high-pitched evening chorus of insects and dawn songs of many birds to the low-frequency, long-distance advertisement calls of mammalian megafauna. The vocal animal community adds a dynamic acoustic layer to the already abundant soundscape of natural habitats, varying in space and time with geology, climate, and seasons.

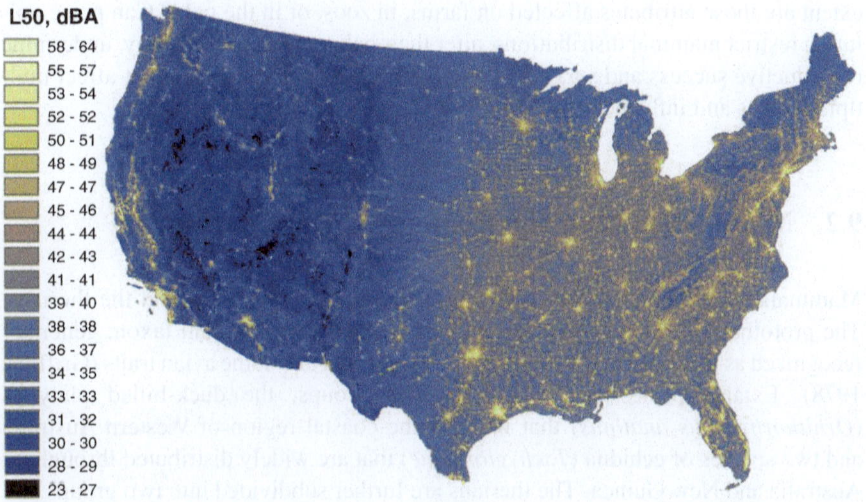

Fig. 9.1 Sound map of the United States of America revealing the noisy network of urbanized areas and roads reflecting the distribution of humans and their activities. L_{50}, A-weighted sound pressure levels predicted by the model that are exceeded 50% of the time. The geographic analyses combined more than 1.5 million hours of sound measurements made at hundreds of sites scattered around the United States and include climatological, geophysical, geographical, and anthropogenic variables (Mennitt et al. 2013).

9.1.2 Noise Pollution and the Human Impact

Man-made sound concerns the fraction of the global soundscape produced by human industry and modern society (Barber et al. 2010; Mennitt et al. 2015). Common man-made sources of noise include traffic noise, heavy construction, and factory operations as well as aviation noise and increasingly noisy recreational vehicles and events. In addition to noisy urban settings, ambient noise can be elevated by resource extraction technologies, cross-country transportation, natural gas developments, and energy production systems such as wind energy installations. Initially, man-made sound was introduced primarily into urban settings, but increasingly, sounds from human activities are more pervasive, extending into rural and remote settings alike. Today, it is difficult to avoid man-made sound (Fig. 9.1), even in areas preserved to maintain their natural qualities, such as national parks (Merchan et al. 2014).

Given the relatively recent but ever expanding presence of man-made sound into diverse habitats, the ecological and evolutionary implications of this new layer of the soundscape is a topic that deserves serious consideration (Katti and Warren 2004; Slabbekoorn and Ripmeester 2008). Specifically, it is important to explore the extent to which man-made sounds can affect hearing performance and interfere with animal communication. In addition, several related questions are addressed in this chapter. How do terrestrial mammals vary in their hearing abilities and to what extent are those attributes affected on farms, in zoos, or in the wild? Can noise pollution restrict mammal distributions, alter their behavior and physiology, undermine reproductive success and survival, and possibly produce outcomes that affect multiple species and influence full ecosystems?

9.2 Mammalian Hearing: Anatomy and Physiology

Mammals are subdivided into two subclasses: the prototherians and the therians. The prototherians, also known as monotremes, represent a small taxon, generally recognized as primitive and exhibiting some reptilian and some avian traits (Griffiths 1978). Extant species are limited to two groups, the duck-billed platypus (*Ornithorhynchus anatinus*) that inhabits the coastal region of Western Australia and two species of echidna (*Tachyglossidae*) that are widely distributed throughout Australia and New Guinea. The therians are further subdivided into two groups, the older marsupials and slightly younger placental mammals. Over the course of the past 200 million years or so, mammalian hearing has been shaped and refined by the forces of natural selection (e.g., Manley 2012; Braga et al. 2015). This process has culminated in the evolution of hearing organs that are sensitive to nanoscale displacements and a dynamic range of 120 dB.

9.2.1 Taxonomic Diversity in Hearing Abilities Among Terrestrial Mammals

Terrestrial mammals range in size from the white-toothed pygmy shrew (*Suncus etruscus*) that weighs approximately 2 g to the African bush elephant (*Loxodonta africana*) that can weigh over 6,000 kg. Therefore, not surprisingly, auditory performance of terrestrial mammals is also remarkably wide ranging (Fig. 9.2). Compared with the hearing thresholds of humans, audiograms of other primates and carnivores often extend into the ultrasonic range (above the human hearing range; >20 kHz), whereas those of elephants extend into the infrasonic range (below the human hearing range; <20 Hz). Rodents are generally more sensitive to higher rather than lower frequencies, although there are exceptions. Fossorial rodents, for example, tend to have poor overall hearing sensitivity but have especially poor hearing at high frequencies. Marsupials, on the other hand, have relatively poor low-frequency hearing, and the more primitive monotremes are the least sensitive and are responsive to only a narrow frequency band somewhat in the middle range for mammals. The hearing of bats is limited to high frequencies, and the upper, high-frequency end of the audiograms of echolocating bats is notably higher than that of nonecholocating bats. Although our understanding of hearing in mammals thus far studied provides

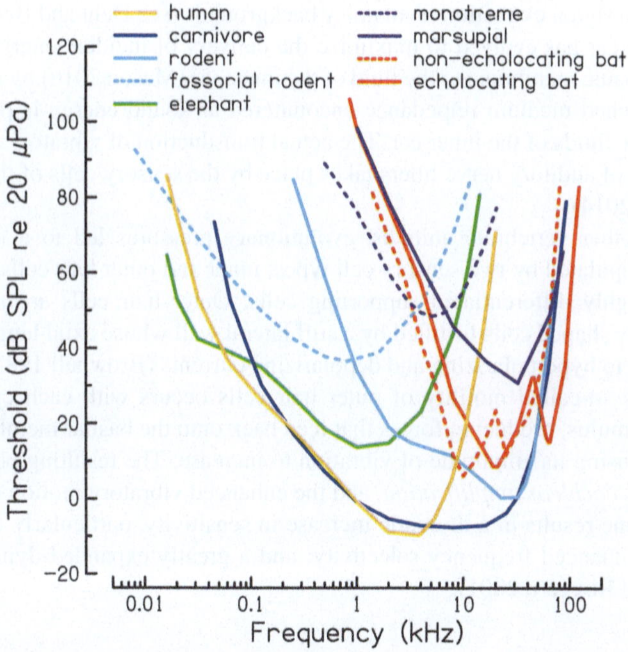

Fig. 9.2 Representation of auditory thresholds as a function of stimulus frequency for selected terrestrial mammals to show the diversity in absolute thresholds and spectral ranges. Curves are polynomial fits to data obtained for humans (averaged from Sivian and White 1933; International Organization for Standardization (1961); Jackson et al. 1999); a carnivore, the domestic cat (*Felis catus*; averaged from Neff and Hind 1955; Heffner and Heffner 1985); a rodent, the Norway rat (Heffner et al. 1994); a fossorial rodent, the blind mole rat (Heffner and Heffner 1992); the Indian elephant (*Elephas maximus*; Heffner and Heffner 1982); a monotreme, the platypus (Gates et al. 1974); and a marsupial, the mouse opossum (*Marmosa elegans*; Frost and Masterton 1994). Raw data are also shown for a nonecholocating bat, the dog-faced fruit bat (*Cynopterus brachyotis*; Heffner et al. 2006), and manually smoothed data are shown for an echolocating bat, the big brown bat (*Eptesicus fuscus*; Koay et al. 1997). All thresholds were derived from behavioral responses, except for those from the platypus, which were based on levels corresponding to cochlear microphonic amplitudes of 1 μV. SPL, sound pressure level.

a good general overview for terrestrial mammals, it should be noted that hearing performance has been studied in a very small fraction of extant species.

Auditory capacity depends on the shape, size, and functionality of the external, middle, and inner ears. In addition to its role in sound source localization, the external ear, which includes the pinna and external ear canal, serves as the first stage in a process that has evolved to minimize degradation of incident acoustic signals by enhancing efficient energy collection. This is accomplished by attenuating low-frequency background sounds and enhancing midfrequency sounds by way of tuned resonance features that promote the amplification of incoming sounds in a species-specific context (Rosowski 1991; Ballachanda 1997). The external ears also contribute to high-frequency filtering, which augments spatial perception and can thereby also serve in signal extraction from noisy backgrounds (see Dent and Bee, Chap. 3). The middle ear has evolved to maximize the delivery of incident energy collected on the tympanic membrane to the fluid of the inner ear (Mason 2016) by minimizing the mismatched medium impedance encountered as sound energy is passed from the air to the fluids of the inner ear. The actual transduction of vibratory energy into stimulation of auditory nerve fibers takes place by the sensory cells of the inner ear (Hudspeth 2014).

Unlike other vertebrate animals, evolutionary pressures led to a mammalian inner ear populated by two sensory cell types; inner and outer hair cells, and a network of highly differentiated supporting cells. Outer hair cells are specialized, cylindrically shaped cells fortified by a stiff lateral wall whose axial length changes in response to hyperpolarizing and depolarizing currents (Brownell 1983; Ashmore 2008). This so-called motility of outer hair cells occurs with each cycle of the acoustic stimulus, producing forces that feed back onto the basilar membrane in the cochlea, causing its amplitude of vibration to increase. The resulting phenomenon is known as *cochlear amplification*, and the enhanced vibratory motion of the basilar membrane results in a dramatic increase in sensitivity, particularly to high frequencies; enhanced frequency selectivity; and a greatly expanded dynamic range (Robles and Ruggero 2001).

9.2.2 Hearing Adaptations for Noisy Environments

Two descending auditory brainstem circuits, the middle ear reflex and the medial olivocochlear reflex, provide feedback to the peripheral auditory system and modulate transduction in noisy conditions (Rosowski 1991; Guinan 2006). As is commonly known, the ossicular chain of the middle ear delivers the incident sound energy collected on the tympanic membrane to the fluid-filled chambers of the inner ear (Rosowski 2013; Mason 2016). Two middle ear muscles, the stapedius and the tensor tympani, insert into the neck of the stapes and onto the malleus, respectively, and contract in response to high-intensity, low-frequency sounds, including chewing sounds as well as the sound of an animal's own voice (Niemeyer 1971; Brask 1978). Contraction of the stapedius limits vibration of the stapes, whereas the tensor tympani draws the malleus medially, increasing tension on the tympanic membrane and attenuating incoming vibrations. The middle ear reflex reduces input to the inner ear in a level-dependent manner up to a maximum of approximately 15 dB and provides a mechanism to avoid discomfort and damage from high-level sounds that are at least 50 ms in duration and persist for no longer than a few seconds (Rosowski 2013).

The second circuit that has a direct impact on the peripheral auditory system during noise exposure is an efferent neural circuit that originates deep in the lower auditory brainstem among neurons of the superior olivary complex. The medial descending division of this projection, named the medial olivocochlear system (MOCS), innervates the cell bodies of outer hair cells and directly alters the

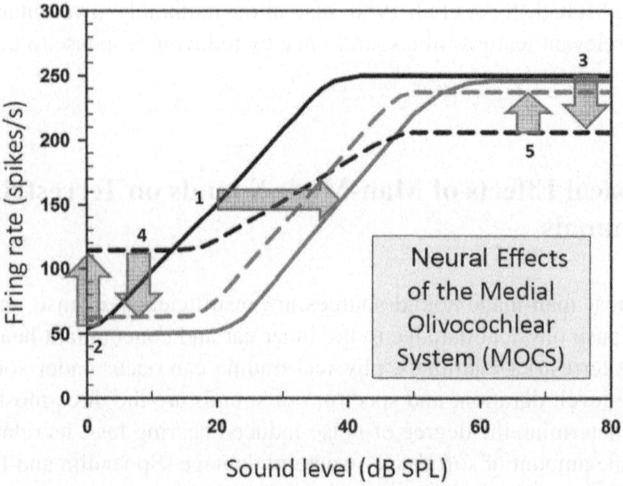

Fig. 9.3 Schematic representation of the effects of the medial olivocochlear system (MOCS) on auditory nerve fiber (ANF) activity. *Solid black line*, quiet conditions; *solid gray line*, quiet conditions plus MOCS activation; *dashed black line*, noisy conditions; *dashed gray line*, noisy conditions plus MOCS activation. Under quiet conditions, MOCS activation shifts the rate-level function elicited by tone bursts horizontally to higher levels (*arrow 1*), resulting in decreased auditory sensitivity. Under noisy conditions, the dynamic range of the ANF responses is compressed due to increased firing rate at low levels (*arrow 2*) and reduced saturation level at high levels due to adaptation (*arrow 3*). Under these circumstances, MOCS activation can partially restore the dynamic range by reducing the response of the ANF to the background noise (*arrow 4*) and elevating the saturation rate (*arrow 5*). MOCS efferents thereby reduce the masking effect of background noise and increase the detection of transient signals. Adapted from Perrot and Collet (2014)

micromechanics of inner ear transduction in response to noise (Warr and Guinan 1979; Perrot and Collet 2014). Potential damage to the inner ear from exposure to intense sound stimulation is counteracted by reducing the electromotile activity of the outer hair cells (reducing the gain of cochlear amplification; Zheng et al. 1997; Maison and Liberman 2000). However, although there is a substantial body of literature validating the protective reflex of the MOCS, Kirk and Smith (2003) argued that the levels of most natural, non-manmade sounds fall well below the levels that induce inner ear trauma and that it is therefore unlikely that this reflex loop evolved by natural selection to protect the inner ear from overexposure to ambient noise.

An alternative biological function attributed to the MOCS that may explain its evolution is improvement in the discrimination of signals in noisy environments (Fig. 9.3) by reducing the masking effects of irrelevant background noise on biologically relevant acoustic signals (Winslow and Sachs 1988; Kawase and Liberman 1993). Although MOCS activation reduces cochlear amplification and shifts the response of auditory nerve fibers in a quiet background toward higher sound levels, when ambient-noise levels are high, the MOCS reflex lessens the response to background noise and enhances the response to the signal, thereby improving the detection of sounds in noisy environments. Furthermore, projections from the cerebral cortex to the MOCS (Syka et al. 1988) also allow mammals to voluntarily concentrate on the relevant features of a soundscape by reducing response to those that are not relevant.

9.3 Physical Effects of Man-Made Sounds on Terrestrial Mammals

Although most man-made sound sources are insufficiently intense under natural settings to cause physical damage to the inner ear and concomitant hearing loss in free-ranging terrestrial mammals, physical trauma can occur under some circumstances. The level, duration, and spectrum of sounds are the three most prominent factors that determine the degree of noise-induced hearing loss, its relative permanence, and the amount of soft tissue structural damage (Spoendlin and Brun 1973). However, different species (Simmons et al. 2016) and strains as well as individuals within a species have varying susceptibilities to noise exposure, which may be a consequence of different hearing sensitivities, age, and prior exposure to noise as well as genetic and metabolic influences (Clark 1991; Lavinsky et al. 2016). In addition, factors such as the intermittency of noise exposure and its predictability are also thought to influence its behavioral effects. The freedom to move about in relation to the location of a sound source adds yet another dimension of complexity to the effort to understand the impact of noise exposure on an animal's behavior. Developmental effects on nonauditory systems and stress resulting from exposure to noise have also been reported (e.g., Freeman et al. 1999; Gao et al. 2009), specifically in laboratory mice (*Mus musculus*) and rats (*Rattus norvegicus*) and a diverse set of captive animals at farms and zoos.

9.3.1 Structural Damage and Hearing Consequences of Noise Overexposure

The soft tissues of the inner ear are vulnerable to noise (Fig. 9.4), and sufficient levels and durations of exposure result in the death of sensory cells and, ultimately, the loss of primary auditory neurons known as spiral ganglion cells. An early indicator of noise overexposure is disarray of sensory cell stereociliary bundles, which may or may not recover over time. The death of outer hair cells is accompanied by a loss of specialized supporting cells and may eventually lead to collapse of the highly structured sensory epithelium within the inner ear that is responsible for sound transduction. The vacuolization of auditory nerve fiber terminals, which innervate inner hair cells, is also an early indicator of serious trauma that can lead to synaptic degeneration and the eventual loss of spiral ganglion cells. Vacuolization is generally regarded as a consequence of overstimulation and the concomitant

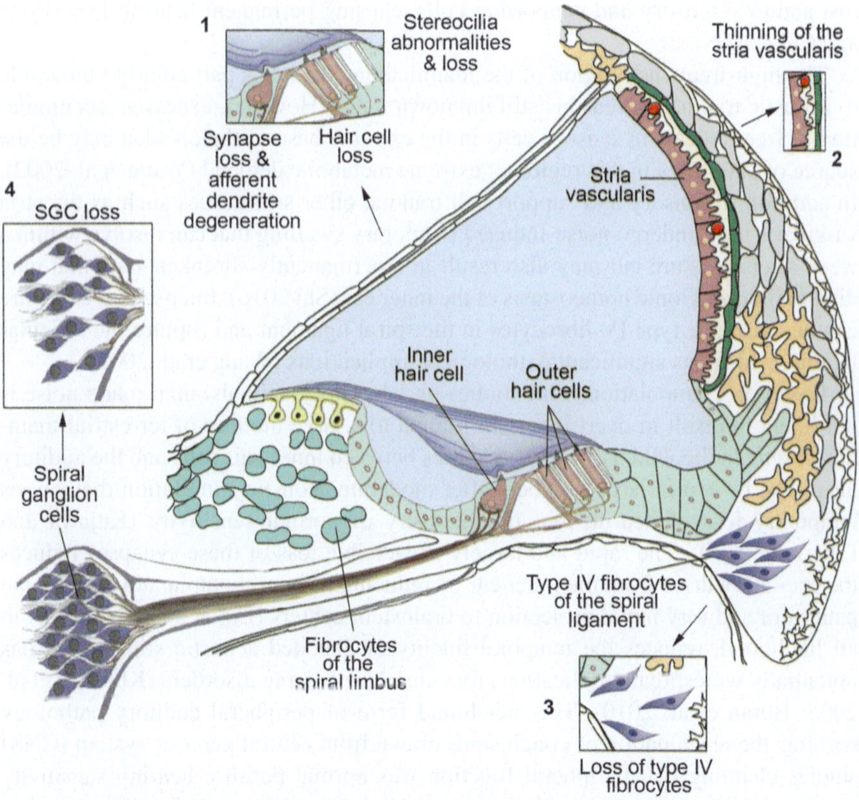

Fig. 9.4 Schematic representation of a cross section of the cochlea illustrating the key structures that are affected by acoustic overexposure. *Inset 1*: injury and loss of sensory cell stereocilia where mechanotransduction channels are located, sensory cell loss and associated supporting cells, loss of inner hair cell ribbon synapses, and degeneration of primary afferent dendrites; *inset 2*: thinning of the stria vascularis, which is a critical source of high K^+ concentrations that carry the mechanoelectrical transduction current; *inset 3*: loss of type IV fibrocytes of the spiral ligament; *inset 4*: loss of primary afferent neurons (spiral ganglion cells [SGC]). Adapted from McGee and Walsh (2015)

release of excessive amounts of the neurotransmitter glutamate, resulting in glutamate excitotoxicity (Pujol and Puel 1999).

The amount of structural damage to the soft tissues of the inner ear determines, to a large degree, not only the amount of hearing loss but also the extent of central processing disorders as well. Loss of sensitivity can be temporary or permanent (see Saunders and Dooling, Chap. 4), and hearing loss degree ranges from slight to profound. Permanent losses between 16 and 25 dB are regarded as slight, between 26 and 40 dB as mild, between 41 and 70 as moderate to moderately severe, between 71 and 90 dB as severe, and of 91 dB and higher are considered profound (Clark 1981). Unlike other vertebrate classes, mammals are generally unable to regenerate lost auditory sensory and supporting cells, causing permanent hearing loss (Edge and Chen 2008).

The high-frequency region of the mammalian cochlea is particularly vulnerable to acoustic trauma for reasons still unknown today. However, excessive accumulation of free radicals in sensory cells in the extreme base of the cochlea may be the source of pathology in this region of extreme metabolic demand (Wang et al. 2002). In addition to sensory and support cell trauma, other soft tissues such as the stria vascularis may undergo noise-induced temporary swelling that can resolve within a week after exposure but may also result in a permanently shrunken state that may disrupt fluid and ionic homeostasis of the inner ear (Shi 2016). Intense overexposure can also damage type IV fibrocytes in the spiral ligament and rupture the reticular lamina, which has significant pathological implications (Wang et al. 2002).

Based on extrapolation from studies on laboratory animals, man-made noise is not likely to result in overt inner ear trauma to a large number of terrestrial mammals living in the wild. However, synapses between inner hair cells and the auditory nerve can be lost in large numbers after moderate acoustic stimulation that causes temporary loss of hearing but full recovery of normal sensitivity (Kujawa and Liberman 2009). The rapid and largely irreversible loss of these synapses reduces the excitatory drive from the inner ear by reducing its neural output and altering the pattern of auditory nerve projection to brainstem centers (Lin et al. 2011). This, in all likelihood, reduces the temporal fidelity of encoded acoustic stimuli and has potentially widespread implications for central processing disorders (Khimich et al. 2005; Buran et al. 2010). This newfound form of peripheral auditory pathology requires the reevaluation of conclusions drawn from central nervous system (CNS) studies claiming that peripheral function was normal because hearing sensitivity was normal. The impact of "auditory synaptopathy" is particularly insidious because traditional measures to assess hearing loss cannot be used to identify the impairment, and, therefore, it is often referred to as "hidden hearing loss."

9.3.2 Noise Exposure Effects on Brain and Brain Development

In addition to cochlear damage, acoustic overstimulation can cause neuronal degeneration in central regions of the auditory system, resulting in decreased cell densities (Coordes et al. 2012). Excessive noise exposure or prolonged stimulation has also been shown to elevate spontaneous discharge rates of central auditory neurons, interrupt the balance between excitation and inhibition, alter the representation of sound in CNS regions, and produce hyperacusis (hypersensitivity to sound) and tinnitus or the perception of phantom sounds (Gourévitch et al. 2014; Eggermont 2017).

Moreover, sound exposure can affect the development and function of nonauditory regions of the CNS without damaging the peripheral auditory system (Recanzone et al. 1993). Although the mechanisms underlying such outcomes are not well understood generally, Guo et al. (2017) recently reported that DNA methylation rates in neurons of several brain regions in Wistar rats were altered in response to exposure to moderate levels of noise over a 3-week period, suggesting that epigenetic factors may underlie at least some forms of indirect noise-induced pathology, including effects on blood pressure and body weight.

In addition, noise exposure has been shown to affect balance (Tamura et al. 2012) and respiration (Oliveira et al. 2001) as well as the cardiovascular, gastrointestinal, and endocrine systems (e.g., Cui et al. 2016) through its influence on the hypothalamic-pituitary-adrenocortical axis and the sympathetic-adrenomedullary system (Burow et al. 2005). Shifts in physiological and behavioral states have also been shown to be directly induced by tissue penetration of low-frequency sounds as well as by vibratory stimulation (Berglund et al. 1996; Mahendra Prashanth and Sridhar 2008). Wysocki (1996) reported that intermittent noise exposure slows wound healing in rats at noise levels commonly experienced in hospitals.

Evidence supporting the notion that exposure to noise may affect nonauditory central processing was also reported by Uran et al. (2012). Two-week-old rats were exposed to wideband noise delivered at 95-97 dB sound pressure level (SPL) for 2 hours/day for 15 days. Exposed animals exhibited mild histological disruption of the hippocampus that correlated with habituation to repeated test sessions in an open field measuring device as well as to impaired novelty preference outcomes in an object recognition task (Fig. 9.5). Uran et al. (2010) speculated that the detrimental impact could be explained by oxidative imbalance in noise-exposed animals resulting from an early increase followed by a decrease in hippocampal levels of reactive oxygen species and an increase in antioxidant enzyme activities.

Given that the underlying mechanisms of noise-induced, indirect outcomes like those reported in these studies have not been determined and in at least some cases the findings have not been replicated, this is a field of inquiry requiring a substantial degree of additional research.

9.3.3 Stress and Sleep Disturbance in Laboratory Rats

It is well-known that environmental noise can act as a stressor in humans, elevating levels of stress hormones (Fig. 9.6), blood pressure, and heart rate (Babisch et al. 2001; Stansfeld and Matheson 2003). Several studies have also revealed that exposure to chronic noise can have an impact on growth in nonhuman mammals (Marti and Armario 1997; Konkle et al. 2003). For example, Alario et al. (1987) showed that chronic exposure to noise can reduce food consumption and thereby affect growth rate in laboratory rats. Similar outcomes were reported in a study in which rats were exposed to 86 dB SPL wideband noise for 15 minutes 8 times a day for 30 days (Michaud et al. 2005). After two to three weeks of exposure body weight was reduced and plasma corticosterone levels were higher in the noise-exposed rats compared with sham-treated control rats.

In a study conducted using laboratory rats, Krebs et al. (1996) reported noise-induced changes in eating and sleeping behavior. Increased eating speed, elevated

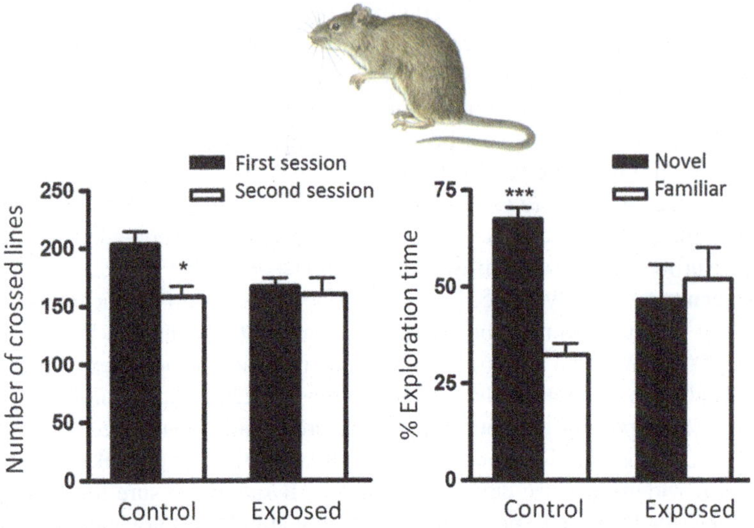

Fig. 9.5 Effects on memory-dependent tasks related to noise-dependent damage to hippocampal development in rats. Results are shown for animals that were exposed between postnatal days 15 and 30 and tested immediately afterward, with 24-hour interval between the two test sessions. *Left*: control rats showed a decrease in the number of lines crossed in an open field test, whereas no difference was observed in noise-exposed rats. *Right*: control rats explored a novel object longer compared with a familiar one, whereas exposed rats explored both objects for the same amount of time. Significance levels for differences between test sessions: $*P < 0.05$; $***P < 0.001$. Findings for both tasks suggest noise-dependent loss of memory. Adapted from Uran et al. (2012)

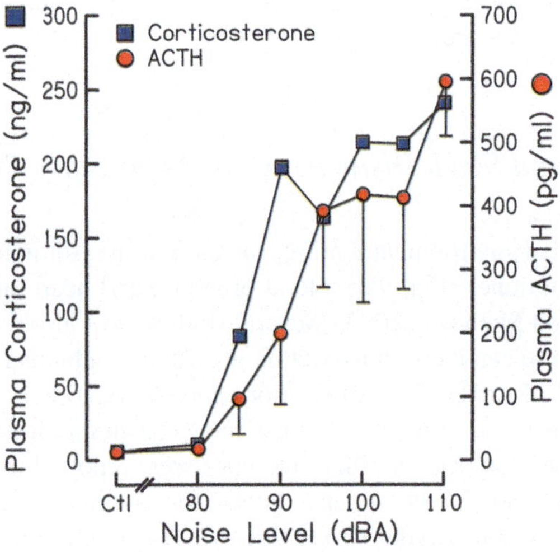

Fig. 9.6 Physiological effects of 30-minute broadband noise exposure on mean plasma corticosterone and adrenocorticotropic hormone (ACTH; corticotropin) levels in adult male rats as a function of environmental noise level. Blood samples were obtained immediately after noise exposure. Control (Ctl) values represent hormone levels in animals maintained under ambient background noise, which was approximately 60 dB(A). Error bars represent standard error of the mean (SEM), which are shown in only one direction for clarity. Adapted from Burow et al. (2005)

vigilance, and increased defecation rates in rats were observed when the level of white noise was elevated 40 dB, to 95 dB SPL, from a background level of 55 dB SPL compared with control rats in which the background noise level was raised 5 dB. In addition, the duration of feeding periods was reduced in treated rats while the durations of exploring, grooming, and resting were longer than normal. The authors interpreted these behavioral changes as stress-evoked adaptations to a threatening environment. The associated physiological changes, directly or indirectly related to noise exposure, could also affect behavior under free-ranging conditions and yield disturbance or deterrence effects.

Rabat et al. (2004, 2005) found that elevated noise levels reduced the amount of time spent in paradoxical sleep (rapid eye movement sleep that plays a role in memory formation) and fragmented the occurrence of slow-wave sleep (deep sleep, potentially influencing daytime performance) in laboratory rats. The authors used noisy airport recordings as the exposure source and showed that the intermittent nature and spectral match with rat hearing sensitivity boosted the magnitude of disturbance. In a follow-up investigation, Rabat et al. (2006) and Rabat (2007) showed that such disturbances of the sleep-wake cycle may affect long-term memory, as revealed by a two-compartment recognition task. In this test, rats were allowed to explore two compartments that they had experienced previously (acquisition trial). In the experimental trial, one compartment was altered in preparation for a second visit (restitution trial). The experimental design was based on the a priori position that only those rats with a memory of the previous compartment layout would more seriously explore the novel environment (Uran et al. 2012). The free exploration scores were significantly more balanced in noise-exposed rats, indicating that their memory of previous experience was degraded relative to that in control rats.

9.3.4 Housing and Transport Noise Affecting Farm Animals

Noisy housing conditions have also been shown to affect stress levels and behavior in farm animals, with consequences for both animal welfare and meat production (e.g., Algers et al. 1978; Schäffer et al. 2001). In an experimental study conducted by Otten et al. (2004), for example, male castrated German Landrace pigs (*Sus scrofa*) were exposed to either 2 hours of broadband noise delivered at 90 dB(Z) daily for 4 weeks or 3 times a week for the same time period. Stress hormone (cortisol and epinephrine/norepinephrine) levels were measured before, during, and after noise exposure. Although changes in stress hormone titers remained normal during and after the first noise exposure, increases in activity were reported during the first 30 min of exposure. Animals exposed to noise on a daily basis spent more time lying down and less time interacting with cohabitants compared with members of the control group that were handled in the same way but were not exposed to noise. Daily weight gain was reduced significantly in the treatment groups on weeks 2 and 4 of sound exposure, while the level of circulating corticotropin and cortisol

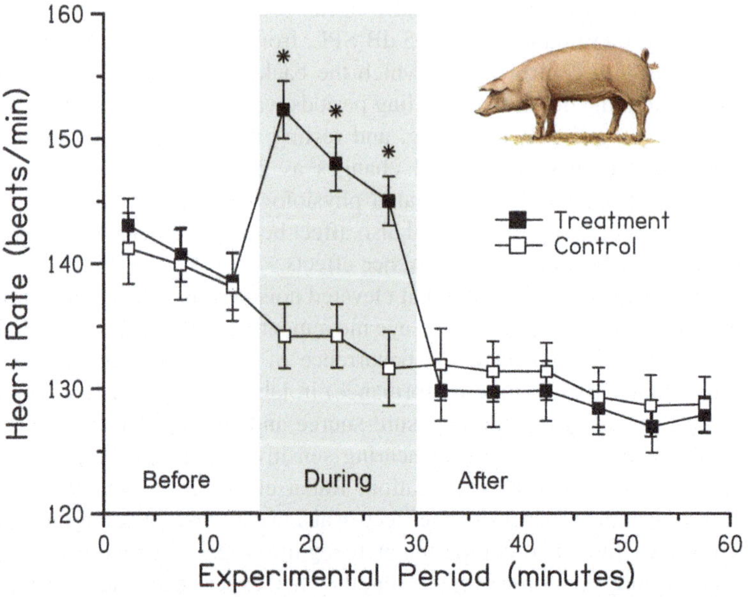

Fig. 9.7 Physiological effects of noise exposure on heart rate in pigs. The mean and standard error for a sample of eight 4-week-old piglets are depicted for the periods before, during, and after sound exposure (combining treatments: farm, transporter, abbatoir, and white noise). Significant difference in sampling times between treatment and control sessions: $^*P < 0.05$. Ambulation scores showed a highly similar pattern for the duration of the experimental period (including significant differences between treatment and control). Adapted from Talling et al. (1996)

hormones diverged between experimental and control groups after day 4 of exposure. Exposed animals exhibited a rapid increase in hormone titer followed by a gradual decrease, whereas hormone levels were relatively constant in the control animals. Otten et al. (2004) concluded that sound levels above 90 dB should be avoided on the grounds of animal welfare and to achieve optimal growth.

Several studies on pigs, cattle (*Bos taurus*), and sheep (*Ovis aries*) have addressed stress-related physiological changes associated with noisy transportation (Agnes et al. 1990; Hall et al. 1998). Experimental studies have, for example, shown that heart rate (Fig. 9.7) and cortisol levels are elevated beyond those observed during loading or handling animals during transportation (Bradshaw et al. 1996; Geverink et al. 1998). Stephens et al. (1985) attempted to separate the perception of different stress factors during pig transportation. Pigs were taught to operate a switch that allowed them to selectively turn off transportation-related vibrations or sounds. Humans that experienced the test conditions expressed some "amusement" by the vibrations initially but found the sounds of a tractor moving over a farm track played at 80-90 dB SPL "uncomfortably loud." If exposed to each factor independently, pigs reliably switched-off the vibratory source but not the sound source, suggesting that the sound alone was not aversive. However, if vibration and sound signals were delivered together, subjects switched off both, presumably reflecting a case of secondary reinforcement. This finding illustrates that noisy conditions, which are not aversive in and of themselves, can become aversive when combined with a known aversive condition by way of learned association.

In addition to housing and transportation conditions, holding pens, designed to rest cattle, pigs, or sheep just before slaughter, are often very noisy (Weeks et al. 2009). Holding pens in England and Wales were found to have peak sound levels up to 110 dB(A), which are usually caused by the indoor resonance of vocal activity produced by the animals themselves and enhanced by noisy machinery or ventilation systems. In an observation made by Geverink et al. (1998), pigs responded to playback of white noise or machinery noise by herding closer together, which is a typical indication of anxiety and risk perception and may be considered as an indicator of reduced welfare.

Studies on cattle revealed variation in acoustic sensitivity among breeds, sexes, and individuals. Relatively high-pitched, intermittent sounds were found to be most potent in triggering a startle response (Talling et al. 1996, 1998), and exposure to sudden, unexpected sounds had a greater effect on cattle behavior than on-going, fluctuating sources of background noise (Waynert et al. 1999). Beef breed cattle were more startle sensitive to sounds than Holstein dairy cattle, and steers and heifers were more sensitive than older bulls and cows (Lanier et al. 2000). Individual responses in the latter study also varied and were rated relative to a temperament score. Sensitivity to sudden acoustic events increased along a four-category scale ranging from animals that walked calmly or stood still in response to a person swinging arms and yelling bids to those that responded with high vigilance, jerky running, and colliding with the fence or other obstacles.

9.4 Effects of Man-Made Noise on Acoustic Communication and Other Behaviors

Although man-made sounds produced at high levels may cause inner ear trauma or chronic stress in mammals, behavioral effects are likely to be more widespread and are, therefore, potentially more of a threat to conservation, health, and welfare. In this light, it is highly relevant that masking can negatively affect the perception of critical sounds in noisy environments (see Dooling and Leek, Chap. 2). Interestingly, senders of acoustic signals for communication have been found to adjust their calls and compensate for masking under noisy conditions. Such counterstrategies in sound production are expected when they benefit both sender and receiver and are not, therefore, useful as acoustic cues in predator-prey situations. Other adverse outcomes of noise exposure that are addressed in this section relate to changes in behavior that signal distress, an especially meaningful consideration in the case of captive animals. Reported changes in the behavior of free-ranging animals are also reviewed and mainly focus on roadside noise and case studies addressing the effects of noisy industrial and recreational activities. Bats are an exceptional group of terrestrial mammals and receive some special attention.

9.4.1 Auditory Masking and Counterstrategies

Masking occurs when a competing noise in the environment decreases the probability that an animal will detect a signal or cue. The most commonly utilized psychoacoustic masking assessment tool is a measurement known as the critical ratio (Fig. 9.8A). The critical ratio is the difference between the masked hearing threshold of a target tone (in dB SPL re 20 μPa) and the spectrum level of a broadband noise masker (i.e., the spectral power density level of the masker in dB re 20 $\mu Pa^2/Hz$). For example, if the masked threshold of a 4-kHz tone is 80 dB SPL and the spectrum level of a broadband masker that is necessary to mask the tone is 50 dB, the critical ratio would be 30 dB. For most mammals studied, the critical ratio increases by approximately 3 dB for every increase of 1 octave, for example, from 2 to 4 kHz.

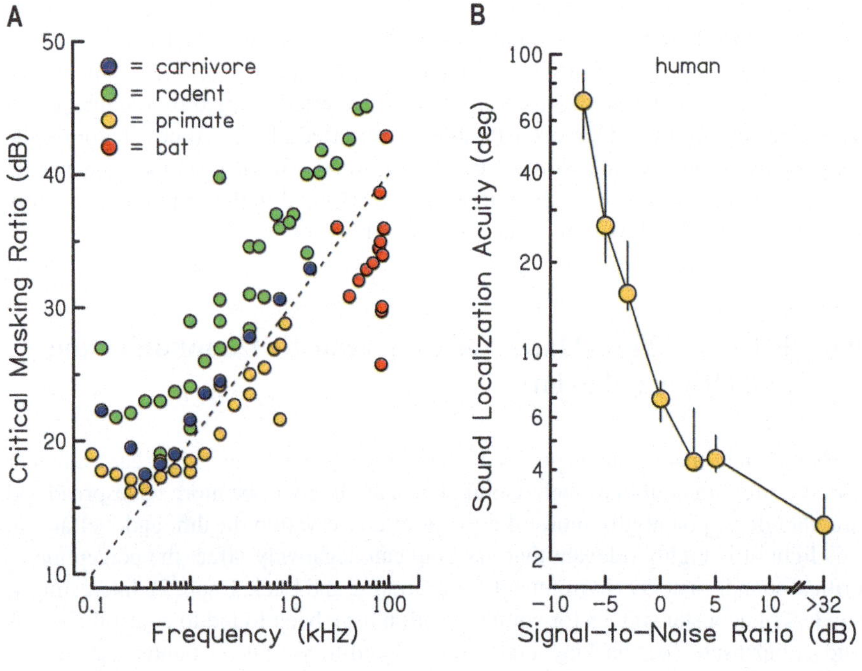

Fig. 9.8 **A:** critical masking ratios as a function of frequency for terrestrial mammalian species from selected taxonomic groups: a carnivore, the domestic cat (*Felis catus*); several laboratory rodents, rats (*Rattus norvegicus*), chinchillas (*Chinchilla lanigera*), and mice (*Mus musculus*); two primates, rhesus monkeys (*Macaca mulatta*) and humans; and greater horseshoe bats (*Rhinolophus ferrumequinum*). Adapted from Fay (1988). **B**: improvement of sound localization acuity as signal-to-noise ratio (SNR) increases in human listeners. Values are medians and error bars represent quartiles. Adapted from Kerber and Seeber (2012)

Vocalizations clearly play a critical role in communication throughout the life span of an animal but can be especially critical at the suckling life stage. Noisy conditions may, therefore, have an impact on the growth and survival of juvenile animals as a direct consequence of masking (see Dooling and Leek, Chap. 2). Algers and Jensen (1985) reported that sow nursing and piglet sucking occur in vocally coordinated bouts. Sow teats synchronously provide milk for limited periods, and piglets must be present at drinking sessions to successfully feed. Shifting from fast to slow grunting by the sow has been shown to signal drinking opportunities that coincide with oxytocin release that is responsible for discrete contraction bouts of the mammary glands. Furthermore, piglets also communicate by grunting and acoustically trigger the sow to expose the udders. Algers and Jensen (1985) showed that elevation of environmental noise level correlated with longer periods of teat massaging and sucking relative to those observed in quiet control groups. They concluded that masking undermined vocal coordination, which resulted in reduced milk transfer and increased piglet energy demand.

Besides interfering with the detection, discrimination, and maintenance of efficient vocal communication, environmental noise can also affect the ability to localize a sound source in a level-dependent manner (Fig. 9.8B). Determining the direction and distance of a sound source is a critical capability that is necessary for locating prey, detecting predators, and orientation and navigation in space. Generally, mammals with larger head sizes and therefore more robust binaural cues are able to localize sound sources in the horizontal plane with higher acuity than animals with smaller head sizes. Humans and elephants can, for example, detect sound source separations as small as 1-2°. Some species, like blind mole rats (*Spalax ehrenbergi*), are notoriously poor at sound localization (Heffner and Heffner 1992), which is likely due to their small head size and a lack of need for localizing sounds in their subterranean habitat. Regardless of head size, localization acuity is diminished when individuals are challenged with noisy backgrounds.

In the field, animals, including terrestrial mammals, take advantage of several counterstrategies to minimize the influence of noise masking. Moving closer to target sound sources relative to sources of background noise, along with head or auricle rotation, provides a mechanism for achieving some degree of masking release. Signaling animals can also compensate for masking by raising vocal output levels (Lombard effect), as has been reported for human speech but which has also been observed in cats (Nonaka et al. 1997), bats (e.g., Habersetzer 1981; Hage et al. 2013), and nonhuman primates (e.g., Sinnot et al. 1975). Brumm et al. (2004) found not only that the sound level of twitter calls of captive common marmosets (*Callithrix jacchus*) increased but also that the duration of syllable production was extended by almost 30% when noise levels were elevated. Cotton-top tamarins (*Saguinus oedipus*) were shown to respond in the very same way by producing more intense and longer syllables in noisy environments (Egnor and Hauser 2006).

In a follow-up study, Egnor et al. (2007) reported that cotton-topped tamarins can also time their calls to occur in silent intervals when presented with a predictable on-off regimen of intense broadband noise (Fig. 9.9). Serial redundancy, another possible counterstrategy for coping with noisy conditions (Brumm and

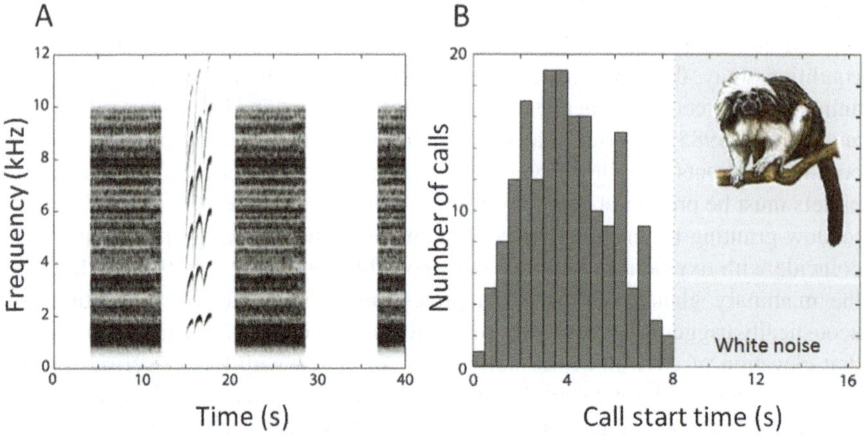

Fig. 9.9 Temporal avoidance of intermittent, broadband noise by calling cotton-top tamarins (Egnor et al. 2007). **A:** spectrogram of a 40-second segment of intermittent, broadband noise with harmonically structured tamarin calls visible in the second quiet interval. **B:** frequency distribution of call occurrence in the time periods without and with the noise masker: all calling activity was restricted to the windows of silence.

Slabbekoorn 2005), was not found in these studies because none of the marmosets or tamarins showed a tendency to increase the number of call syllables in response to increased noise levels. However, vocal adjustments in amplitude and the timing of calls within silent windows were found and facilitated the capacity of calling monkeys to retain their active space (the area in which communicative signals can be heard by conspecifics; see Larsen and Radford, Chap. 5).

9.4.2 Noise-Induced Behavioral Changes at Zoos

Zoo environments are generally noisy and complex environments. Objective studies to determine the influence of environmental noise on the well-being of resident animals are typically challenging. Opportunities to replicate findings, separating visual versus acoustic effects, habituation of animals to recurring acoustic events, and variation in coping style or personality among and within species are a few of the challenges related to data acquisition and interpretation in zoo settings. Nevertheless, man-made noise may adversely affect the behavior of animals in these captive environments. For example, two groups of white-handed gibbons (*Hylobates lar*) housed in two different Canadian zoos appeared to become more active with elevated noise levels and visitor numbers (Cooke and Shillaci 2007). The animals were commonly engaged in self-directed scratching and more frequently initiated interactive behaviors, such as making eye contact with visitors, open mouth displays, and increased locomotive behaviors such as brachiating, hanging, and bipedal walking, than under quiet conditions.

Owen et al. (2004) and Powell et al. (2006) studied the effects of ambient noise on pairs of giant pandas *(Ailuropoda melanoleuca)* at two US zoos. Under noisy conditions, both groups reported a general increase in activity, heightened alertness, increased locomotion, and, in rare instances, the production of honking vocalizations that are thought to be an indicator of distress. Noisy periods were also associated with more anxiety-related behaviors, such as scent marking, grooming, bleating, urination, and defecation. In addition, urinary corticoid metabolite levels were elevated on noisy days relative to quiet days in three of the four animals studied. One of the male pandas exhibited more stereotypical behavior during noisy demolition activities, while one of the females appeared especially sensitive to sounds during estrus and lactation. However, Owen et al. (2004) reported that they "found no compelling evidence that these adjustments indicate substantive detrimental effects on well-being or reproduction." In another study, Owen et al. (2014) reported that ambient-noise levels affected maternal behavior in Bornean sun bears (*Helarctos malayanus*); the mother attended her cub more on noisy than quiet days. Obviously, more data are needed before such anecdotal observations can be confirmed or interpreted.

Two investigations of the impact that noise has on zoo animals are of particular interest given their experimental, albeit qualitative, nature. Birke (2002) asked visiting groups to be more or less noisy while standing in front of an orangutan (*Pongo pygmaeus*) exhibit and observed the behavioral responses. Adult animals confronted by noisy groups, independent of group size, made visual contact with visitors, while infants approached and sought comfort (held onto adults) more frequently under noisier conditions. Larsen et al. (2014) investigated the effect of visitor noise on koala (*Phascolarctos cinereus*) behavior. Using audio playbacks of recordings taken in the absence of visitors versus the presence of quiet or noisy visitors, they found that vigilance time increased with rising noise levels.

Quadros et al. (2014) also studied the responsiveness to crowd noise for a range of mammals housed at the Belo Horizonte Zoo, Brazil, that included elephants and giraffes (*Giraffa camelopardalis*) as well as apes, monkeys, big cats, canids, and deer species. Typical reactions across taxa included increased vigilance and movement. They also reported that species of higher popularity that attract larger crowd sizes were consequently exposed to the highest sound levels. These findings, minimal as they are, suggest that zoo designers should consider the impact of noise disturbance when developing exhibits. In addition, efforts to retrofit particularly noisy facilities with mitigating devices should be a high priority when there are signs that man-made sounds are affecting the welfare of animals on exhibit.

Animal shelters for dogs (*Canis familiaris*) can also be very noisy depending on the building layout and materials used for construction and on the number and size of animals occupying the facility (Coppola et al. 2006). This may be particularly problematic because dogs are known to be highly sensitive to sounds, a characteristic that is exploited in dog training activities in which dogs can, for example, be taught to respond to a whistle (McConnell 1990). It has also been reported that dogs are stressed by sound and acquire digestion disorders as a response to noisy housing conditions (Gue et al. 1987).

The findings reviewed here suggest that zoo noise can affect the health and welfare of some captive animals, but large-scale conclusions should be made cautiously given the small number of facilities and species thus far assessed in an objective manner.

9.4.3 Acoustic Roadside Ecology: Vigilance Behavior

Noise-dependent avoidance is thought to be a dominant causal factor affecting animal distributions and population connectivity along roads and urban areas (Forman and Alexander 1998; Baker and Harris 2007). Avoidance of large roads and highways, in which exposure to chemical, light, and noise pollution all play a role, has been reported for diverse mammal species including hedgehogs (*Erinaceus europaeus*; Rondini and Doncaster 2002), moose (*Alces alces*; Laurian et al. 2008), and grizzly bears (*Ursus arctos*; McLellan and Shackleton 1988).

Francis et al. (2015) reported noise-dependent loss and degradation of habitat for the squirrel glider (*Petaurus norfolcensis*), a nocturnal marsupial from eastern Australia. They studied squirrel glider activity using infrared motion sensor cameras in urban and rural areas, taking artificial light and traffic noise levels into account as well as tree height, hollow-bearing tree density, vegetation cover, and road and housing density. The negative effects of traffic noise on site occupancy were also found in a telemetric monitoring study on red squirrels (*Tamiasciurus hudsonicus*), a placental counterpart from Arizona, US (Chen and Koprowski 2015).

Shannon et al. (2014) conducted a rare experimental noise-exposure study using prairie dogs (*Cynomys ludovicianus*). Data were collected with a single traffic-noise recording in two different colonies located at sites removed from actual roads and traffic. Results revealed a 21% decline in the number of animals observed aboveground, and those at the surface exhibited significantly increased vigilance behavior at the expense of foraging, resting, and social activities. Playback noise was delivered from a speaker positioned approximately 100 meters from colonies at levels comparable to previously analyzed roadside traffic noise. However, the impact on behavior in this study may not simply be attributed to an inherent effect of traffic noise; vigilant behavior may also increase in response to novel sounds, especially in the case of visual absence of a logical source.

Kern and Radford (2016) also reported increased vigilance among a group of dwarf mongoose (*Helogale parvula*) in South Africa. This species exhibits sentinel behavior; members of social groups as large as 30 individuals adopt a raised position and scan for predators, warning others of potential danger. Sentinels produce high-amplitude threat-specific alarm calls that trigger an escape response by group members in addition to low-amplitude surveillance calls, which allow foragers to spend valuable time foraging rather than scanning for predators. A playback experiment showed that road noise undermined the benefits of the surveillance calls and triggered a rise in the number of scans and the overall duration of vigilance behavior.

In a second experiment in which wild dwarf mongooses were study subjects (Morris-Drake et al. 2016), road-noise playback was combined with fecal presentations to demonstrate that man-made noise could disrupt information processing across sensory modalities (also see Halfwerk and Slabbekoorn 2015). In the absence of road noise, habituated dwarf mongooses exhibited greater vigilance after detecting predator feces relative to control, herbivore feces. However, in the presence of road noise, the animals did not exhibit variation in vigilance as expected and did not show the adaptive response after fecal source discrimination. These two experimental studies suggest that man-made sounds can affect mongoose fitness through a reduction in foraging efficiency through masking of sentinel calls and an increased predation risk as a result of undermining the effects on adaptive responsiveness to fecal discrimination.

Duarte et al. (2011) published the results of a study on the effects of man-made noise on spatial behavior of an urban-dwelling mammal species, the black-tufted marmoset (*Callithrix penicillata*). The authors monitored a single group of nine individuals living in an urban park in the Brazilian city of Belo Horizonte. Their data indicated that the monkeys spent more time in the central and quiet areas of the park than in areas close to the park edges that were noisy due to road traffic, especially during weekdays. The animals were shown to adjust this spatial pattern during weekends when the roads were relatively quiet and the central areas were noisier due to the increased presence of park visitors. The authors argued that traffic noise led the animals to occupy nonoptimal, less than desirable spatial patterns. A behavior that may to some extent mitigate this noise-induced occupation of undesirable spatial locations for some species is to shift the time of occupation. Radio-collared grizzly bears were reported to avoid roads during daylight hours but to take advantage of the roadside terrain of their home range at night in a 7-year study in the Rocky Mountains (McLellan and Shackleton 1988).

It is relevant to note that there are also reports indicating that some animals may be relatively unresponsive to man-made sounds. Walther (1969) described the flight and avoidance behavior of Thomson's gazelles *(Gazella thomsoni)* in the Serengeti National Park, Tanzania. He reported that the animals reacted only to man-made sounds, such as human voices, the ringing of an alarm clock, or the slamming of a car door, when the source was within 30 meters or if they were already alarmed by visual cues. This relative insensitivity may have been due to habituation to the frequent presence of harmless tourists and their sounds, which may be an even more common phenomenon today. Elephants, for example, were recently reported to respond selectively to the ethnicity, sex, and age of human voices and appeared to rate the relative danger of human subcategories. Defensive bunching and investigative smelling by the elephants was most prominent in response to the voices of adult male Masai, who traditionally hunt elephants (McComb et al. 2014).

9.4.4 Industrial and Recreational Noise Pollution

Other relevant studies have concentrated on the effects of noise generated by industrial and recreational activities on terrestrial mammals. Rabin et al. (2006), for example, reported on antipredation behavior in California ground squirrels (*Spermophilus beecheyi*) at the Altamont Pass Wind Resource Area in California where squirrels live near wind turbines. The turbines generated average sound levels of 110 dB SPL at heights of 0.25 meters above the ground, whereas the average ambient-noise level was 76 dB, with turbines producing peak levels of 118 dB SPL. Conspecific alarm calls were recorded at nonturbine sites, were filtered to remove ambient noise, and subsequently played back at a level of 105 dB SPL at a distance of 0.25 meters from the speaker. Playback of conspecific alarm calls was associated with increased vigilance and excessive cautionary behavior among animals at the turbine farm compared with those at control sites. The authors concluded that the wind turbine noise caused enhanced antipredator behavior to the alarm calls.

Bradshaw et al. (1997) examined the potential effects of noise generated as a product of petroleum exploration on movement and behavior of woodland caribou (*Rangifer tarandus caribou*) in boreal mixed woods and peatland of northeastern Alberta, Canada. A propane cannon was used to simulate extraction blasts. The signal level varied from 90 to 110 dB SPL at 2 meters from the source. Caribou were exposed to 1 blast per minute for 1 hour per trial at a distance of 331 meters on average. Animals were stationary, feeding, or moving slowly in a random manner before the test started. Immediately after a blast, the animals stopped feeding and moved away from the source, crossing more distinct habitat transitions than did control animals, possibly resulting in higher energy expenditure. There also was a nonsignificant trend for reduced foraging time. The conclusion was that blast sounds may cause a temporary or permanent habitat loss for caribou as well as a reduction in the time available for foraging that may negatively affect acquisition of daily energy requirements.

Similar outcomes were reported in the case of a study of seismic activity associated with the search for oil in the Loango National Park, Gabon (Rabanal et al. 2010). The investigation was conducted before, during, and after low-impact seismic explorations, and transect count data focused on the indirect signs of animal presence. The results revealed that large, wide-ranging rainforest mammals (elephants and apes) avoided areas of seismic exploration, whereas the territories of smaller animals with more restricted ranges (duikers and monkeys) were unaffected. Another passive monitoring study of animal calls and seismic blasting in the same area reported that forest elephants (*Loxodonta cyclotis*) became more nocturnal, which was not directly linked to the intensity or frequency of dynamite detonation but was attributed to the increased daytime presence of oil workers (Wrege et al. 2010). This finding underscores the complexity of such studies and reinforces the need for more research in this area.

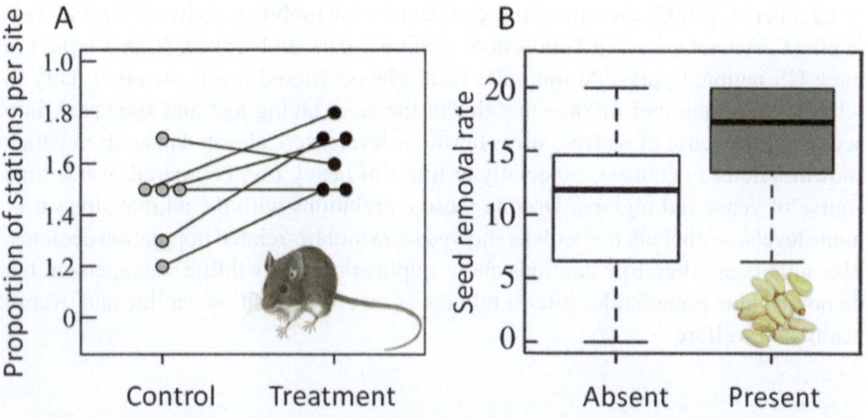

Fig. 9.10 **A:** impact of generator noise on mice presence. Control sites were six quiet well pads and treatment sites were six active gas wells with noisy compressors in New Mexico. *Peromyscus* mice were detected more frequently at treatment sites than at control sites and showed an apparent preference for noisy conditions (see text for potential explanation). **B:** impact of mice presence on seed removal. Pine seed removal rates (per 24 hours) were higher in a seed removal experiment (20 seeds scattered on the ground) when *Peromyscus* mice were present (contributing to a noise-dependent stand of pine seedlings). Adapted from Francis et al. (2012)

Francis et al. (2012) studied the effects of noise pollution on terrestrial mammals inhabiting the surroundings of sound-generating gas extraction stations in New Mexico. *Peromyscus* mice were more abundant at noisy sites compared with quiet control sites [approximately 14 dB(A) difference in levels], a finding that correlated positively with higher seed predation rates observed in a pine seed removal experiment (Fig. 9.10). The impact on a mammalian seed predator influenced the composition and rejuvenation of vegetation by altering seed survival and dispersal. A reasonable explanation for the apparent "preference" of mice for noisy conditions may be a relative lack of hearing sensitivity for low-frequency compressor noise combined with reduced competition with a seed collecting jay (*Aphelocoma californica*) that tends to avoid noisy gas extraction sites. It is also possible that the risk of predation may be reduced due to the deterrence or masking effects of the man-made sounds on owls (Mason et al. 2016; Senzaki et al. 2016).

Weissenberger et al. (1996) reported increased heart rate, alertness, and alarm behavior in mule deer (*Odocoileus hemionus crooki*) and mountain sheep (*Ovis canadensis mexicana*) in response to simulated low-altitude jet aircraft overflights (sound levels ranging from 92 to 112 dB). However, the behavioral and physiological measures quickly returned to predisturbance levels within 3 min postexposure. The results of another study on the effects of real F-16 aircraft overflights at an altitude of approximately 125 meters were in line with these moderate and rapidly fading effects and revealed no significant behavioral or spatial effects on mountain sheep (*Ovis canadensis nelsoni*) in the Desert National Wildlife Refuge (Krausman et al. 1998). Five mountain sheep fitted with heart rate monitors revealed a significant increase in heart rate in only 21 of 149 overflights, and heart rates returned to preflight levels within 2 min.

Creel et al. (2002) investigated the effect of snowmobile activity on stress levels in elk (*Cervus elaphus*) in Yellowstone National Park and wolves (*Canis lupus*) in three US national parks. Among elk, fecal glucocorticoid levels covaried daily in relation to the number of snowmobiles in the area, taking age and snowpack into account. In the case of wolves, stress hormone levels were elevated in areas in which snowmobile use occurred, especially at times of heavy use, compared over a time course of years and regions. Despite these correlations with the animal stress hormone levels, neither elk nor wolves showed snowmobile-related population declines. The authors are therefore cautious about implications for wildlife management but do not exclude potential long-term influences on elk population decline and overall health and welfare.

9.4.5 Bat Life in Noisy Environments

Bats are a special group of terrestrial mammals with respect to the potential impact of man-made sounds on hearing physiology and behavior. They are aerial predators, typically exhibiting nocturnal activity peaks, and use vocal signals to communicate, detect, and capture prey. Depending on the species, bats attend to relatively low-frequency sounds made by active prey or the high-frequency reflections of echolocation calls. An anecdotal report made by Shirley et al. (2001) suggested that disturbance from a nearby music festival significantly delayed the departure time of bats leaving their roosts; however, the delay may have been due to light levels at the festival rather than acoustic interference because sunset time is correlated with the time that bats leave their roost.

In a more controlled laboratory investigation, playbacks of traffic and gas compressor noise at a variety of levels and distances resulted in a two- to threefold increase in the amount of time required for pallid bats (*Antrozous pallidus*) to locate the sounds of prey (Bunkley and Barber 2015), reducing foraging success. Interestingly, in a similar study, foraging efficiency in Daubenton's bats (*Myotis daubentonii*) was reduced during traffic noise playbacks; however, the effects appeared to be due to noise avoidance and not the result of acoustic masking or reduced attention (Luo et al. 2015). Additional correlational and experimental studies on the impact of man-made noise on bats have also shown clear effects on local diversity and density, call production features, spatial preferences, and foraging efficiency (e.g., Bunkley et al. 2015; Luo et al. 2015).

A correlational study on the effect of compressor noise on aerial foraging in the San Juan Basin of New Mexico revealed significantly reduced bat activity near sound-generating gas-extraction stations in comparison to quiet well pads (Bunkley et al. 2015). This pattern was observed in species using relatively low-frequency echolocation calls (<35 kHz), whereas species using relatively high-frequency calls remained equally abundant around stations, independent of local noise levels. This frequency-dependent effect suggests that sound is a critical factor underlying the reduction of suitable bat habitat.

Schaub et al. (2008) conducted an experimental study on the greater mouse-eared bat (*Myotis myotis*), a gleaning species that listens passively for rustling sounds to locate prey. They used a choice experiment to test whether background noise affects spatial preference and performance during foraging. Two separate compartments containing food (mealworms), one associated with a noisy background and the other not, served as the experimental setup. Bats avoided compartments exhibiting traffic noise as well as "natural" noise (wind in vegetation) or computer-generated broadband noise. When bats did forage in the noisy compartment, they were less efficient than in the quiet compartment in terms of detecting and successfully capturing prey. In a follow-up experiment with the same bat species, Siemers and Schaub (2011) also showed that foraging effort and success declined with noise level, suggesting that foraging conditions decline with proximity to highways.

A number of studies have reported acoustic changes in bat calls under noisy conditions. Brazilian free-tailed bats (*Tadarida brasiliensis*) produce relatively low-frequency but longer calls with a narrower frequency bandwidth when occupying noisier sites compared with quiet sites associated with gas-extraction stations (Bunkley et al. 2015). Hage et al. (2013) and Hage and Mezner (2013) also reported upward shifts in call amplitude and frequency during noisy conditions in greater horseshoe bats (*Rhinolophus ferrumequinum*). Spectral shifts were also observed when noise bands were in the 10- to 30-kHz range, well below the 65- to 75-kHz range that is used for echolocation in this species.

A recent study also revealed the multimodal complexity and flexibility of cues used in bat hunting strategies (Gomes et al. 2016). Fringe-lipped bats (*Trachops cirrhosis*) were shown to hunt frogs by using low-frequency acoustic as well as visual cues that were layered on top of the high-frequency reflections of their echolocation calls. The animals were shown to bias perceptual weights depending on the variation in signal-to-noise ratios across available information channels and shifted their hunting strategy to vision and echolocation cues in the presence of low-frequency masking noise.

The papers reviewed above demonstrate that both gleaning and aerial foraging bat species avoid man-made sounds, a behavior that reduces the size of habitats that are suitable for foraging. Those species that persist in noisy areas may experience lower feeding rates even when expending an equivalent effort to those hunting in lower noise habitats, raising conservation concern in an energy expenditure context (Schaub et al. 2008; Siemers and Schaub 2011). Although at least some bat species appear to be flexible in terms of reliance on acoustic cues and echolocation call production, it is not clear whether such species are negatively affected by man-made sounds. It may well be that temporal or spectral shifts in echolocation calls (e.g., Hage et al. 2013; Bunkley et al. 2015) or a redundancy reduction associated with information processing across modalities (Gomes et al. 2016) may still have negative effects on bat foraging efficiency. Consequently, despite their relatively high-frequency hearing range, noise exposure should also be a serious concern when considering bat conservation protocols.

9.5 Conclusions

After reviewing the literature addressing questions related to the impact of man-made sound on terrestrial mammals, it is tempting to catalog the impact into simple high-risk, medium-risk, and low-risk categories. By all accounts, high-risk conditions generally refer to exposure conditions producing physical inner ear trauma and permanent loss of acoustic sensitivity. Medium-risk conditions can be thought of as those that produce temporary hearing loss without apparent inner ear trauma, although recent findings suggest that auditory synaptopathy should be taken into consideration. Low-risk conditions might refer to man-made sounds that produce a more moderate physiological or behavioral response or that mask biologically relevant sounds.

However, low-risk events may also have dramatic outcomes in terms of reproductive success and survival. Missed mating or foraging opportunities can have large and lasting consequences, and failing to detect the sound of a predator has clear and evident survival implications. Furthermore, the widespread nature of moderate and more subtle effects of noise exposure in time and space add to conservation concerns because of the possible accumulation of potentially undesirable outcomes. It is therefore important to not only recognize and assess the impact of duration, intensity, and spectral properties of man-made sounds in efforts to evaluate their potentially trauma- and stress-inducing power, but it is equally necessary to include measures of species-specific ecology and behavioral sensitivity. Furthermore, it may well be that there are critical life stages or seasonal conditions that determine the vulnerability of a species to man-made acoustic disturbances. Finally, when the variability of hearing thresholds and ranges represented among terrestrial mammals are combined with the probability of exposure and the timing of exposure, the cumulative impact on the health and welfare of individuals, populations, and communities must be factored into realistic risk management assessments.

These concluding remarks may apply to all taxa, but terrestrial mammals stand out due to their high sensitivity to airborne sounds and the spectral diversity in hearing ranges across species. Their exquisite acoustic sensitivity makes them especially vulnerable to inner ear damage and disrupted behavioral patterns in response to man-made noise. Crepuscular and nocturnal habits are also specific to many terrestrial mammal species and complicate observation and measurement strategies, which should encourage the development of better detection technology. However, in at least some rural and urban settings, terrestrial mammals may be at least partially released from man-made noise pollution pressures, decoupling elements of behavior, like foraging, from noisy human activity. Unfortunately, human activities are encroaching into increasingly larger areas of natural habitats and noise produced by activities such as gas-extraction equipment is generated continuously regardless of time of day or season. Future studies of the effects of man-made noise on terrestrial mammals will therefore remain and likely grow in interest due to our concern for animal welfare and conservation.

Compliance with Ethics Requirements
Hans Slabbekoorn declares that he has no conflict of interest.
JoAnn McGee declares that she has no conflict of interest.
Edward J. Walsh declares that he has no conflict of interest.

References

Agnes, F., Sartorelli, P., Abdi, B. H., & Locatelli, A. (1990). Effect of transport loading or noise on blood biochemical variables in calves. *American Journal of Veterinary Research*, 51, 1679-1681.

Alario, P., Gamallo, A., Beato, M. J., & Trancho, G. (1987). Body weight gain, food intake and adrenal development in chronic noise stressed rats. *Physiology & Behavior*, 40, 29-32.

Algers, B., & Jensen, P. (1985). Communication during suckling in the domestic pig. Effects of continuous noise. *Applied Animal Behaviour Science*, 14, 49-61.

Algers, B., Ekesbo, I., & Strömberg, S. (1978). The impact of continuous noise on animal health. *Acta Veterinaria Scandinavica Supplementum*, 68, 1-26.

Ashmore, J. (2008). Cochlear outer hair cell motility. *Physiological Reviews*, 88, 173-210.

Babisch, W., Fromme, H., Beyer, A., & Ising, H. (2001). Increased catecholamine levels in urine in subjects exposed to road traffic noise: The role of stress hormones in noise research. *Environment International*, 26, 475-481.

Baker, P. J., & Harris, S. (2007). Urban mammals: What does the future hold? An analysis of the factors affecting patterns of use of residential gardens in Great Britain. *Mammal Reviews*, 37, 297-315.

Ballachanda, B. B. (1997). Theoretical and applied external ear acoustics. *Journal of the American Academy of Audiology*, 8, 411-420.

Barber, J. R., Crooks, K. R., & Fristrup, K. M. (2010). The costs of chronic noise exposure for terrestrial organisms. *Trends in Ecology and Evolution*, 25, 180-189.

Berglund, B., Hassmén, P., & Job, R. F. (1996). Sources and effects of low-frequency noise. *The Journal of the Acoustical Society of America*, 99, 2985-3002.

Birke, L. (2002). Effects of browse, human visitors and noise on the behaviour of captive orang utans. *Animal Welfare*, 11, 189-202.

Bradshaw, C. J. A., Boutin, S., & Hebert, D. M. (1997). Effects of petroleum exploration on woodland caribou in northeastern Alberta. *The Journal of Wildlife Management*, 61, 1127-1133.

Bradshaw, R. H., Parrott, R. F., Goode, J. A., Lloyd, D. M., Rodway, R. G., & Broom, D. M. (1996). Behavioural and hormonal responses of pigs during transport: Effect of mixing and duration of journey. *Animal Science* 62, 547-554.

Braga, J., Loubes J.-M., Descouens, D., Dumoncel, J., Thackeray, J. F., Kahn, J.-L., de Beer, F., Riberon, A., Hoffman, K., Balaresque, P., & Gilissen, E. (2015). Disproportionate cochlear length in genus *Homo* shows a high phylogenetic signal during apes' hearing evolution. *PLoS ONE*, 10(6), e0127780.

Brask, T. (1978). The noise protection effect of the stapedius reflex. *Acta Oto-Laryngologica*, 86, 116-117.

Brownell, W. E. (1983). Observations on a motile response in isolated outer hair cells. In W. R. Webster & L. M. Aitken (Eds.), *Mechanisms of Hearing* (pp. 5-10). Clayton, Victoria, Australia: Monash University Press.

Brumm, H., & Slabbekoorn, H. (2005). Acoustic communication in noise. *Advances in the Study of Behavior*, 35, 151-209.

Brumm, H., Voss, K., Köllmer, I., & Todt, D. (2004). Acoustic communication in noise: Regulation of call characteristics in a New World monkey. *Journal of Experimental Biology*, 207, 443-448.

Bunkley, J. P., & Barber, J. R. (2015). Noise reduces foraging efficiency in pallid bats (*Antozous pallidus*). *Ethology*, 121, 1116-1121.

Bunkley, J. P., McClure, C. J. W., Kleist, N. J., Francis, C. D., & Barber, J. R. (2015). Anthropogenic noise alters bat activity levels and echolocation calls. *Global Ecology and Conservation*, 3, 62-71.

Buran, B. N., Strenzke, N., Neef, A., Gundelfinger, E. D., Moser, T., & Liberman, M. C. (2010). Onset coding is degraded in auditory nerve fibers from mutant mice lacking synaptic ribbons. *The Journal of Neuroscience*, 30, 7587-7597.

Burow, A., Day, H. E., & Campeau, S. (2005). A detailed characterization of loud noise stress, intensity analysis of hypothalamo-pituitary-adrenocortical axis and brain activation. *Brain Research*, 1062, 63-73.

Chen, H. L., & Koprowski, J. L. (2015). Animal occurrence and space use change in the landscape of anthropogenic noise. *Biological Conservation*, 192, 315-322.

Clark, J. G. (1981). Uses and abuses of hearing loss classification. *American Speech-Language-Hearing Association*, 23, 493-500.

Clark, W. W. (1991). Recent studies of temporary threshold shift (TTS) and permanent threshold shift (PTS) in animals. *The Journal of the Acoustical Society of America*, 90, 155-163.

Cooke, C. M., & Schillaci, M. A. (2007). Behavioral responses to the zoo environment by white handed gibbons. *Applied Animal Behaviour Science*, 106, 125-133.

Coordes, A., Gröschel, M., Ernst, A., & Basta, D. (2012). Apoptotic cascades in the central auditory pathway after noise exposure. *Journal of Neurotrauma*, 29, 1249-1254.

Coppola, C. L., Enns, R. M., & Grandin, T. (2006). Noise in the animal shelter environment: Building design and the effects of daily noise exposure. *Journal of Applied Animal Welfare Science*, 9, 1-7.

Creel, S., Fox, J. E., Hardy, A., Sands, J., Garrott, B., & Peterson, R. O. (2002). Snowmobile activity and glucocorticoid stress responses in wolves and elk. *Conservation Biology*, 16, 809-814.

Cui, B., Gai, Z., She, X., Wang, R., & Xi, Z. (2016). Effects of chronic noise on glucose metabolism and gut microbiota-host inflammatory homeostasis in rats. *Scientific Reports*, 6, 36693.

Duarte, M. H. L., Vecci, M. A., Hirsch, A., & Young, R. Y. (2011). Noisy human neighbours affect where urban monkeys live. *Biology Letters*, 7, 840-842.

Edge, A. S., & Chen, Z. Y. (2008). Hair cell regeneration. *Current Opinion in Neurobiology*, 18, 377-382.

Eggermont, J. J. (2017). Acquired hearing loss and brain plasticity. *Hearing Research*, 343, 176-190.

Egnor, S. E., & Hauser, M. D. (2006). Noise-induced vocal modulation in cotton-top tamarins (*Saguinus oedipus*). *American Journal of Primatology*, 68, 1183-1190.

Egnor, S. E., Wickelgren, J. G., & Hauser, M. D. (2007). Tracking silence: Adjusting vocal production to avoid acoustic interference. *Journal of Comparative Physiology A: Neuroethology, Sensory, Neural, and Behavioral Physiology*, 193, 477-483.

Fay, R. R. (1988). *Hearing in Vertebrates: A Psychophysics Databook*. Winnetka, IL: Hill-Fay Associates.

Forman, R. T. T., & Alexander, L. E. (1998). Roads and their major ecological effects. *Annual Review of Ecology and Systematics*, 29, 207-231.

Francis, C. D., Keist, N. J., Ortega, C. P., & Cruz, A. (2012). Noise pollution alters ecological services: Enhanced pollination and disrupted seed dispersal. *Proceedings of the Royal Society B: Biological Sciences*, 279, 2727-2735.

Francis, M. J., Spooner, P. G., & Matthews, A. (2015). The influence of urban encroachment on squirrel gliders (*Petaurus norfolcensis*): Effects of road density, light and noise pollution. *Wildlife Research*, 42, 324-333.

Freeman, S., Khvoles, R., Cherny, L., & Sohmer, H. (1999). Effect of long-term noise exposure on the developing and developed ear in the rat. *Audiology & Neurotology*, 4, 207-218.

Frost, S. B., & Masterton, R. B. (1994). Hearing in primitive mammals: *Monodelphis domestica* and *Marmosa elegans*. *Hearing Research*, 76, 67-72.

Gao, F., Zhang, J., Sun, X., & Chen, L. (2009). The effect of postnatal exposure to noise on sound level processing by auditory cortex neurons of rats in adulthood. *Physiology & Behavior*, 97, 369-373.

Gates, G. R., Saunders, J. C., Bock, G. R., Aitkin, L. M., & Elliott, M. A. (1974). Peripheral auditory function in the platypus, *Ornithorhynchus anatinus*. *The Journal of the Acoustical Society of America*, 56, 152-156.

Geverink, N. A., Bühnemann, A., van de Burgwal, J. A., Lambooij, E., Blokhuis, H. J., & Wiegant, V. M. (1998). Responses of slaughter pigs to transport and lairage sounds. *Physiology & Behavior*, 63, 667-673.

Gomes, D. G. E., Page, R. A., Geipel, I., Taylor, R. C., Ryan, M. J., & Halfwerk, W. (2016). Bats perceptually weight prey cues across sensory systems when hunting in noise. *Science*, 6305, 1277-1280.

Gourévitch, B., Edeline, J. M., Occelli, F., & Eggermont, J. J. (2014). Is the din really harmless? Long-term effects of non-traumatic noise on the adult auditory system. *Nature Reviews Neuroscience*, 15, 483-491.

Griffiths, M. (1978). *The Biology of the Monotremes*. New York: Academic Press.

Gue, M., Fioramonti, J., Frexinos, J., Alvinerie, M., & Bueno, L. (1987). Influence of acoustic stress by noise on gastrointestinal motility in dogs. *Digestive Diseases and Sciences*, 32, 1411-1417

Guinan, J., J., Jr. (2006). Olivocochlear efferents: Anatomy, physiology, function, and the measurement of efferent effects in humans. *Ear and Hearing*, 27, 589-607.

Guo, L., Li, P. H., Li, H., Colicino, E., Colicino, S., Wen, Y., Zhang, R., Feng, X., Barrow, T. M., Cayir, A., & Baccarelli, A. A. (2017). Effects of environmental noise exposure on DNA methylation in the brain and metabolic health. *Environmental Research*, 153, 73-82.

Habersetzer, J. (1981). Adaptive echolocation sounds in the bat *Rhinopoma hardwickei*. A field study. *Journal of Comparative Physiology A: Neuroethology, Sensory, Neural, and Behavioral Physiology*, 144, 559-566.

Hage, S. R., & Metzner, W. (2013). Potential effects of anthropogenic noise on echolocation behavior in horseshoe bats. *Communicative Integrative Biology* 6, e24753.

Hage, S. R., Jiang, T., Berquist, S. W., Feng, J., & Metzner, W. (2013). Ambient noise induces independent shifts in call frequency and amplitude within the Lombard effect in echolocating bats. *Proceedings of the National Academy of Sciences of the United States of America*, 110, 4063-4068.

Halfwerk, W., & Slabbekoorn, H. (2015). Pollution going multimodal: The complex impact of the human-altered sensory environment on animal perception and performance. *Biology Letters*, 11, 20141051.

Hall, S. J. G., Kirkpatrick, S. M., Lloyd, D. M., & Broom, D. M. (1998). Noise and vehicular motion as potential stressors during the transport of sheep. *Animal Science*, 67, 467-473.

Heffner, H. E., Heffner, R. S., Contos, C., & Ott, T. (1994). Audiogram of the hooded Norway rat. *Hearing Research*, 73, 244-248.

Heffner, R. S., & Heffner, H. E. (1982). Hearing in the elephant (*Elephas maximus*): Absolute sensitivity, frequency discrimination, and sound localization. *Journal of Comparative Physiology and Psychology*, 96, 926-944.

Heffner, R. S., & Heffner, H. E. (1985). Hearing range of the domestic cat. *Hearing Research*, 19, 85-88.

Heffner, R. S., & Heffner, H. E. (1992). Hearing and sound localization in blind mole rats, *Spalax ehrenbergi*. *Hearing Research*, 62, 206-216.

Heffner, R. S., Koay, G., & Heffner, H. E. (2006). Hearing in large (*Eidolon helvum*) and small (*Cynopterus brachyotis*) non-echolocating fruit bats. *Hearing Research*, 221, 17-25.

Hudspeth, A. J. (2014). Integrating the active process of hair cells with cochlear function. *Nature Reviews Neuroscience*, 15, 600-614.

International Organization for Standardization. (1961). *Normal Equal-Loudness Level Contours for Pure Tones and Normal Threshold of Fearing Under Free Field Listening Conditions*. ISO R/226:1961, International Organization for Standardization, Geneva.

Jackson, L. L., Heffner, R. S., & Heffner, H. E. (1999). Free-field audiogram of the Japanese macaque (*Macaca fuscata*). *The Journal of the Acoustical Society of America*, 106, 3017-3023.

Katti, M., & Warren, P. S. (2004). Tits, noise and urban bioacoustics. *Trends in Ecology and Evolution*, 19, 109-110.

Kawase, T., & Liberman, M. C. (1993). Antimasking effects of the olivocochlear reflex. I. Enhancement of compound action potentials to masked tones. *Journal of Neurophysiology*, 70, 2519-2532.

Kerber, S., & Seeber, B. U. (2012). Sound localization in noise by normal-hearing listeners and cochlear implant users. *Ear and Hearing*, 33, 445-457.

Kern, J. M., & Radford, A. N. (2016). Anthropogenic noise disrupts use of vocal information about predation risk. *Environmental Pollution*, 218, 988-995.

Khimich, D., Nouvian, R., Pujol, R., tom Dieck, S., Egner, A., Gundelfinger, E. D., & Moser, T. (2005). Hair cell synaptic ribbons are essential for synchronous auditory signalling. *Nature*, 434, 889-894.

Kirk, E. C., & Smith, D. W. (2003). Protection from acoustic trauma is not a primary function of the medial olivocochlear efferent system. *Journal of the Association for Research in Otolaryngology*, 4, 445-465.

Koay, G., Heffner, H. E., & Heffner, R. S. (1997). Audiogram of the big brown bat (*Eptesicus fuscus*). *Hearing Research*, 105, 202-210.

Konkle, A. T., Baker, S. L., Kentner, A. C., Barbagallo, L. S., Merali, Z., & Bielajew, C. (2003). Evaluation of the effects of chronic mild stressors on hedonic and physiological responses: Sex and strain compared. *Brain Research*, 992, 227-238.

Krausman, P. R., Wallace, M. C., Hayes, C. L., & DeYoung, D. W. (1998). Effects of jet aircraft on mountain sheep. *Journal of Wildlife Management*, 62, 1246-1254.

Krebs, H., Macht, M., Weyers, P., Weijers, H. G., & Janke, W. (1996). Effects of stressful noise on eating and non-eating behavior in rats. *Appetite*, 26, 193-202.

Kujawa, S. G., & Liberman, M. C. (2009). Adding insult to injury: Cochlear nerve degeneration after "temporary" noise-induced hearing loss. *The Journal of Neuroscience*, 29, 14077-14085.

Lanier, J. L., Grandin, T., Green, R. D., Avery, D., & McGee, K. (2000). The relationship between reaction to sudden, intermittent movements and sounds and temperament. *Journal of Animal Science*, 78, 1467-1474.

Larsen, M. J., Sherwen, S. L., & Rault, J.-L. (2014). Number of nearby visitors and noise level affect vigilance in captive koalas. *Applied Animal Behaviour Science*, 154, 76-82.

Laurian, C., Dussault, C., Ouellet, J.-P., Courtois, R., Poulin, M., & Breton, L. (2008). Behavior of moose relative to a road network. *Journal of Wildlife Management*, 72, 1550-1557.

Lavinsky, J., Ge, M., Crow, A. L., Pan, C., Wang, J., Salehi, P., Myint, A., Eskin, E., Allayee, H., Lusis, A. J., & Friedman, R. A. (2016). The genetic architecture of noise-induced hearing loss: Evidence for a gene-by-environment interaction. *G3: Genes, Genomes, Genetics*, *6*, 3219-3228.

Lin, H. W., Furman, A. C., Kujawa, S. G., & Liberman, M. C. (2011). Primary neural degeneration in the guinea pig cochlea after reversible noise-induced threshold shift. *Journal of the Association for Research in Otolaryngology*, 12, 605-616.

Luo, J., Siemers, B. M., & Koselj, K. (2015). How anthropogenic noise affects foraging. *Global Change Biology*, 21, 3278-3289.

Mahendra Prashanth, K. V., & Sridhar, V. (2008). The relationship between noise frequency components and physical, physiological and psychological effects of industrial workers. *Noise and Health*, 10, 90-98.

Maison, S. F., & Liberman, M. C. (2000). Predicting vulnerability to acoustic injury with a noninvasive assay of olivocochlear reflex strength. *The Journal of Neuroscience*, 20, 4701-4707.

Manley, G. A. (2012). Evolutionary paths to mammalian cochleae. *Journal of the Association for Research in Otolaryngology*, 13, 733-743.

Marti, O., & Armario, A. (1997). Influence of regularity of exposure to chronic stress on the pattern of habituation of pituitary-adrenal hormones, prolactin and glucose. *Stress*, 1, 179-189.

Mason, J. T., McClure, C. J. W., & Barber, J. R. (2016). Anthropogenic noise impairs owl hunting behavior. *Biological Conservation*, 199, 29-32.

Mason, M. J. (2016). Structure and function of the mammalian middle ear. II: Inferring function from structure. *Journal of Anatomy*, 228, 284-299.

McComb, K., Shannon, G., Sayialel, K. N., & Moss, C. (2014). Elephants can determine ethnicity, gender, and age from acoustic cues in human voices. *Proceedings of the National Academy of Sciences of the United States of America*, 111, 5433-5438.

McConnell, J. C. (1990). Acoustic structure and receiver response in domestic dogs (*Canis familiaris*). *Animal Behaviour*, 39, 897-904.

McGee, J., & Walsh, E. J. (2015). Cochlear transduction and the molecular basis of peripheral auditory pathology. In P. W. Flint, B. H. Haughey, V. J. Lund, J. Niparko, M. A. Richardson, K. T. Robbins, & J. R. Thomas (Eds.), *Cummings Otolaryngology Head and Neck Surgery*, 6th ed. (pp. 2234-2274). Philadelphia: Mosby, Elsevier, Inc.

McLellan, B. N., & Shackleton, D. M. (1988). Grizzly bears and resource extraction industries: Effects of roads on behavior, habitat use and demography. *Journal of Applied Ecology*, 25, 451-460.

Mennitt, D., Fristrup, K., Sherrill, K., & Nelson, L. (2013). Mapping sound pressure levels on continental scales using a geospatial sound model. *Proceedings of 42nd International Congress and Exposition on Noise Control Engineering 2013 (INTER-NOISE 2013): Noise Control for Quality of Life*, Innsbruck, Austria, September 15-18, 2013, pp. 41-51.

Mennitt, D. J., Fristrup, K. M., & Nelson, L. (2015). A spatially explicit estimate of environmental noise exposure in the contiguous United States. *The Journal of the Acoustical Society of America*, 137, 2339-2340.

Merchan, C. I., Diaz-Balteiro, L., & Soliño, M. (2014). Noise pollution in national parks: Soundscape and economic valuation. *Landscape and Urban Planning*, 123, 1-9.

Michaud, D. S., Miller, S. M., Ferrarotto, C., Keith, S. E., Bowers, W. J., Kumarathsan, P., Marro, L., & Trivedi, A. (2005). Exposure to chronic noise and fractionated X-ray radiation elicits biochemical changes and disrupts body weight gain in rats. *International Journal of Radiation Biology*, 81, 299-307.

Morris-Drake, A., Kern, J. M., & Radford, A. N. (2016). Cross-modal impacts of anthropogenic noise on information use. *Current Biology*, 26, R903-R912.

Neff, W. D., & Hind, J. E. (1955). Auditory thresholds of the cat. *The Journal of the Acoustical Society of America*, 27, 480-483.

Niemeyer, W. (1971). Relations between the discomfort level and the reflex threshold of the middle ear muscles. *Audiology*, 10, 172-176.

Nonaka, S., Takahashi, R., Enomoto, K., Katada, A., & Unno, T. (1997). Lombard reflex during PAG-induced vocalization in decerebrate cats. *Neuroscience Research*, 29, 283-289.

Oliveira, M. J., Pereira, A. S., Castelo Branco, N. A., Grande, N. R., & Aguas, A. P. (2001). In utero and postnatal exposure of Wistar rats to low frequency/high intensity noise depletes the tracheal epithelium of ciliated cells. *Lung*, 179, 225-232.

Otten, W., Kanitz, E., Puppe, B., Tuchscherer, M., Brüssow, K. P., Nürnberg, G., & Stabenow, B. (2004). Acute and long term effects of chronic intermittent noise stress on hypothalamic-pituitary-adrenocortical and sympatho-adrenomedullary axis in pigs. *Animal Science*, 78, 271-283.

Owen, M. A., Swaisgood, R. R., Czekala, N. M., Steinman, K., & Lindburg, D. G. (2004). Monitoring stress in captive giant pandas (*Ailuropoda melanoleuca*), behavioral and hormonal responses to ambient noise. *Zoo Biology*, 23, 147-164.

Owen, M. A., Hall, S., Bryant, L., & Swaisgood, R. R. (2014). The influence of ambient noise on maternal behavior in a Bornean sun bear (*Helarctos malayanus euryspilus*). *Zoo Biology*, 33, 49-53.

Perrot, X., & Collet, L. (2014). Function and plasticity of the medial olivocochlear system in musicians: A review. *Hearing Research*, 308, 27-40.

Pijanowski, B. C., Farina, A., Gage, S. H., Dumyahn, S. L., & Krause, B. L. (2011). What is soundscape ecology? An introduction and overview of an emerging new science. *Landscape Ecology*, 26, 1213-1232.

Powell, D. M., Carlstead, K., Tarou, L. R., Brown, J. L., & Monfort, S. L. (2006). Effects of construction noise on behavior and cortisol levels in a pair of captive giant pandas (*Ailuropoda melanoleuca*). *Zoo Biology*, 25, 391-408.

Pujol, R., & Puel, J. L. (1999). Excitotoxicity, synaptic repair, and functional recovery in the mammalian cochlea: A review of recent findings. *Annals of the New York Academy of Sciences*, 884, 249-254.

Quadros, S., Goulart, V. D. L., Passos, L., Vecci, M. A. M., & Young, R. J. (2014). Zoo visitor effect on mammal behaviour: Does noise matter? *Applied Animal Behaviour Science*, 156, 78-84.

Rabanal, L. I., Kuehl, H. S., Mundry, R., Robbins, M. M., & Boesch, C. (2010). Oil prospecting and its impact on large rainforest mammals in Loango National Park, Gabon. *Biological Conservation*, 143, 1017-1024.

Rabat, A. (2007). Extra-auditory effects of noise in laboratory animals: The relationship between noise and sleep. *Journal of the American Association of Laboratory Animal Science*, 46, 35-41.

Rabat, A., Bouyer, J. J., Aran, J. M., Courtiere, A., Mayo, W., & Le Moal, M. (2004). Deleterious effects of an environmental noise on sleep and contribution of its physical components in a rat model. *Brain Research*, 1009, 88-97.

Rabat, A., Bouyer, J. J., Aran, J. M., Le Moal, M., & Mayo, W. (2005). Chronic exposure to an environmental noise permanently disturbs sleep in rats: Inter-individual vulnerability. *Brain Research*, 1059, 72-82.

Rabat, A., Bouyer, J. J., George, O., Le Moal, M., & Mayo, W. (2006). Chronic exposure of rats to noise: Relationship between long-term memory deficits and slow wave sleep disturbances. *Behavioural Brain Research*, 171, 303-312.

Rabin, L. A., Coss, R. C., & Owings, D. H. (2006). The effects of wind turbines on antipredator behavior in California ground squirrels (*Spermophilus beecheyi*). *Biological Conservation*, 131, 410-420.

Recanzone, G. H., Schreiner, C. E., & Merzenich, M. M. (1993). Plasticity in the frequency representation of primary auditory cortex following discrimination training in adult owl monkeys. *The Journal of Neuroscience*, 13, 87-103.

Robles, L., & Ruggero, M. A. (2001). Mechanics of the mammalian cochlea. *Physiological Reviews*, 81, 1305-1352.

Rondinini, C., & Doncaster, C. (2002). Roads as barriers to movement for hedgehogs. *Functional Ecology*, 16, 504-509.

Rosowski, J. J. (1991). The effects of external- and middle-ear filtering on auditory threshold and noise-induced hearing loss. *The Journal of the Acoustical Society of America*, 90, 124-135.

Rosowski, J. J. (2013). Comparative middle ear structure and function in vertebrates. In S. Puria, R. R. Fay, & A. N. Popper (Eds.), *The Middle Ear* (pp. 31-65). New York: Springer-Verlag.

Schäffer, D., Marquardt, V., Marx, G., & von Borell, E. (2001). Noise in animal housing: A review with emphasis on pig housing. *Deutsche Tierärztliche Wochenschrift*, 108, 60-66.

Schaub, A., Ostwald, J., & Siemers, B. M. (2008). Foraging bats avoid noise. *Journal of Experimental Biology*, 211, 3174-3180.

Senzaki, M., Yamaura, Y, Francis, C. D., & Nakamura, F. (2016). Traffic noise reduces foraging efficiency in wild owls. *Scientific Reports*, 6, 30602.

Shannon, G., Angeloni, L. M., Wittemyer, G., Fristrup, K. M., & Crooks, K. R. (2014). Road traffic noise modifies behaviour of a keystone species. *Animal Behaviour*, 94, 135-141.

Shi, X. (2016). Pathophysiology of the cochlear intrastrial fluid-blood barrier (review). *Hearing Research*, 338, 52-63.

Shirley, M. D. F., Armitage, V. L., Barden, T. L., Gough, M., Lurz, P. W. W., Oatway, D. E., & Rushton, S. P. (2001). Assessing the impact of a music festival on the emergence behaviour of a breeding colony of Daubenton's bats (*Myotis daubentonii*). *Journal of Zoology*, 254, 367-373.

Siemers, B. M., & Schaub, A. (2011). Hunting at the highway: Traffic noise reduces foraging efficiency in acoustic predators. *Proceedings of the Royal Society B: Biological Sciences*, 278, 1646-1652.

Simmons, A. M., Hom, K. N., Warnecke, M., & Simmons, J. A. (2016). Broadband noise exposure does not affect hearing sensitivity in big brown bats (*Eptesicus fuscus*). *Journal of Experimental Biology*, 209, 1031-1040.

Singh, P., Price, T. D., & Hartley, I. (2015). Causes of the latitudinal gradient in birdsong complexity assessed from geographical variation within two Himalayan warbler species. *Ibis*, 157, 511-527.

Sinnot, J. M., Stebbins, W. C., & Moody, D. B. (1975). Regulation of voice amplitude by the monkey. *The Journal of the Acoustical Society of America*, 58, 412-414.

Sivian, L. J., & White, S. D. (1933). On minimum audible sound fields. *The Journal of the Acoustical Society of America*, 4, 288-321.

Slabbekoorn, H., & Ripmeester, E. A. P. (2008). Birdsong and anthropogenic noise: Implications and applications for conservation. *Molecular Ecology*, 17, 72-83.

Spoendlin, H., & Brun, J. P. (1973). Relation of structural damage to exposure time and intensity in acoustic trauma. *Acta Otolaryngology*, 75, 220-226.

Stansfeld, S. A., & Matheson, M. P. (2003). Noise pollution: Non-auditory effects on health. *British Medical Bulletin*, 68, 243-257.

Stephens, D. B., Bailey, K. J., Sharman, D. F., & Ingram, D. L. (1985). An analysis of some behavioural effects of the vibration and noise components of transport in pigs. *Quarterly Journal of Experimental Physiology*, 70, 211-217.

Syka, J., Popelář, J., Druga, R., & Vlkova, A. (1988). Descending central auditory pathway—Structure and function. In J. Syka & R. B. Masterton (Eds.), *Auditory Pathway* (pp. 279-292). Boston: Springer.

Talling, J. C., Waran, N. K., Wathes, C. M., & Lines, J. A. (1996). Behavioural and physiological responses of pigs to sound. *Applied Animal Behaviour Science*, 48, 187-202.

Talling, J. C., Waran, N. K., Wathes, C. M., & Lines, J. A. (1998). Sound avoidance by domestic pigs depends upon characteristics of the signal. *Applied Animal Behaviour Science*, 58, 255-266.

Tamura, H., Ohgami, N., Yajima, I., Iida, M., Ohgami, K., Fujii, N., Itabe, H., Kusudo, T., Yamashita, H., & Kato, M. (2012). Chronic exposure to low frequency noise at moderate levels causes impaired balance in mice. *PLoS ONE*, 7, e39807.

Uran, S. L., Caceres, L. G., & Guelman, L. R. (2010). Effects of loud noise on hippocampal and cerebellar-related behaviors. Role of oxidative state. *Brain Research*, 1361, 102-114.

Uran, S. L., Aon-Bertolino, M. L., Caceres, L. G., Capani, F., & Guelman, L. R. (2012). Rat hippocampal alterations could underlie behavioral abnormalities induced by exposure to moderate noise levels. *Brain Research*, 1471, 1-12.

Walther, F. R. (1969). Flight behaviour and avoidance of predators in Thomson's gazelle *(Gazella thomsoni)*. *Behaviour*, 34, 184-219.

Wang, Y., Hirose, K., & Liberman, M. C. (2002). Dynamics of noise-induced cellular injury and repair in the mouse cochlea. *Journal of the Association for Research in Otolaryngology*, 3, 248-268.

Warr, W. B., & Guinan, J. J. (1979). Efferent innervation of the organ of Corti: Two separate systems. *Brain Research*, 173, 152-155.

Waynert, D. F., Stookey, J. M., Schwartzkopf-Genwein, K. S., Watts, J. M., & Waltz, C. S. (1999). Response of beef cattle to noise during handling. *Applied Animal Behaviour Science*, 62, 27-42.

Weeks, C. A., Brown, S. N., Lane, S., Haesman, L., Benson, T. , & Warriss, P. D. (2009). Noise levels in lairages for cattle, sheep and pigs in abattoirs in England and Wales. *Veterinary Record*, 165, 308-314.

Weir, J. T., Wheatcroft, D. J., & Price, T. D. (2012). The role of ecological constraint in driving the evolution of avian song frequency across a latitudinal gradient. *Evolution*, 66, 2773-2783.

Weisenberger, M. E., Krausman, P. R., Wallace, M. C., DeYoung, D. W., & Maughan, O. E. (1996). Effects of simulated jet aircraft noise on heart rates and behavior of desert ungulates. *Journal of Wildlife Management*, 60, 52-61.

Wilson, D. E., & Reeder, D. A. M. (Eds.). (2005). *Mammal Species of the World. A Taxonomic and Geographic Reference,* 3rd ed. Baltimore, MD: Johns Hopkins University Press.

Winslow, R. L., & Sachs, M. B. (1988). Single-tone intensity discrimination based on auditory-nerve rate responses in backgrounds of quiet, noise, and with stimulation of the crossed olivocochlear bundle. *Hearing Research*, 35, 165-189.

Wrege, P. H., Rowland, E. D., Thompson, B. G., & Batruch, N. (2010). Use of acoustic tools to reveal otherwise cryptic responses of forest elephants to oil exploration. *Conservation Biology*, 24, 1578-1585.

Wysocki, A. B. (1996). The effect of intermittent noise on wound healing. *Advances in Skin and Wound Care*, 9, 35-39.

Zheng, X. Y., Henderson, D., McFadden S. L., & Hu, B. H. (1997). The role of the cochlear efferent system in acquired resistance to noise-induced hearing loss. *Hearing Research*, 104, 191-203.

Chapter 10
Effects of Noise on Marine Mammals

Christine Erbe, Rebecca Dunlop, and Sarah Dolman

Abstract Marine mammals (whales, dolphins, seals, sea lions, sea cows) use sound both actively and passively to communicate and sense their environment, covering frequencies from a few hertz to greater than 100 kHz, differing with species. Although a few documents on marine mammal sound production and reception date back 200 years, concern about the effects of man-made noise on marine mammals has only been documented since the 1970s. Underwater noise can interfere with key life functions of marine mammals (e.g., foraging, mating, nursing, resting, migrating) by impairing hearing sensitivity, masking acoustic signals, eliciting behavioral responses, or causing physiological stress. Many countries are developing and updating guidelines and regulations for underwater noise management in relation to marine mammal conservation. In the United States, the Marine Mammal Protection Act, enacted in 1972, is increasingly being applied to underwater noise emission. Common mitigation methods include (1) time/area closures, (2) the establishment of safety zones that are monitored by visual observers or passive acoustics and that lead to shut-down or low-power operations if animals enter these zones, (3) noise reduction gear like bubble curtains around pile driving, and (4) noise source modifications or operational parameters like soft starts. Mitigation management mostly deals with single operations (like a one-month seismic survey). Key questions that remain are how noise impacts accumulate over time and multiple exposures, how multiple acoustic and nonacoustic stressors interact, and how effects on individuals affect a population as a whole.

C. Erbe (✉)
Centre for Marine Science and Technology, Curtin University, Perth, WA, Australia
e-mail: c.erbe@curtin.edu.au

R. Dunlop
Cetacean Ecology and Acoustics Laboratory, School of Veterinary Science, University of Queensland, Gatton, QLD, Australia
e-mail: r.dunlop@uq.edu.au

S. Dolman
Whale and Dolphin Conservation, Chippenham, Wiltshire, UK
e-mail: sarah.dolman@whales.org

H. Slabbekoorn et al. (eds.), *Effects of Anthropogenic Noise on Animals*, Springer Handbook of Auditory Research 66,
https://doi.org/10.1007/978-1-4939-8574-6_10

Keywords Behavioral response · Bioacoustic impact · Environmental management · Marine Mammal Protection Act · Marine mammals · Masking · Population consequences of acoustic disturbance · Population consequences of disturbance · Safety zone · Stress · Temporary threshold shift · Underwater noise

10.1 Introduction

There are about 130 species of marine mammals taxonomically grouped into 21 families (Table 10.1). Cetaceans (whales, dolphins, and porpoises) and sirenians (sea cows) are fully aquatic. The marine carnivores (seals, sea lions, and otters), however, split their time between land and water. Marine mammals inhabit all of the world's oceans, from the deep offshore waters (with sperm whales [*Physeter macrocephalus*], elephant seals [*Mirounga* sp.], and Cuvier's beaked whales [*Ziphius*

Table 10.1 Marine mammal taxonomy

Latin name	Common name
Order Cetacea	Whales, dolphins & porpoises
Suborder Mysticeti	Baleen whales
Family Balaenidae	Right and bowhead whales
Family Neobalaenidae	Pygmy right whale
Family Balaenopteridae	Rorquals
Family Eschrichtiidae	Gray whale
Suborder Odontoceti	Toothed whales
Family Delphinidae	Oceanic dolphins
Family Platanistidae	South Asian river dolphins
Family Iniidae	Amazon river dolphin, boto
Family Lipotidae	Chinese river dolphin, baiji
Family Pontoporiidae	Franciscana
Family Phocoenidae	Porpoises
Family Monodontidae	Narwhal and beluga
Family Physeteridae	Sperm whale
Family Kogiidae	Pygmy and dwarf sperm whales
Family Ziphiidae	Beaked whales
Order Sirenia	Sea cows
Family Trichechidae	Manatees
Family Dugongidae	Dugongs
Order Carnivora	Carnivores
Family Mustelidae	Marine otters
Family Ursidae	Polar bear
Suborder Pinnipedia	Seals, sea lions, and walrus
Family Phocidae	True seals
Family Otariidae	Eared seals and sea lions
Family Odobenidae	Walrus

cavirostris] diving down to 2–3 km; e.g., Schorr et al. 2014) to the shallow coastal waters, and a few species, such as river dolphins, are in rivers.

Marine mammals live in a medium through which sound propagates better than potential cues or signals of any other sensory modality, such as light. They have therefore evolved to use sound both actively and passively in all biologically important behaviors (Tyack 2000), including socializing, traveling, hunting, breeding, and parental care. Examples of marine mammal sounds are the behavior-specific and signature whistles of dolphins (Caldwell and Caldwell 1965; Herzing 1996) and the song of humpback whales (*Megaptera novaeangliae*; Payne and McVay 1971). Cultural transmission of sound structure is evident in killer whales (*Orcinus orca*; Ford 1991) who have dialects that can be used to distinguish between populations living in the same area. Odontocetes (toothed whales) also emit sound to echolocate during navigation and foraging (Au 1993). Examples of passive sound usage include listening to acoustic cues from the environment, predators, and prey (e.g., Deecke et al. 2002; Gannon et al. 2005).

Knowledge of the auditory capabilities of marine mammals is important to understand their acoustic ecology, how they sense their environment, over what ranges they remain in acoustic contact, whether they can detect predators and prey, and how they receive ambient and man-made noise. Studies examining the hearing of marine mammals date back two centuries (e.g., Home 1812). However, it was not until the 1970s that underwater sound emitted by human activities in the oceans was first recognized to sometimes be in conflict with marine mammals. Payne and Webb (1971) concluded that ship noise decreased the communication range of baleen whales, a concern still echoing 40 years later (Clark et al. 2009). Impacts documented in the 1970s also include hauled-out walrus (*Odobenus rosmarus*) disturbance by aircraft associated with Arctic petroleum exploration (Salter 1979) and, opportunistically, a beaked whale mass stranding coincident with naval maneuvers (van Bree and Kristensen 1974). The Marine Mammal Protection Act (MMPA; passed in 1972) and the Endangered Species Act (ESA; passed in 1973) set the legal framework for conservation (including marine mammals) in the United States. A symposium on the effects of sound on wildlife held in Spain in 1977 included discussions of the impacts of man-made sound on marine biological systems and resulted in a book on the effects of man-made noise on wildlife (Fletcher and Busnel 1978).

Since then, dedicated research rather than opportunistic observations has grown (Williams et al. 2015), leading to the landmark book *Marine Mammals and Noise* (Richardson et al. 1995). In the 1990s, the Heard Island Feasibility Test and the Acoustic Thermometry of Ocean Climate (ATOC) experiments caused widespread public concern, which resulted in a large-scale marine mammal research program (National Research Council 1994, 2000). Much of the research on the sound impacts on marine mammals over the past two decades has been driven by "take" authorizations under the MMPA[1] that require baseline and in situ monitoring. In fact, the MMPA has increasingly been applied to sound sources so that nearly all "incidental

[1] The MMPA defines "take" as "hunt, harass, capture, or kill."

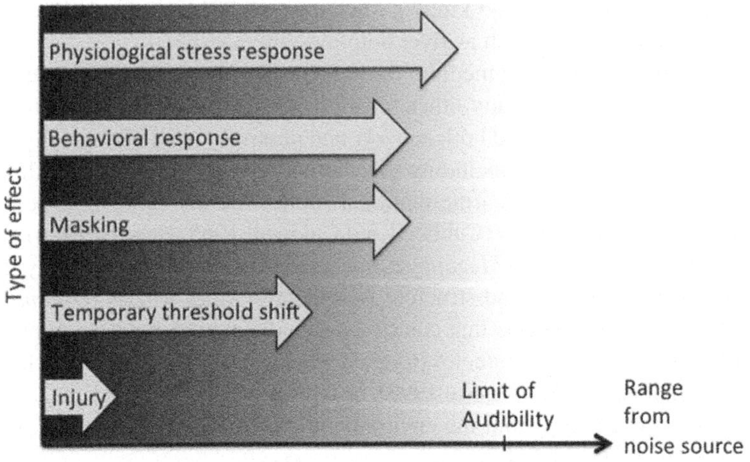

Fig. 10.1 Assuming a source of sound is located on the left side, its received level decreases with range. Near the source, a variety of bioacoustic impacts may be possible. Some effects such as stress, behavioral responses, or masking of communication may extend to long ranges where the sound is just audible. The ranges over which the above effects happen and the order of effects by range may depend on the type of sound, its spectral and temporal characteristics, the local sound propagation environment, ambient-noise conditions, the characteristics of the auditory system of the receiving animal, its current behavioral state, and/or past experience

take" authorizations issued under the Act today are at least partly, and in many cases primarily, focused on acoustic impacts (Roman et al. 2013). Public concern in the United States has culminated in law suits under the MMPA and ESA, specifically criticizing the US Navy's use of active sonar (Zirbel et al. 2011).

Underwater sound from human activities can have a variety of immediate effects on marine mammals, including injury, temporary loss of hearing, behavioral responses, masking, and stress (Fig. 10.1). Severity of the impacts typically decreases with the range from the sound source and depends on the specific scenario consisting of the type of sound, the acoustic environment, and the receiving individual. At the longest ranges, the sound might barely be audible or discernible above the ambient noise. The animal's hearing abilities and the level of ambient noise determine the range of audibility.

In extreme cases, close to the source, injuries such as tissue or organ damage (e.g., a permanent loss of hearing called permanent threshold shift [PTS]; see Southall et al. 2007) may be found (see Saunders and Dooling, Chap. 4). If hearing loss recovers with time, it is termed a temporary threshold shift (TTS). TTS has been demonstrated in a number of odontocetes and pinnipeds (walruses, seals, and sea lions) in controlled sound exposure experiments (e.g., Kastelein et al. 2013). Severe to profound hearing loss has been measured in some wild, stranded odontocetes (Mann et al. 2010), but the cause and whether this was TTS or PTS is unknown. Less extreme behavioral responses might be seen both near and far from the source. Beluga whales (*Delphinapterus leucas*), for example, responded to faraway (tens of kilometers) icebreakers that were expected to be barely audible (Finley et al. 1990). Acoustic masking occurs when noise interferes with the detection of acoustic signals important to animals. This can also happen at long ranges, such as when the call of a faraway conspecific is masked by similarly faint man-made noise. Such "extreme" scenarios were modeled for icebreakers and beluga whales based on behavioral masked hearing experiments with a captive beluga whale involving beluga calls and different types of icebreaker sound at different levels (Erbe and Farmer 1998, 2000). Stress is a physiological response and might be a direct result of exposure to man-made sounds that are unknown or resemble the sounds of predators or are an indirect result of exposure when injury or masking cause stress (Wright et al. 2007). Therefore, stress can occur at various ranges. The concept of impact ranges or zones, as illustrated in Fig. 10.1, applies to the immediate impacts on individual animals near an active source, and most evidence of sound impacts on marine mammals is related to short-term, individual responses. Figure 10.1 does not capture extreme responses like mass strandings (Cox et al. 2006), where whales were likely subjected to only moderate received levels not expected to cause physical damage and yet stranded and died due to perhaps more complex processes.

The National Research Council (2005) defined an effect as "biologically significant" if it keeps an animal from growing, surviving, and reproducing, thereby potentially affecting the survival of its population. The challenge is to figure out how temporary responses accumulate over space, time, and individuals to ultimately lead to population-level effects. Behavioral effects might accumulate over many years before such impacts are realized. However, in the case of sound-related mass strandings, a single instance of behavioral disturbance can affect the local population. A framework to develop the progression from immediate, individual impacts to population impacts is provided by the population consequences of disturbance (PCoD) model, and this chapter is organized along the stages of the PCoD model.

10.2 Underwater Sound

In this chapter, the focus is on waterborne sound. Pinnipeds, polar bears (*Ursus maritimus*), and otters (*Lutrinae*) spend time both on land and in water and are hence subject to sound impacts in both media. Responses to airborne sound are not reviewed here. Instead, the reader is referred to the comprehensive review work by Richardson et al. (1995).

Understanding the ambient sound conditions in marine mammal habitats is important because ambient sound limits the detection of and likely response to man-made sound (see Larsen and Radford, Chap. 5). The ocean is naturally noisy. Wind, rain, breaking waves, cracking polar ice, and subsea earthquakes and volcanoes all contribute to the ambient noise in certain geographic regions. Some of these natural sounds propagate over hundreds to thousands of kilometers so that Antarctic ice breakup is recorded on hydrophones near Australia (Gavrilov and Li 2007). Wenz (1962) summarized the spectral characteristics of typical ambient-noise sources,

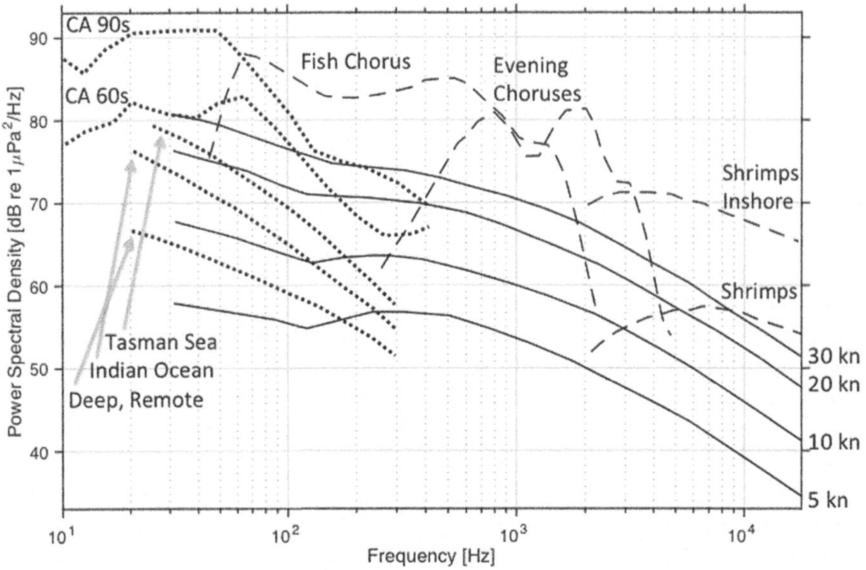

Fig. 10.2 Typical source spectra of ambient noise: wind, biological choruses, and distant shipping. Distant shipping sound was recorded at five locations: off California in the late 1990s (CA 90s; Andrew et al. 2002) and early 1960s (CA 60s; Wenz 1969); in the Tasman Sea, Australia; in the southeast Indian Ocean; and in Australian deep water remote from shipping lanes. Wind-dependent noise is shown at four different wind speeds. The tropical biological choruses vary with location, time of day, and season (Cato 1978). Shrimp noise typically only exists in shallow (<40 m) water. Based on Cato (2008)

yielding the widely used *Wenz curves*. Other significant contributors to underwater sound are, of course, marine animals, including mammals, fishes, crustaceans, and urchins, many of which create biological choruses (Cato 1978). Under conditions where many animals call at the same time, they can raise the ambient level in a characteristic frequency band for several hours. Typical spectra of such choruses along with wind-dependent ambient noise and distant shipping are shown in Fig. 10.2.

All marine operations produce underwater sound: shipping, transport, oil and gas, defense, tourism, fishing, offshore minerals, offshore wind and water energy, and on- and near-shore construction (Richardson et al. 1995; Wyatt 2008). Sound produced in air, such as by airplanes and helicopters, transmits into the water at incidence angles less than 13° from the vertical. Similarly, sound produced in air on ship decks or oil platforms enters the water by radiation through the hull or support legs. Figure 10.3 shows smoothed and simplified example source spectra of underwater noise emitted by human activities. Such source spectra are typically used in conjunction with sound propagation models (e.g., Jensen et al. 2011) to predict received levels at some range for the purpose of environmental impact assessment.

The nature of the sound propagation environment plays an important role because it changes the spectral and temporal characteristics of a sound as it travels from the

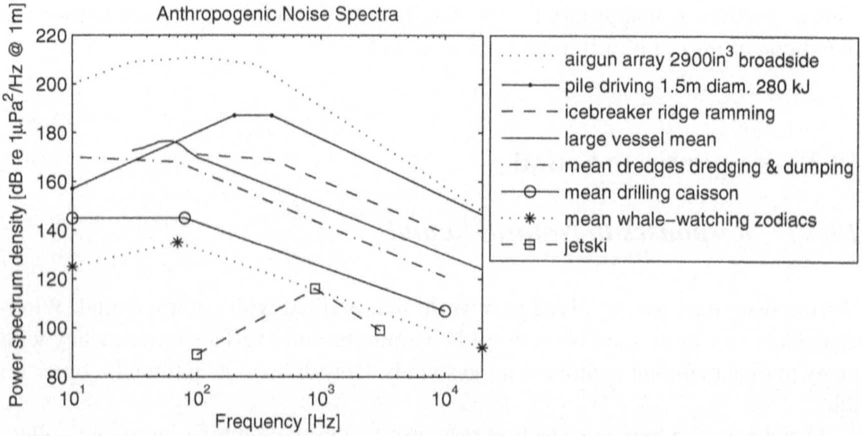

Fig. 10.3 Stylized far-field equivalent source spectra of example anthropogenic operations (reprinted from Erbe 2012). All sounds were recorded at some range and back-propagated to a nominal distance of 1 m. In the case of the pulsed sources (airgun array and pile driving), the power spectrum density was computed over the 90% pulse length, which is the duration from the 5% to the 95% point on the cumulative energy curve (see, e.g., Erbe 2011)

source to the receiver (see Larsen and Radford, Chap. 5), in this case, a marine mammal. Hence, propagation affects the potential for bioacoustic impact. Overall, the broadband received sound level attenuates with range, but the rate of attenuation depends on the bathymetry, the hydroacoustic profile of the water column, and the geoacoustic parameters of the upper seafloor. The spectral characteristics change with range because energy at different frequencies is attenuated at different rates. In deep water, energy at low frequencies (<100 Hz) can travel over very long ranges, which is why ship noise has the potential to mask the calls of baleen whales over many tens of kilometers. In the case of pulsed sound, the duration of the pulse typically increases with range. Thus, sound from a seismic airgun array might consist of 100-ms pulses every few seconds and marine mammals close to the source likely detect the calls of conspecifics through the quiet gaps in the seismic sound pattern. At a 100-km range, however, each pulse might be several seconds long (Guerra et al. 2011), forming a continuous (albeit band-limited) sound.

As the waveform of the sound changes during propagation, the various acoustic quantities, which might be responsible for different types of effects in different animal species, also change. Obviously, source level alone is no indicator for impact. The received root-mean-square sound pressure level (SPL_{rms}), the received sound exposure level (SEL; weighted or not), and the received peak SPL (SPL_{peak}) have most commonly been investigated as potential indicators for impact (e.g., Southall et al. 2007). Other parameters might play a role, e.g., the signal-to-noise ratio, kurtosis, duty cycle, and/or pulse rise time. Different acoustic quantities, either alone or in combination, are likely linked to different types of effect, and this link might be different in different species. Comparing sound sources merely by source level or source spectrum is inappropriate. As such, Fig. 10.3 should not be used to rank the likelihood of impact of different types of sound.

10.3 Responses to Sound

10.3.1 Responses to Natural Sound

Marine mammals have evolved in a world that is filled with natural sound. Wind-dependent elevation of ambient noise is ubiquitous and overlaps in frequency with many marine mammal communication sounds. How do marine mammals cope with this?

The changes in human speech in response to elevated ambient noise are collectively known as the Lombard effect, where signalers modify vocal characteristics such as level, pitch, and/or rate of signal production in a noisy environment by which they may improve signal detection probability at the receiver (Lombard 1911). Humpback whales were found to increase the source level of their social vocalizations by 0.9 dB for every 1-dB increase in wind-elevated ambient noise (Dunlop et al. 2014), maintaining about 60-dB signal excess above the ambient-noise level in medium wind conditions. There was evidence, however, of an upward limit to this response, perhaps due to anatomical constraints. When the ceiling is reached, a change in spectral characteristics or call type might be an alternative option by which to communicate in noisy conditions. Another study found that humpback whales switched communication signal type from primarily vocal signals to mechanical signals generated at the surface (breaches, slaps) in the same spread of ambient-noise levels as in the Lombard study (Dunlop et al. 2010). It is unclear whether the use of different signal types changed the message sent or maintained the original communication.

Vocalizing conspecifics, such as singing humpback whales, also raise the background noise in which animals must continue to communicate with one another. The "cocktail party effect" (Cherry 1953) is experienced by receivers due to acoustic interference from multiple vocalizing conspecifics (akin to the challenge humans face when communicating with each other at a noisy party). To some extent, the receiver is able to focus on the signaler and filter out the background noise of conspecific sounds. Most of the research on how animals communicate in noisy social aggregations has been carried out in birds and frogs (see Bee and Micheyl 2008 for a review; see also Simmons and Narins, Chap. 7). Many marine mammals live in large groups too, making a cacophony of calls. How one group member is able to communicate successfully with another in among the chatter has not been studied.

Currently, there is no information in the literature on behavioral changes (e.g., in diving behavior or movement patterns) or physiological changes (e.g., TTS or stress) in response to natural fluctuations in ambient noise. Understanding the

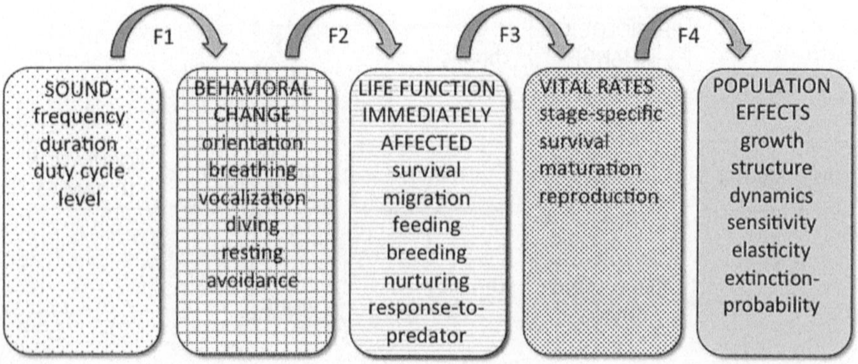

Fig. 10.4 Population consequences of acoustic disturbance (PCAD) model breaking down the link between sound and population-level impact into a set of stages connected by transfer functions (F1-F4). Modified from the National Research Council (2005)

natural repertoire of responses and their frequency of occurrence might aid in assessing the biological significance of responses to man-made noise.

10.3.2 The Population Consequences of Disturbance Framework

The biggest challenge in bioacoustic impact assessments and in the management of underwater sound is how to progress from short-term observations of individual responses to predictions of population-level consequences. The population consequences of acoustic disturbance (PCAD) model (National Research Council 2005) was developed as a conceptual framework linking behavioral and some physiological responses to man-made sound with biologically significant, population-level effects (Fig. 10.4). The PCAD model breaks the causal relationship between individual behavior change and population effects into a set of more manageable stages connected by transfer functions. The model starts with measurements of sound characteristics, such as the spectral characteristics and the duration, and links these via transfer function 1 to short-term, individual behavior change, such as a change in dive pattern or vocalization rate. A sudden change in diving might affect an animal's foraging activity. An onset of avoidance might disrupt resting or nursing. Disruption of vocalization might interfere with breeding. Transfer function 2 makes these links between behavioral change and the life functions immediately affected. If feeding is repeatedly disrupted, an animal might suffer caloric and nutritional deficiencies affecting its survival. Interrupted breeding comes at a cost to reproduction. Transfer function 3 links life functions to vital rates. Transfer function 4 yields population effects, such as a reduced population growth rate and changes in population structure. Unfortunately, the paucity of data underlying the various stages and

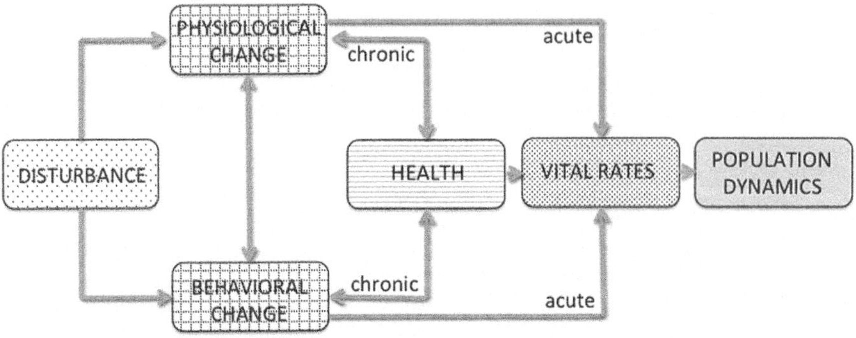

Fig. 10.5 Population consequences of disturbance (PCoD) model linking disturbance of individuals to population-level effects (Harwood et al. 2014)

transfer functions limits the PCAD model to a conceptual rather than predictive model.

The PCAD framework was broadened to include disturbance other than man-made noise and to account for the impact of disturbance on physiology in addition to behavior (Harwood et al. 2014; New et al. 2014). The result is the PCoD model (Fig. 10.5). PCoD begins with a disturbance (either acoustic or not), which results in a behavioral or physiological response. In the acute case, these responses immediately affect vital rates (e.g., survival or reproduction). For chronic disturbance, the animal's health is impaired, eventually impacting vital rates. Changes in vital rates lead to changes in population dynamics.

The PCoD model has been translated into a formal, mathematical model that can be parameterized with data from case studies. The data needed to implement the PCoD model for the case of acoustic disturbance include the sound field around the source, the sound parameters and their levels that cause behavioral or physiological responses (ideally as dose-response curves), the number of animals that are likely going to be exposed to these levels, the relationship between physiological impacts and vital rates (ideally by age and gender), the relationship between the number of behavioral disturbances and vital rates, the population size, and demographic parameters. Uncertainty in all of these input parameters can be included in the model (Harwood et al. 2014).

10.3.3 Disturbance

Disturbance in the PCoD model can be any interruption of "normal" functioning and leads to behavioral or physiological changes in an animal. The disturbance might be some form of alteration of the environment such as climate change, artificial light at nighttime, chemical discharge, the mere presence of an oilrig or vessel, or the sound emitted by industrial operations. Within the legal framework of the US

MMPA, disturbance is considered Level B harassment. For the purpose of this chapter, disturbance is deemed acoustic disturbance as a result of underwater sound from anthropogenic activities. For brevity, acoustic disturbance to hauled-out pinnipeds by airborne sound, such as that from overflying aircraft, is excluded in this overview.

10.3.4 Behavioral Change

Behavioral response study (BRS) designs are often followed to assess whether or not there is a significant behavioral change in an animal in response to an acoustic stimulus. BRSs in marine mammals have focused on five main research areas (Deecke 2006): (1) to determine the function of conspecific vocalizations, (2) as a method of wildlife management (e.g., using heterospecific sounds to deter animals from specific areas), (3) to study predator-prey interactions, (4) to study individual and kin recognition, and (5) to determine the response to anthropogenic noise, the focus of this chapter.

In the literature, BRSs using an anthropogenic stimulus are sometimes called "controlled exposure experiments" (CEEs), although this implies the anthropogenic stimulus is given in carefully controlled doses, which may not always be true. The experimental design is a "before, during, and after" (BDA) procedure, where the behavior of the animals is measured before, during, and after the stimulus is given. An appropriate before period provides one type of control. The before behavior is compared with the during behavior to look for a significant change. The after period allows the assessment of the animals' behavioral "recovery" and to determine if the behavioral change was short term (only in the during phase) or long term (i.e., the animals continue to display a change in behavior after the stimulus has ended). The during phase can be classified according to the "treatment" given: usually either an "active" treatment (where the sound stimulus is presented) or a "control" treatment (where no stimulus is given but everything else remains the same). The control treatment helps determine other factors that may have contributed to the behavioral response (e.g., a response to the tow vessel rather than to the towed airguns, as studied in the Behavioural Response of Australian Humpback whales to Seismic Surveys (BRAHSS) experiment; Dunlop et al. 2015, 2016; Fig. 10.6). Treatments could also be sounds from other cetaceans. Sometimes the calls of killer whales, the apex predator, are used (e.g., Allen et al. 2014). One can then compare the behavioral response to the anthropogenic stimulus with the response to a "biologically meaningful" stimulus.

Although the majority of literature on "marine mammals and man-made sounds" reports behavioral responses, carrying out a scientifically robust BRS is not easy, resulting in common errors (Campbell and Stanley 1966), which make interpretation of results and comparison among studies difficult. The experimenter might wrongly attribute an observed behavioral change to the acoustic stimulus, when, in fact, it was due to some other environmental parameter (internal validity error). A

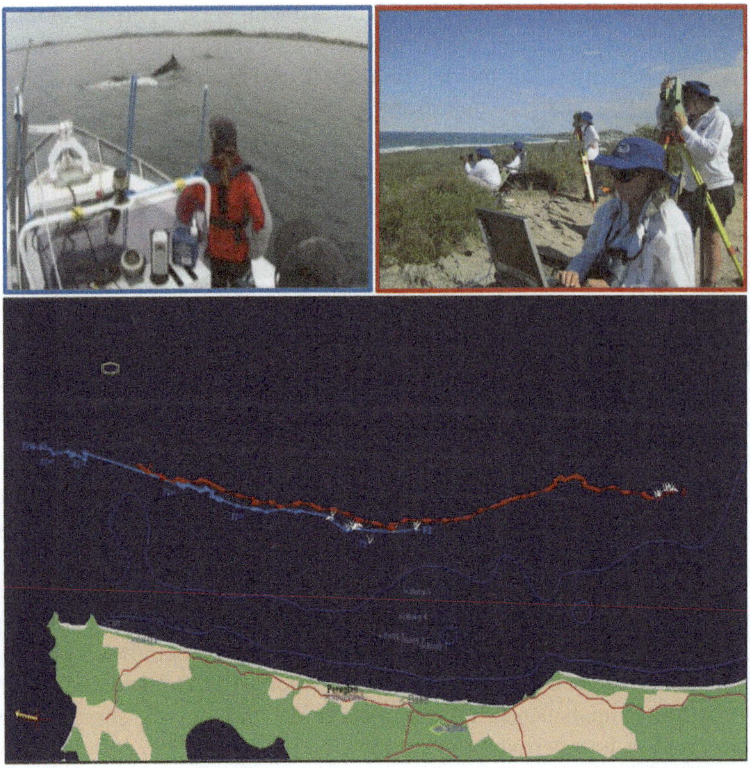

Fig. 10.6 In the Behavioural Response of Australian Humpback whales to Seismic Surveys (BRAHSS) experiment, migrating humpback whales were tracked by boat (*top left*) and from shore (*top right*), yielding tracks (*bottom*; *blue line*, boat based; *red line*, land based of the same group) that were compared between noise exposure and control conditions

common mistake is that replicates are either spatially or temporally segregated. Furthermore, conclusions are commonly generalized (e.g., to other man-made sounds, entire populations, or other species) beyond the validity of the experiment (external validity error). Exposing animals to more exemplars of the stimulus in multiple geographic regions or ecological settings and using more species will overcome this problem, although this will often require a larger number of experiments and will have cost and ethical implications.

Summaries of behavioral responses of marine mammals to man-made noise show a large variability in the received levels (differing by many tens of decibels) and the severity in the response from minor to severe (Richardson et al. 1995; Southall et al. 2007; Gomez et al. 2016). These differences are partly due to different populations, sound sources, contexts, and environments (Ellison et al. 2012; Dunlop et al. 2013). The large within-species variability might be explained by individual differences such as prior exposure (habituation versus sensitization), motivation, age, gender, and health. One would not expect all animals in a population to

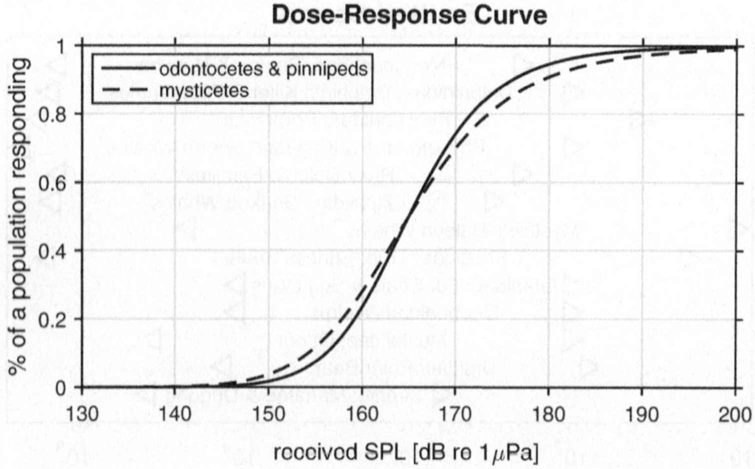

Fig. 10.7 Dose-response relationship used by the US Navy to estimate the percentage of a population of marine mammals responding to naval sonar during the Atlantic Fleet active sonar training exercises and the Gulf of Alaska Navy training activities (US Department of the Navy 2008, 2009). *SPL* sound pressure level

respond at the same received level all the time. Rather, the response of a population can be represented as a dose-response curve (Fig. 10.7), showing the range in sound levels over which a certain percentage might react (e.g., Miller et al. 2014). The usefulness of the received level as a predictor for the behavioral response remains questionable (Gomez et al. 2016), and the criteria to determine whether or not an animal responds can be difficult to define. Movement and avoidance metrics (e.g., a deviation in course, speed, or dive profile), or a change in behavioral state (e.g., from feeding to traveling) might be too broad scale. Animals may be exhibiting more subtle reactions like changes in vocal signals or fine-scale movement. The use of a multisensor digital acoustic recording tag (DTAG; Johnson and Tyack 2003), which, along with the acoustic data, simultaneously records orientation and movement of the whales, has advanced these studies, finding changes in fluke rate, duration and rate of descent and ascent (DeRuiter et al. 2013), and changes in acoustic behavior (Miller et al. 2009).

When exposed to naval low-frequency sonar, humpback whales increased the length of song (Miller et al. 2000; Fristrup et al. 2003), beaked whales ceased echolocation (Tyack et al. 2011; DeRuiter et al. 2013), and long-finned pilot whales (*Globicephala melas*) increased their call rate (Rendell and Gordon 1999). In the presence of boat noise, killer whales increased their call duration (Foote et al. 2004) and level (Holt et al. 2009); beluga whales increased their call level, reduced their call rate, and shifted the mean frequency up (Lesage et al. 1998; Scheifele et al. 2005); bottlenose dolphins (*Tursiops truncatus*) increased their whistle rate (Buckstaff 2004); and fin whales (*Balaenoptera physalus*) decreased their call duration and bandwidth (Castellote et al. 2012). These acoustic responses could be due

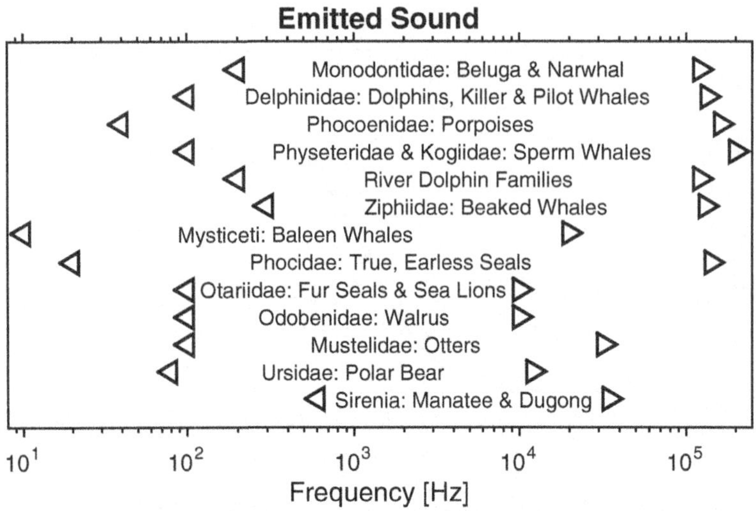

Fig. 10.8 Rough bandwidths of sound emitted by several marine mammal families. Modified from Erbe (2012)

to the boat disturbance per se, changes in context due to the presence of the boat, changes in social behavior, a response to experienced masking, or any combination of these.

10.3.5 Physiological Change

Masking

Masking is the interference of ambient noise with the detection or recognition of signals (e.g., whale communication sounds or dolphin echolocation clicks). The frequencies emitted by various groupings of marine mammals are sketched in Fig. 10.8, covering a range from 10 Hz to 200 kHz. Underwater sound of abiotic, biotic, or anthropogenic origin covers a similar range (see Figs. 10.2 and 10.3), likely making masking a common and ubiquitous phenomenon.

Various parameters relating to an animal's hearing capabilities play a role in masking (Erbe et al. 2016a). Any sound within the hearing range of an animal can be masked. The audiograms (i.e., hearing thresholds as a function of frequency) of marine mammals are summarized by Erbe et al. (2016a). The minimum thresholds recorded from individuals belonging to several species grouped by family are shown in Fig. 10.9. No audiogram exists for any of the mysticete species, sperm whales, and polar bears under water.

Masking depends on the spectral characteristics of both signal and noise at the receiver (see Dooling and Leek, Chap. 2). At a low signal-to-noise ratio, the signal

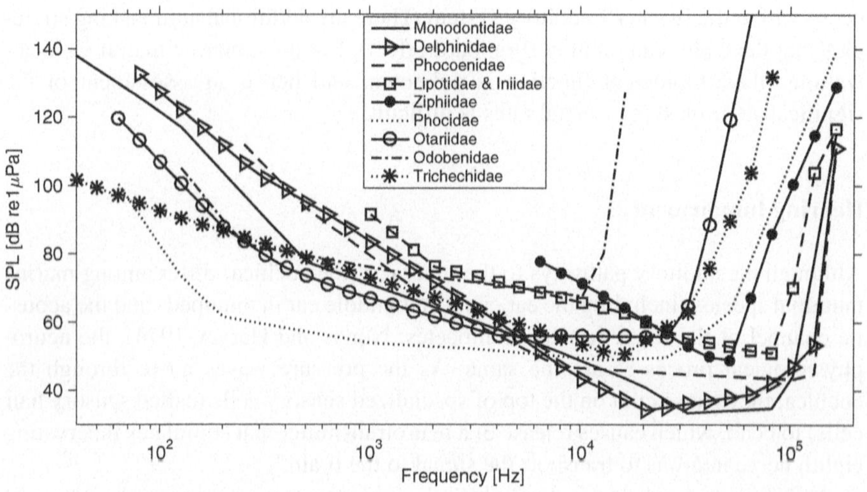

Fig. 10.9 Audiograms of several marine mammal families. The minimum thresholds recorded from any individual within these groups were interpolated to one-third octave frequencies. Updated from Erbe (2012)

might merely be detectable but not recognizable. A higher signal-to-noise ratio is needed for the animal to recognize or discriminate the signal, as known from studies with birds (Dooling et al. 2009; see also Halfwerk, Lohr, and Slabbekoorn, Chap. 8). The critical ratio (CR) is defined as the difference in the signal (tone) intensity level and the power spectrum density level of masking (white) noise at the detection threshold. CRs have been measured in a dozen marine mammal species (Erbe et al. 2016a). The CR has proven to be a strong predictor for masking in birds (Dooling and Blumenrath 2014) when the noise is continuous and broadband and the signal has strong tonal character. The CR was also a good predictor for the masking of a tonal beluga call in broadband ship noise (Erbe and Farmer 1998; Erbe 2008).

In realistic listening scenarios, signal and noise have complex spectral and temporal structures and likely arrive at the listener from different directions. If the ambient noise is amplitude modulated across a wide band of frequencies, the animal can use information from outside the band of the signal to determine when the signal occurs, simply as a difference in correlation between bands. This is called a comodulation masking release and has been demonstrated with beluga whales, bottlenose dolphins, California sea lions (*Zalophus californianus*), and harbor seals (e.g., Branstetter and Finneran 2008; Erbe 2008). If the ambient noise has quieter gaps (as in the case of strongly amplitude-modulated ship noise and natural ice-cracking noise), and if the signal is long or repetitive, the animal might detect the signal from the pieces that emerge through the intermittent noise pattern by gap listening, as shown in beluga whales (Erbe 2008). If the signal and the noise arrive from different directions, a spatial release from masking occurs based on directional hearing capabilities, as measured in bottlenose dolphins, California sea lions, and harbor seals (e.g., Turnbull 1994; Holt and Schusterman 2007). The above processes occur within the listener's auditory system. There are additional antimasking strategies that the caller can employ (Lombard effect). For most marine mammal vocalizations, their biological function is unknown, and hence an assessment of the significance of masking to vital rates is difficult.

Hearing Impairment

Although the auditory pathways to the inner ear (the cochlea) differ among marine mammal species (including the ear canal and middle ear in pinnipeds and the acoustic channel of the lower jaw in odontocetes; Norris and Harvey 1974), the neurophysiological processes are the same. As the pressure waves move through the cochlea, they cause cilia on the top of specialized sensory cells (called sensory hair cells) to bend, which causes release of a neurotransmitter that stimulates innervating eighth nerve neurons to transport the signal to the brain.

A PTS occurs when the neurophysiological process is permanently damaged (see Saunders and Dooling, Chap. 4). One of the most common ways is damage to the sensory hair cells from overexposure to sound, causing hair cell death and/or damage to the innervating neurons of the eighth nerve. A PTS is measured as a permanent increase in the hearing threshold (audiogram) at various frequencies. A TTS occurs when there is temporary impairment of the sensory hair cells; in other words, the animal's hearing threshold recovers to the normal audiogram after acoustic exposure (see Saunders and Dooling, Chap. 4). However, recent studies have shown that even a TTS may not be completely recoverable in that the nerves that transport the electrical signal to the brain may be irreversibly damaged, a damage that does not affect the audiogram but affects hearing in noisy conditions (Kujawa and Liberman 2009; Liberman 2016).

There are no data on the sound characteristics that could cause PTS in any marine mammal because, for ethical reasons, PTS has not been intentionally induced in controlled experiments. Rather, small amounts of TTS have been induced with pure tones, sonar signals, band-limited white noise, or airguns in beluga whales, bottlenose dolphins, harbor porpoises (*Phocoena phocoena*), Yangtze finless porpoises (*Neophocaena phocaenoides asiaeorientalis*), California sea lions, harbor seals, and elephant seals. The level of TTS depends on a number of factors that may include sound level, pressure rise time, duration, duty cycle, and spectral characteristics. Maximum TTS is typically seen at frequencies higher than the stimulus frequency (Kastak et al. 2008), and this difference was shown to increase with the sound level (Kastelein et al. 2014a). The relationship between exposure level and frequency agrees with equal loudness contours (Finneran and Schlundt 2013). Pinnipeds seem equally susceptible to airborne and underwater sound if exposure levels are given in terms of sensation levels (relative to the audiogram; Kastak et al. 2006). Exposures with equal cumulative SELs but different interpulse intervals produced different amounts of TTS (Kastelein et al. 2014b). TTS recovery has followed a $-10\log$ (minute) slope in some individuals (Kastak et al. 2006).

There is interesting evidence of a conditioned hearing sensitivity reduction in false killer whales (*Pseudorca crassidens*) and bottlenose dolphins whereby a brief and loud "warning" sound reduced the sensitivity to a subsequent sound (Nachtigall and Supin 2013, 2014). This mechanism might reduce the potential for hearing damage in certain circumstances.

Stress

The stress response in animals involves two different but interconnected systems (Hall 2011). The first is the sympathetic nervous system response in which the release of epinephrine and norepinephrine triggers fast physiological changes. These include an increase in heart rate, blood pressure, and gas exchange as well as a redistribution of blood to the brain and muscles, away from the stomach and other organs that are nonessential for fight or flight responses. These short-term stress responses act as adaptive countermeasures to potentially life-threatening events and can co-occur with a range of fight-or-flight behavioral responses. The second type of stress response, the hypothalamic-pituitary-adrenal (HPA) axis, is a chain of endocrine reactions, with the goal of restoring homeostasis. The whole HPA process usually begins between 3 and 5 min after the stress event and can last up to several hours after the event has ceased.

Studies with land and marine vertebrates have shown that acute stress responses can lead to a number of detrimental effects including poor body condition, poor immune function and disease resistance, decreased reproductive rates, and, in some animals, increased mortality rates (Romero and Butler 2007). Chronic (i.e., lasting days or longer) stress responses may become maladaptive if there is a prolonged activation of the stress response. For example, if animals are in a constant state of stress, particular behaviors such as the ability to find food, escape from predators, and socialize with conspecifics may be hindered (reviewed by Chrousos and Gold 1992).

Anthropogenic sources of underwater sound have the potential to cause a stress response in marine mammals. Cetaceans are subject to physiological challenges such as those associated with deep diving, prolonged fasting, thermoregulation, and osmoregulation. These processes are under endocrine control, and the breakdown of such systems may dramatically impact on the survival of an individual, especially one that lives near mammalian physiological limits (Wright et al. 2011). Acute or chronic stress in animals is quite difficult to measure given that there is potential stress associated with sampling (e.g., Ortiz and Worthy 2000; Lanyon et al. 2012). Normal diurnal (e.g., Suzuki et al. 2003) and seasonal fluctuations (e.g., Mashburn and Atkinson 2004; Myers et al. 2010) should also be taken into account. An increase in cortisol in the blood is commonly used as an indicator of stress. One of the few available studies on the physiological response to a sound stimulus involved a captive beluga whale and a captive bottlenose dolphin. Both were blood sampled before and after exposure to various levels of a seismic water gun as well as a pure tone resembling a sonar ping (Romano et al. 2004). Several physiological parameters were measured, indicating an increase in stress after exposure to high-level sounds. However, this type of study would be extremely difficult to carry out in the wild. Measuring levels of glucocorticoids from fecal (Rolland et al. 2012), blubber (Trana et al. 2015), or blow (Hogg et al. 2009) samples may be more practicable in wild animals; however, such studies carry other risks and uncertainties.

Other Physiological Effects

Sound exposure may also induce other physiological effects that are more subtle or hard to measure unless they are extensive enough to materialize in the form of increased levels of stress hormones or reduced fitness over long periods of time; it is possible that marine mammals may, in at least some cases, suffer from sound-induced neurological disorders that go undetected (Tougaard et al. 2015).

Beaked whales may be particularly susceptible to other physiological impacts. After a review of recent findings (e.g., Jepson et al. 2003; Fernandez et al. 2005) and of the anatomy and physiology of beaked whales (Rommel et al. 2006), Cox et al. (2006) suggested that rapid surfacing on sound exposure might cause gas-bubble disease in deep-diving beaked whales and explain the morbidity and mortality seen after sonar trials. Tyack et al. (2006) calculated that decompression problems are more likely to result from an abnormal behavioral response at the surface, such as repeated shallow dives, and ruled out a direct acoustic effect that triggers bubble growth. The mechanism(s) by which intense sound may lead to stranding and sometimes the death of beaked whales remains undetermined.

10.3.6 Changes in Health, Vital Rates, and Population Dynamics

Relating a change in physiology and/or behavior to a change in the animal's health (if chronic) or vital rates (if acute) is difficult and requires targeted work on the biological significance of the change. A short-term change in behavior or physiology may not necessarily be biologically significant, and therefore, it cannot be assumed that every change in behavior in response to an acoustic disturbance will lead to a change in an animal's health or vital rates. However, it cannot be assumed that because an animal or population shows little or no response, they are not vulnerable (Beale and Monaghan 2004). Even prolonged changes in behavior might not have long-term population impacts. Prolonged seismic surveys did not lead to permanent or broad-scale displacement of harbor porpoises into a suboptimal habitat (Thompson et al. 2013). However, steady increases in ambient shipping noise might have led to permanent changes in the vocalization parameters of right whales (Parks et al. 2007). The population dynamics of bottlenose dolphins were modeled and found to be unaffected by large increases in disturbance from vessels (New et al. 2013). Gray whales (*Eschrichtius robustus*) and killer whales returned after multiyear abandonment of their habitat due to anthropogenic disturbance (Bryant et al. 1984; Morton and Symonds 2002). Therefore, a species may be capable of short- and long-term modifications at the population level in response to changes in background noise conditions. Such long-term studies show that marine mammals have the ability to cope, to some extent, with changes in their acoustic environment. However, the question remains as to whether or not there is an upper limit to these changes as well as whether or not these changes have an associated cost.

It is easy to conceive different pathways from disturbance to population consequences through the PCoD model. Underwater sound might mask the song of whales, impacting mating success and ultimately population survival. Loud sound might cause TTS, putting animals at temporarily increased risk of ship strike or predation because they cannot detect the threat. Although these pathways are conceptually simple, determining the biological significance of the initial disturbance and quantifying the various transfer functions are extremely difficult.

The most tangible approach to populating the PCoD model is a bioenergetics pathway. The idea is that underwater sound disrupts foraging, leading to reduced energy intake and perhaps additional energy expenditure in avoidance, impacting maternal fitness and resulting in reduced birth rate and pup health, potentially leading to pup or adult death (Costa 2012). To fully parameterize the PCoD model, years of baseline data on foraging behavior, general health, and vital rates of individuals within that population as well as background information on the demographics and dynamics of the population are needed. Perhaps the only species for which a full PCoD model can be established at this stage is the elephant seal, for which good data on at-sea movement patterns, foraging behavior, reproductive biology, and demography are available, and the link between maternal mass and pup mass and survival is understood. In addition, this species is an ideal PCAD/PCoD candidate because all vital behaviors happen on land, with only foraging occurring at sea and hence being subject to disturbance by underwater sound (Costa et al. 2016).

Another data-rich species is the bottlenose dolphin, where some links between an acoustic stimulus and behavioral change, between health and vital rates, and between vital rates and population dynamics have been made (for a review, see New et al. 2013). The lack of data to parameterize the transfer functions for other species leads to the development of models that are based on expert opinion and simulated data. An agent-based model gives each "agent" (animal) various behavioral and/or physiological rules (including movement and dive parameters) based on a combination of observations and expert opinion. Simulations on how the agents respond to a sound stimulus are then carried out to assess the potential for impacts on the population (e.g., Nabe-Nielsen et al. 2014).

10.4 Mitigation

Reducing SELs is the most effective available means of reducing actual and potential impacts on both individuals and populations of marine mammals. Mechanisms to achieve this include reducing sound levels at the source, reducing sound propagation, or avoiding noisy activities at times and in places where sensitive species are present.

Figure 10.10 illustrates mitigation methods that involve the source (e.g., using the lowest practical power for all operations, vibratory pile driving, or alternative foundations like pile screwing instead of impact pile driving), additional sound level reduction gear installed near the source (e.g., bubble curtains or cofferdams around piles being driven), location/timing of operations (e.g., time/area closures), operational parameters (e.g., reducing ship speed and hence cavitation noise; soft start during seismic surveying and pile driving intended as a warning to marine mammals; this also includes acoustic deterrent devices), and mitigation procedures (e.g., the observation of a safety zone and reducing power or shutting down if animals

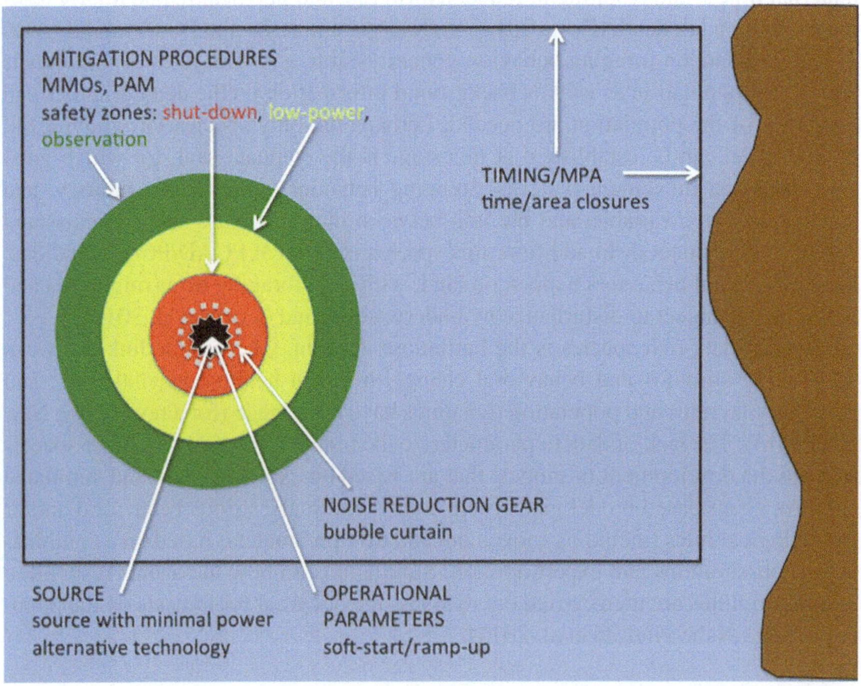

Fig. 10.10 Mitigation at the source (e.g., by using alternative, quieter technology or by modifying operational parameters), immediately near the source (e.g., by installing noise absorption gear), around the source (e.g., by using marine mammal observers [MMOs] or passive acoustic monitoring [PAM] to detect animals within certain safety zones), or over larger areas and times of year by establishing time/area closures (e.g., in marine protected areas [MPA])

enter the zone). Mitigation options for differing operations including seismic airgun surveys, naval sonar, pile driving, shipping, and explosions have been reviewed and their effectiveness and practicality have been discussed (Wright 2014). There are still many remaining questions regarding the effectiveness of the various mitigation methods.

The most commonly applied mitigation is the use of safety zones. During operations, these zones are monitored for animal presence, and if animals are sighted, often the operation switches to low power or shuts down to reduce injury to individuals. Safety zones are mostly monitored by marine mammal observers (MMOs) using binoculars. This is only practical in daylight and during good visibility. Sometimes passive acoustic monitoring is used, but it only works for vocalizing animals (Erbe 2013). Infrared, sonar, and other tools have been used to improve monitoring in certain circumstances. Common criticisms are that the size of safety zones is often determined by practicality and not (just) impact and the risk of not detecting animals. Wider impacts might happen at longer ranges and lower levels. Furthermore, these mitigation methods consider a single operation. Animals, however, are potentially exposed to multiple operations over considerable space and time. It is therefore difficult to assess, manage, and mitigate for these long-term, cumulative, and cross-border effects. A combination of wider marine spatial planning and effective mitigation measures around the source as well as collaboration among stakeholders and consistency in mitigation and regulation across jurisdictions and political borders is needed to achieve adequate management.

10.5 Regulation

Although research on underwater sound impacts on marine mammals has grown steadily over recent decades, there continue to be pressing data needs for conservation management. Furthermore, there is a significant delay in science transfer, meaning that guidance and policy lag behind the current state of scientific knowledge.

The existing legal mechanisms and guidance available to managers for reducing the impacts of individual sound sources have been reviewed (e.g., Weir and Dolman 2007; Dolman et al. 2009). Details of the limitations in existing management and mitigation, including their effectiveness, have been summarized for various jurisdictions (Parsons et al. 2009; Herschel et al. 2014).

The Joint Nature Conservation Committee (JNCC) first produced seismic guidelines in 1995. Thresholds and guidance were replicated, to various degrees, by numerous countries around the world. Guidance for a wider range of sound sources, including pile driving and explosives use, has since been developed (Joint Nature Conservation Committee 2010a, b). Currently, shipping remains unregulated with regard to sound pollution globally, but the International Maritime Organization (IMO) has issued voluntary guidelines for quieting underwater radiated sound from commercial ships (International Maritime Organization 2012). The "state of the art" in mitigation and monitoring has been described for seismic surveys (Nowacek et al. 2013).

The United States set the first thresholds for levels of sound beyond which marine mammals should not be exposed to prevent injury and disturbance under the US MMPA. The MMPA regulates Level A and B harassment (i.e., injury and disturbance respectively). Specifically, the 1994 amendments defined Level A harassment as “any act of pursuit, torment or annoyance which has the potential to injure a marine mammal or marine mammal stock in the wild” and Level B harassment as “any act of pursuit, torment, or annoyance which has the potential to disturb a marine mammal or marine mammal stock in the wild by causing disruption of behavioral patterns including, but not limited to, migration, breathing, nursing, breeding, feeding, or sheltering.”

The United States recently published an Ocean Noise Strategy (Gedamke et al. 2016) and a technical guidance providing thresholds for the onset of TTS and PTS (National Marine Fisheries Service 2016), which involved a complex review process and has taken a decade to complete. Thresholds for the onset of observable behavioral impacts have been slower, largely due to considerable variability and lack of supporting field data, although a requirement for criteria has been identified and a matrix framework that incorporates contextual factors by categorizing species, activities, and geographic areas to develop a series of step functions based on available literature documenting behavioral links was suggested (Fitch et al. 2011). This expert panel also stated that injury and behavioral harassment criteria neglect physiological stress, masking, and other factors (Fitch et al. 2011). The auditory impact criteria are now under review pursuant to Trump’s 2017 Executive Order (13795) entitled “America First Offshore Energy Strategy.”

Although the number and scale of field studies on underwater sound impacts have increased dramatically, policy is still based on studies with a few individuals of a few species, and management mostly addresses one event at a time. Mitigating immediate impacts on individuals is important, as is monitoring for long-term effects. Detecting any declines in populations, especially cryptic ones such as beaked whales, will require a large increase in monitoring effort and collaboration among countries and jurisdictions.

Conservation management is completely lacking throughout large parts of the world. Sound regulation in Antarctic and Arctic waters continues to be managed by individual nations and varies accordingly (Scott and Dolman 2006). Sound-related resolutions and statements of concern issued by various international bodies, such as the Convention on Migratory Species (CMS), have been reviewed elsewhere (Dolman et al. 2011; Simmonds et al. 2014).

The European Union (EU) first formally enshrined underwater sound in law for the determination of good environmental status (GES) under the Marine Strategy Framework Directive (2008/56/EC; Dekeling et al. 2016). Member states are required to monitor and may need to limit the amount of anthropogenic noise in European waters (van der Graaf et al. 2012). Two sound-related indicators are being defined under the Directive: one for intense sounds of short duration such as sonar, seismic surveys, and pile driving (Indicator 11.1.1) and one for low-frequency ambient noise associated primarily with shipping. No thresholds have been set and no impact indicator exists currently. Dekeling et al. (2013) outline monitoring guidance with respect to these MSFD indicators, including establishing registers of most intense sound sources and monitoring programs for ambient noise.

Fortunately, there seems to be a gradual shift from management that focuses on near-field source mitigation to prevent injury to wider, more holistic management that begins early in the planning process and is based on an effective reduction of a wider range of possible impacts. Improved early and transparent planning will help reduce the overlap between marine mammals and human activities. In addition to wide, often national-level spatial measures, habitat-based solutions such as marine protected areas can provide an effective method of reducing impacts in known areas of importance during sensitive periods (Dolman et al. 2009; Hoyt 2011). More holistic, habitat-based, multisectoral management also allows that cumulative stressors (acoustic and nonacoustic, e.g., bycatch, prey depletion, and contaminants) from different human activities be addressed. Regulators face the considerable challenge of managing these cumulative and interacting impacts with little scientific guidance.

A number of new tools are being proposed and developed to help assess the overall impact of multiple threat exposures. The United States has developed a product called CetSound (http://cetsound.noaa.gov/) to aid in the assessment and management of cumulative impacts. CetSound provides best available distribution and density maps for every cetacean species and maps of additional, biologically important areas for small resident populations and migratory species across the entire US territorial sea and exclusive economic zone. Through the CetMap process, the National Marine Fisheries Service is mapping sound levels from major chronic and intermittent sources across entire US waters.

In a pivotal case, the mass stranding of Cuvier's beaked whales was linked to naval sonar operations in the Bahamas. A prominent lawsuit followed in 2008, when a Los Angeles federal court ruled in favor of the defendant (Natural Resources Defense Council) that the US Navy should adopt specific safety measures during active sonar use to protect marine mammals (Zirbel et al. 2011). The mitigation measures included a ban on the use of sonar within 12 nautical miles of the California coast, shutdown when marine mammals entered within 2200 yards of the source, and power down during surface ducting conditions. The US Navy appealed, and the case ended up in the Supreme Court, where two of the six mitigation measures were overturned (Parsons et al. 2008). In September 2015, a US federal court settled a case that included, for the first time, spatial-temporal restrictions during active sonar and explosive use off Hawaii and California (Case No. 1:13-CV-00684-SOM-RLP).

10.6 Summary and Conclusions

Research is active on all aspects of marine mammal bioacoustics and sound impacts, including hearing and sound perception, sound production and call repertoires, behavioral responses to sound, masking, TTS, and stress. Studies are increasing in complexity, becoming multivariate, addressing complex questions in acoustic ecology, and considering cumulative exposures, potentially long-term impacts, and population consequences. As the complexity of studies grows, it is essential that researchers with diverse backgrounds collaborate. Studying marine mammals can be difficult, time consuming, and expensive, in particular in the wild. As a result, the sample size is often small, and variability and uncertainty are poorly understood. Pooling data from multiple studies is nearly impossible because of differences in measurement and analysis methodology, and reporting. Having agreed guidelines for best practice or standards would be invaluable but require dedicated effort and time to develop (Erbe et al. 2016b).

In behavioral-response experiments in the field, the experimental condition typically exposes animals to sound from an anthropogenic source, and in the control condition, animals are observed with the source present but off. The baseline study should observe the same animals in the absence of the source and its sound, assumed quiet. However, in the field, the baseline is hardly ever quiet. In many regions on Earth, the baseline and the control include ambient anthropogenic and nonanthropogenic noise. So really, these projects study the effect of additional anthropogenic noise to an already noisy ocean. How can one work out the "additional response" to the "additional noise"? This question was considered within the framework of the International Quiet Ocean Experiment (Boyd et al. 2011), which also included an interesting thought experiment: What if one could treat animals with silence? What if one could temporarily switch off all sound in a restricted habitat? What behavioral and other responses would be observed?

In order to "judge" animal responses to anthropogenic noise, it would be sensible to examine their responses to natural sounds (e.g., from wind or biological choruses). Such studies could put observed behaviors into "perspective." It is surprising how little attention this research question has received. Another important field of research is the effectiveness of common as well as novel mitigation methods.

The big questions remain. What are the population consequences of acoustic disturbance? How do impacts accumulate over multiple exposures as well as with acoustic and nonacoustic stressors? The PCoD model provides a framework within which these questions can begin to be addressed. A combination of long-term surveillance and well-replicated and controlled experiments, including behavioral-response studies, is needed to assess population-level effects with any confidence (Nowacek et al. 2007). Implementing solutions will require innovative approaches.

As legislation and public profile have become more focused on marine sound issues in recent years, our understanding of the range of potential impacts has advanced, monitoring technologies have become more sophisticated, mitigation methods have been developed, and research funding has increased. Although the translation of science into policy is slow and somewhat convoluted, all of these advances have directed the research focus, influenced policy, and, as a result, have improved our knowledge and management of marine sound pollution in marine mammal habitats.

Compliance with Ethics Requirements
Christine Erbe declares that she has no conflict of interest.
Rebecca Dunlop declares that she has no conflict of interest.
Sarah Dolman declares that she has no conflict of interest.

References

Allen, A. N., Schanze, J. J., Solow, A. R., & Tyack, P. L. (2014). Analysis of a Blainville's beaked whale's movement response to playback of killer whale vocalizations. *Marine Mammal Science,* 30(1), 154–168.

Andrew, R., Bruce, M. H., & James, A. M. (2002). Ocean ambient sound: Comparing the 1960s with the 1990s for a receiver off the California coast. *Acoustics Research Letters Online,* 3(2), 65–70.

Au, W. W. L. (1993). *The Sonar of Dolphins*. New York: Springer-Verlag.

Beale, C. M., & Monaghan, P. (2004). Behavioural responses to human disturbance: A matter of choice? *Animal Behaviour,* 68(5), 1065–1069.

Bee, M. A., & Micheyl, C. (2008). The cocktail party problem: What is it? How can it be solved? And why should animal behaviorists study it? *Journal of Comparative Psychology,* 122(3), 235–251.

Boyd, I. L., Frisk, G., Urban, E., Tyack, P., Ausubel, J., Seeyave, S., Cato, D., Southall, B., Weise, M., Andrew, R., Akamatsu, R., Dekeling, R., Erbe, C., Farmer, D. M., Gentry, R., Gross, T., Hawkins, A. D., Li, F. C., Metcalf, K., Miller, J. H., Moretti, D., Rodrigo, C., & Shinke, T. (2011). An International Quiet Ocean Experiment. *Oceanography,* 24(2), 174–181.

Branstetter, B. K., & Finneran, J. J. (2008). Comodulation masking release in bottlenose dolphins (*Tursiops truncatus*). *The Journal of the Acoustical Society of America,* 124(1), 625–633.

Bryant, P. J., Lafferty, C. M., & Lafferty, S. K. (1984). Reoccupation of Laguna Guerrero Negro, Baja California, Mexico, by gray whales. In M. L. Jones, S. L. Swartz, & S. Leatherwood (Eds.), *The Gray Whale: Eschrichtius robustus* (pp. 375–387). Orlando, FL: Academic Press.

Buckstaff, K. C. (2004). Effects of watercraft noise on the acoustic behavior of bottlenose dolphins, *Tursiops truncatus*, in Sarasota Bay, Florida. *Marine Mammal Science,* 20(4), 709–725.

Caldwell, M. C., & Caldwell, D. K. (1965). Individualized whistle contours in bottle-nosed dolphins (*Tursiops truncatus*). *Nature,* 207(4995), 434–435.

Campbell, D. T., & Stanley, J. C. (1966). *Experimental and Quasi-Experimental Designs for Research*. Chicago, IL: Rand McNally.

Castellote, M., Clark, C., & Lammers, M. (2012). Acoustic and behavioural changes by fin whales (*Balaenoptera physalus*) in response to shipping and airgun noise. *Biological Conservation,* 147(1), 115–122.

Cato, D. H. (1978). Marine biological choruses observed in tropical waters near Australia. *The Journal of the Acoustical Society of America,* 64(3), 736–743.

Cato, D. H. (2008). Ocean ambient noise: Its measurement and its significance to marine animals. Paper presented at the Conference on Underwater Noise Measurement, Impact and Mitigation, Southampton, UK, October 14-15, 2008. *Proceedings of the Institute of Acoustics*, 30, 1-9.

Cherry, E. C. (1953). Some experiments on the recognition of speech, with one and with two ears. *The Journal of the Acoustical Society of America,* 25, 975–979.

Chrousos, G. P., & Gold, P. W. (1992). The concepts of stress and stress system disorders. Overview of physical and behavioral homeostasis. *Journal of the American Medical Association, 267*(9), 1244–1252.

Clark, C. W., Ellison, W. T., Southall, B. L., Hatch, L., Van Parijs, S. M., Frankel, A., & Ponirakis, D. (2009). Acoustic masking in marine ecosystems: Intuitions, analysis, and implication. *Marine Ecology Progress Series, 395*, 201–222.

Costa, D. P. (2012). A bioenergetics approach to developing the PCAD model. In A. N. Popper & A. Hawkins (Eds.), *The Effects of Noise on Aquatic Life* (pp. 423–426). New York: Springer-Verlag.

Costa, D. P., Schwarz, L., Robinson, P., Schick, R. S., Morris, P. A., Condit, R., Crocker, D. E., & Kilpatrick, A. M. (2016). A bioenergetics approach to understanding the population consequences of disturbance: Elephant seals as a model system. In A. N. Popper & A. Hawkins (Eds.), *The Effects of Noise on Aquatic Life II* (pp. 161–169). New York: Springer-Verlag.

Cox, T. M., Ragen, T. J., Read, A. J., Vos, E., Baird, R. W., Balcomb, K., Barlow, J., Caldwell, J., Cranford, T., Crum, L., Amico, A. D., Spain, G. D., Fernandez, A., Finneran, J., Gentry, R., Gerth, W., Gulland, F., Hidebrand, J., Houser, D., Hullar, T., Jepson, P. D., Ketten, D., MacLeod, C. D., Miller, P., Moore, S., Mountain, D. C., Palka, D., Ponganis, P., Rommel, S., Rowles, T., Taylor, B., Tyack, P., Wartzok, D., Gisiner, R., Mead, J., & Benner, L. (2006). Understanding the impacts of anthropogenic sound on beaked whales. *Journal of Cetacean Research and Management, 7*(3), 177–187.

Deecke, V. B. (2006). Studying marine mammal cognition in the wild: A review of four decades of playback experiments. *Aquatic Mammals, 32*(4), 461–482.

Deecke, V. B., Slater, P. J. B., & Ford, J. K. B. (2002). Selective habituation shapes acoustic predator recognition in harbour seals. *Nature, 420*(6912), 171–173.

Dekeling, R., Tasker, M., Ainslie, M., Andersson, M., André, M., Borsani, F., Brensing, K., Castellote, M., Dalen, J., Folegot, T., van der Graaf, S., Leaper, R., Liebschner, A., Pajala, J., Robinson, S., Sigray, P., Sutton, G., Thomsen, F., Werner, S., Wittekind, D., & Young, J. V. (2016). The European Marine Strategy: Noise monitoring in European marine waters from 2014. In A. N. Popper & A. Hawkins (Eds.), *The Effects of Noise on Aquatic Life II* (pp. 205–215). New York: Springer-Verlag.

Dekeling, R. P. A., Tasker, M. L., van der Graaf, A. J., Ainslie, M. A., Andersson, M. H., André, M., Borsani, J. F., Brensing, K., Castellote, M., Cronin, D., Dalen, J., Folegot, T., Leaper, R., Pajala, J., Redman, P., Robinson, S. P., Sigray, P., Sutton, G., Thomsen, F., Werner, S., Wittekind, D., & Young, J. V. (2013). *Monitoring Guidance for Underwater Noise in European Seas: Background Information and Annexes*. Guidance Report, Second Report of the Technical Subgroup on Underwater Noise (TSG Noise), November 2013.

DeRuiter, S. L., Southall, B. L., Calambokidis, J., Zimmer, W. M. X., Sadykova, D., Falcone, E. A., Friedlaender, A. S., Joseph, J. E., Moretti, D., Schorr, G. S., Thomas, L., & Tyack, P. L. (2013). First direct measurements of behavioural responses by Cuvier's beaked whales to mid-frequency active sonar. *Biology Letters, 9*(4), 20130223. https://doi.org/10.1098/rsbl.2013.0223.

Dolman, S. J., Weir, C. R., & Jasny, M. (2009). Comparative review of marine mammal guidance implemented during naval exercises. *Marine Pollution Bulletin, 58*, 465–477.

Dolman, S. J., Evans, P. G. H., Notarbartolo-di-Sciara, G., & Frisch, H. (2011). Active sonar, beaked whales and European regional policy. *Marine Pollution Bulletin, 63*(1–4), 27–34.

Dooling, R. J., & Blumenrath, S. H. (2014). Avian sound perception in noise. In H. Brumm (Ed.), *Animal Communication in Noise* (pp. 229–250). Berlin: Springer-Verlag.

Dooling, R. J., West, E. W., & Leek, M. R. (2009). Conceptual and computational models of the effects of anthropogenic noise on birds. Paper presented at the 5th International Conference on Bioacoustics 2009, Holywell Park, UK, March 31 to April 2, 2009. *Proceedings of the Institute of Acoustics, 31*(1), 99–106.

Dunlop, R. A., Cato, D. H., & Noad, M. J. (2010). Your attention please: Increasing ambient noise levels elicits a change in communication behaviour in humpback whales (*Megaptera novaeangliae*). *Proceedings of the Royal Society B: Biological Sciences, 277*(1693), 2521–2529.

Dunlop, R. A., Noad, M. J., Cato, D. H., Kniest, E., Miller, P., Smith, J. N., & Stokes, D. M. (2013). Multivariate analysis of behavioural response experiments in humpback whales (*Megaptera novaeangliae*). *Journal of Experimental Biology, 216*, 759–770.

Dunlop, R. A., Cato, D. H., & Noad, M. J. (2014). Evidence of a Lombard response in migrating humpback whales (*Megaptera novaeangliae*). *The Journal of the Acoustical Society of America, 136*(1), 430–437.

Dunlop, R. A., Noad, M. J., McCauley, R. D., Kniest, E., Paton, D., & Cato, D. H. (2015). The behavioural response of humpback whales (*Megaptera novaeangliae*) to a 20 cubic inch air gun. *Aquatic Mammals, 41*(4), 412–433.

Dunlop, R. A., Noad, M. J., McCauley, R. D., Kniest, E., Slade, R., Paton, D., & Cato, D. H. (2016). Response of humpback whales (*Megaptera novaeangliae*) to ramp-up of a small experimental air gun array. *Marine Pollution Bulletin, 103*(1–2), 72–83.

Ellison, W., Southall, B., Clark, C., & Frankel, A. (2012). A new context-based approach to assess marine mammal behavioral responses to anthropogenic sounds. *Conservation Biology, 26*(1), 21–28.

Erbe, C. (2008). Critical ratios of beluga whales (*Delphinapterus leucas*) and masked signal duration. *The Journal of the Acoustical Society of America, 124*(4), 2216–2223.

Erbe, C. (2011). *Underwater Acoustics: Noise and the Effects on Marine Mammals, A Pocket Handbook*, 3rd ed. Brisbane, Australia: JASCO Applied Sciences. Available at http://oalib.hlsresearch.com/PocketBook 3rd ed.pdf.

Erbe, C. (2012). The effects of underwater noise on marine mammals. In A. N. Popper & A. Hawkins (Eds.), *The Effects of Noise on Aquatic Life* (pp. 17–22). New York: Springer-Verlag.

Erbe, C. (2013). Underwater passive acoustic monitoring and noise impacts on marine fauna—A workshop report. *Acoustics Australia, 41*(1), 113–119.

Erbe, C., & Farmer, D. M. (1998). Masked hearing thresholds of a beluga whale (*Delphinapterus leucas*) in icebreaker noise. *Deep Sea Research Part II: Topical Studies in Oceanography, 45*(7), 1373–1388.

Erbe, C., & Farmer, D. M. (2000). Zones of impact around icebreakers affecting beluga whales in the Beaufort Sea. *The Journal of the Acoustical Society of America, 108*(3), 1332–1340.

Erbe, C., Reichmuth, C., Cunningham, K. C., Lucke, K., & Dooling, R. J. (2016a). Communication masking in marine mammals: A review and research strategy. *Marine Pollution Bulletin, 103*, 15–38.

Erbe, C., Ainslie, M. A., de Jong, C. A. F., Racca, R., & Stocker, M. (2016b). The need for protocols and standards in research on underwater noise impacts on marine life. In A. N. Popper & A. Hawkins (Eds.), *The Effects of Noise on Aquatic Life II* (pp. 1265–1271). New York: Springer-Verlag.

Fernandez, A., Edwards, J. F., Rodriguez, F., de los Monteros, A. E., Herraez, P., Castro, P., Jaber, J. R., Martin, V., & Arbelo, M. (2005). "Gas and fat embolic syndrome" involving a mass stranding of beaked whales (family *Ziphiidae*) exposed to anthropogenic sonar signals. *Veterinary Pathology, 42*(4), 446–457.

Finley, K. J., Miller, G. W., Davis, R. A., & Greene, C. R. (1990). Reactions of belugas, *Delphinapterus leucas*, and narwhals, *Monodon monoceros*, to ice-breaking ships in the Canadian high arctic. *Canadian Bulletin of Fisheries and Aquatic Sciences, 224*, 97–117.

Finneran, J. J., & Schlundt, C. E. (2013). Effects of fatiguing tone frequency on temporary threshold shift in bottlenose dolphins (*Tursiops truncatus*). *The Journal of the Acoustical Society of America, 133*(3), 1819–1826.

Fitch, R., Harrison, J., & Lewandowski, J. (2011). *Marine Mammal and Sound Workshop July 13 and 14, 2010*. Report to the National Ocean Council Ocean Science and Technology Interagency Policy Committee, Washington, DC.

Fletcher, J. L., & Busnel, R.-G. (Eds.). (1978). *Effects of Noise on Wildlife*. New York: Academic Press.

Foote, A. D., Osborne, R. W., & Hoelzel, A. R. (2004). Environment: Whale-call response to masking boat noise. *Nature,* 428(6986), 910. https://doi.org/10.1038/428910a.

Ford, J. K. B. (1991). Vocal traditions among resident killer whales (*Orcinus orca*) in coastal waters of British Columbia. *Canadian Journal of Zoology,* 69(6), 1454–1483.

Fristrup, K. M., Hatch, L. T., & Clark, C. W. (2003). Variation in humpback whale (*Megaptera novaeangliae*) song length in relation to low-frequency sound broadcasts. *The Journal of the Acoustical Society of America,* 113(6), 3411–3424.

Gannon, D. P., Barros, N. B., Nowacek, D. P., Read, A. J., Waples, D. M., & Wells, R. S. (2005). Prey detection by bottlenose dolphins, *Tursiops truncatus*: An experimental test of the passive listening hypothesis. *Animal Behaviour,* 69, 709–720.

Gavrilov, A., & Li, B. (2007). *Antarctica as One of the Major Sources of Noise in the Ocean*. Paper presented at the 2nd International Conference and Exhibition, Underwater Acoustic Measurements: Technologies and Results, Heraklion, Crete, June 25-29, 2007.

Gedamke, J., Harrison, J., Hatch, L., Angliss, R., Barlow, J., Berchok, C., Caldow, C., Castellote, M., Cholewiak, D., DeAngelis, M. L., Dziak, R. P., Garland, E. C., Guan, S., Hastings, S., Holt, M. M., Laws, B., Mellinger, D. K., Moore, S. E., Moore, T. J., Oleson, E., Pearson-Meyer, J., Piniak, W., Redfern, J. V., Rowles, T., Scholik-Schlomer, A., Smith, A. B., Soldevilla, M., Stadler, J., Van Parijs, S. M., & Wahle, C. (2016). *Ocean Noise Strategy Roadmap*. Boston, MA: National Oceanic and Atmospheric Administration.

Gomez, C., Lawson, J., Wright, A. J., Buren, A., Tollit, D., & Lesage, V. (2016). A systematic review on the behavioural responses of wild marine mammals to noise: The disparity between science and policy. *Canadian Journal of Zoology,* 94(12), 801–819.

Guerra, M., Thode, A., Blackwell, S., & Macrander, M. (2011). Quantifying seismic survey reverberation off the Alaskan North Slope. *The Journal of the Acoustical Society of America,* 130(5), 3046–3058.

Hall, J. E. (2011). *Guyton and Hall Textbook of Medical Physiology*, 12th ed. Philadelphia, PA: Saunders/Elsevier.

Harwood, J., King, S., Schick, R., Donovan, C., & Booth, C. (2014). A protocol for implementing the interim population consequences of disturbance (PCoD) approach: Quantifying and assessing the effects of UK offshore renewable energy developments on marine mammal populations. Report Number SMRUL-TCE-2013-014, *Scottish Marine and Freshwater Science,* 5(2). Available at http://www.gov.scot/Resource/0044/00443360.pdf.

Herschel, A., Stephenson, S., Sparling, C., Sams, C., & Monnington, J. (2014). *ORJIP Project 4, Phase 1 Use of Deterrent Devices and Improvements to Standard Mitigation During Piling*. Research Summary 300100 (2013): S00, Offshore Renewables Joint Industry Programme (ORJIP).

Herzing, D. L. (1996). Vocalizations and associated underwater behavior of free-ranging Atlantic spotted dolphins, *Stenella frontalis* and bottlenose dolphins, *Tursiops truncatus*. *Aquatic Mammals,* 22(2), 61–79.

Hogg, C. J., Rogers, T. L., Shorter, A., Barton, K., Miller, P. J. O., & Nowacek, D. P. (2009). Determination of steroid hormones in whale blow: It is possible. *Marine Mammal Science,* 25(3), 605–618.

Holt, M. M., & Schusterman, R. J. (2007). Spatial release from masking of aerial tones in pinnipeds. *The Journal of the Acoustical Society of America,* 121(2), 1219–1225.

Holt, M. M., Noren, D. P., Veirs, V., Emmons, C. K., & Veirs, S. (2009). Speaking up: Killer whales (*Orcinus orca*) increase their call amplitude in response to vessel noise. *The Journal of the Acoustical Society of America,* 125(1), EL27-EL32.

Home, E. (1812). An account of some peculiarities in the structure of the organ of hearing in the *Balaena mysticetus* of *Linnaeus*. *Philosophical Transactions of the Royal Society of London,* 102(1), 83–89.

Hoyt, E. (2011). *Marine Protected Areas for Whales, Dolphins and Porpoises*, 2nd ed. London: Earthscan.

International Maritime Organization (IMO). (2012). *Provisions for Reduction of Noise from Commercial Shipping and Its Adverse Impacts on Marine Life*. Document DE 57/17, IMO Sub-Committee on Ship Design and Equipment.

Jensen, F. B., Kuperman, W. A., Porter, M. B., & Schmidt, H. (2011). *Computational Ocean Acoustics*, 2nd ed. New York: Springer-Verlag.

Jepson, P. D., Arbelo, M., Deaville, R., Patterson, I. A. P., Castro, P., Baker, J. R., Degollada, E., Ross, H. M., Herraez, P., Pocknell, A. M., Rodriguez, F., Howie, F. E., Espinosa, A., Reid, R. J., Jaber, J. R., Martin, V., Cunningham, A. A., & Fernandez, A. (2003). Gas-bubble lesions in stranded cetaceans. *Nature,* 425(6958), 575–576. https://doi.org/10.1038/425575a.

Johnson, M. P., & Tyack, P. L. (2003). A digital acoustic recording tag for measuring the response of wild marine mammals to sound. *IEEE Journal of Oceanic Engineering,* 28(1), 3–12.

Joint Nature Conservation Committee (JNCC). (2010a). *JNCC Guidelines for Minimising the Risk of Injury and Disturbance to Marine Mammals from Seismic Surveys*. Aberdeen, UK: Joint Nature Conservation Committee.

Joint Nature Conservation Committee (JNCC). (2010b). *Statutory Nature Conservation Agency Protocol for Minimising the Risk of Injury to Marine Mammals from Piling Noise*. Aberdeen, UK: Joint Nature Conservation Committee.

Kastak, D., Holt, M. M., Mulsow, J., Kastak, C. J. R., Schusterman, R. J., & Southall, B. L. (2006). Towards a predictive model of noise-induced temporary threshold shift for an amphibious marine mammal, the California sea lion (*Zalophus californianus*). *The Journal of the Acoustical Society of America,* 120(5), 3226.

Kastak, D., Mulsow, J., Ghoul, A., & Reichmuth, C. (2008). Noise-induced permanent threshold shift in a harbor seal. *The Journal of the Acoustical Society of America,* 123(5), 2986.

Kastelein, R. A., Gransier, R., & Hoek, L. (2013). Comparative temporary threshold shifts in a harbor porpoise and harbor seal, and severe shift in a seal. *The Journal of the Acoustical Society of America,* 134(1), 13–16.

Kastelein, R. A., Schop, J., Gransier, R., & Hoek, L. (2014a). Frequency of greatest temporary hearing threshold shift in harbor porpoises (*Phocoena phocoena*) depends on the noise level. *The Journal of the Acoustical Society of America,* 136(3), 1410–1418.

Kastelein, R. A., Hoek, L., Gransier, R., Rambags, M., & Claeys, N. (2014b). Effect of level, duration, and inter-pulse interval of 1-2 kHz sonar signal exposures on harbor porpoise hearing. *The Journal of the Acoustical Society of America,* 136(1), 412–422.

Kujawa, S. G., & Liberman, M. C. (2009). Adding insult to injury: Cochlear nerve degeneration after "temporary" noise-induced hearing loss. *The Journal of Neuroscience,* 29(45), 14077–14085.

Lanyon, J. M., Sneath, H. L., & Long, T. (2012). Evaluation of exertion and capture stress in serum of wild dugongs (*Dugong dugon*). *Journal of Zoo and Wildlife Medicine,* 43(1), 20–32.

Lesage, V., Barrette, C., Kingsley, M. C. S., & Sjare, B. (1998). The effect of vessel noise on the vocal behavior of belugas in the St. Lawrence River estuary, Canada. *Marine Mammal Science,* 15(1), 65–84.

Liberman, M. C. (2016). Noise-induced hearing loss: Permanent versus temporary threshold shifts and the effects of hair cell versus neuronal degeneration. In A. N. Popper & A. Hawkins (Eds.), *The Effects of Noise on Aquatic Life II* (pp. 1–7). New York: Springer-Verlag.

Lombard, É. (1911). Le signe de l'élévation de la voix. *Annales des Maladies de L'Oreille et du Larynx,* XXXVII(2), 101-109.

Mann, D. A., Cook-Hill, M., Manire, C., Greenhow, D., Montie, E., Powell, J., Wells, R., Bauer, G., Cunningham-Smith, P., Lingenfelser, R., Digiovanni Jr., R., Stone, A., Brodsky, M., Stevens, R., Kieffer, G., & Hoetjes, P. (2010). Hearing loss in stranded odontocete dolphins and whales. *PLoS ONE,* 5(11), e13824.

Mashburn, K. L., & Atkinson, S. (2004). Evaluation of adrenal function in serum and feces of Steller sea lions (*Eumetopias jubatus*): Influences of molt, gender, sample storage, and age on glucocorticoid metabolism. *General and Comparative Endocrinology,* 136(3), 371–381.

Miller, P. J. O., Biassoni, N., Samuels, A., & Tyack, P. L. (2000). Whale songs lengthen in response to sonar. *Nature,* 405(6789), 903.

Miller, P. J. O., Johnson, M. P., Madsen, P. T., Biassoni, N., Quero, M., & Tyack, P. L. (2009). Using at-sea experiments to study the effects of airguns on the foraging behavior of sperm whales in the Gulf of Mexico. *Deep Sea Research Part I: Oceanographic Research Papers,* 56(7), 1168–1181.

Miller, P. J. O., Antunes, R. N., Wensveen, P. J., Samarra, F. I. P., Alves, A. C., Tyack, P. L., Kvadsheim, P. H., Kleivane, L., Lam, F.-P. A., Ainslie, M. A., & Thomas, L. (2014). Dose-response relationships for the onset of avoidance of sonar by free-ranging killer whales. *The Journal of the Acoustical Society of America,* 135(1), 975–993.

Morton, A. B., & Symonds, H. K. (2002). Displacement of *Orcinus orca* (L.) by high amplitude sound in British Columbia, Canada. *ICES Journal of Marine Science,* 59(1), 71–80.

Myers, M. J., Litz, B., & Atkinson, S. (2010). The effects of age, sex, season and geographic region on circulating serum cortisol concentrations in threatened and endangered Steller sea lions (*Eumetopias jubatus*). *General and Comparative Endocrinology,* 165(1), 72–77.

Nabe-Nielsen, J., Sibly, R. M., Tougaard, J., Teilmann, J., & Sveegaard, S. (2014). Effects of noise and by-catch on a Danish harbour porpoise population. *Ecological Modelling,* 272, 242–251.

Nachtigall, P. E., & Supin, A. Y. (2013). A false killer whale reduces its hearing sensitivity when a loud sound is preceded by a warning. *Journal of Experimental Biology,* 216(12), 3062–3070.

Nachtigall, P. E., & Supin, A. Y. (2014). Conditioned hearing sensitivity reduction in a bottlenose dolphin (*Tursiops truncatus*). *Journal of Experimental Biology,* 217(16), 2806–2813.

National Marine Fisheries Service. (2016). *Technical Guidance for Assessing the Effects of Anthropogenic Sound on Marine Mammal Hearing: Underwater Acoustic Thresholds for Onset of Permanent and Temporary Threshold Shifts.* Report by the National Oceanic and Atmospheric Administration (NOAA) and US Department of Commerce (DOC), Silver Spring, MD.

National Research Council. (1994). *Low-Frequency Sound and Marine Mammals: Current Knowledge and Research Needs.* Washington, DC: National Academies Press.

National Research Council. (2000). *Marine Mammals and Low-Frequency Sound: Progress Since 1994.* Washington, DC: National Academies Press.

National Research Council. (2005). *Marine Mammal Populations and Ocean Noise: Determining When Noise Causes Biologically Significant Effects.* Washington, DC: National Academies Press.

New, L. F., Harwood, J., Thomas, L., Donovan, C., Clark, J. S., Hastie, G., & Lusseau, D. (2013). Modelling the biological significance of behavioural change in coastal bottlenose dolphins in response to disturbance. *Functional Ecology,* 27, 314–322.

New, L. F., Clark, J. S., Costa, D. P., Fleishman, E., Hindell, M. A., Klanjcek, T., Lusseau, D., Kraus, S., McMahon, C. R., Robinson, P. W., Schick, R. S., Schwarz, L. K., Simmons, S. E., Thomas, L., Tyack, P. L., & Harwood, J. (2014). Using short-term measures of behaviour to estimate long-term fitness of southern elephant seals. *Marine Ecology Progress Series,* 496, 99–108.

Norris, K., & Harvey, G. (1974). Sound transmission in the porpoise head. *The Journal of the Acoustical Society of America,* 56(2), 659–664.

Nowacek, D. P., Thorne, L. H., Johnston, D. W., & Tyack, P. L. (2007). Responses of cetaceans to anthropogenic noise. *Mammal Review,* 37(2), 81–115.

Nowacek, D. P., Bröker, K., Donovan, G., Gailey, G., Racca, R., & Reeves, R. R. (2013). Responsible practices for minimizing and monitoring environmental impacts of marine seismic surveys with an emphasis on marine mammals. *Aquatic Mammals,* 39(4), 356–377.

Ortiz, R. M., & Worthy, G. A. J. (2000). Effects of capture on adrenal steroid and vasopressin concentrations in free-ranging bottlenose dolphins (*Tursiops truncatus*). *Comparative Biochemistry and Physiology Part A: Molecular & Integrative Physiology,* 125(3), 317–324.

Parks, S. E., Clark, C. W., & Tyack, P. L. (2007). Short- and long-term changes in right whale calling behavior: The potential effects of noise on acoustic communication. *The Journal of the Acoustical Society of America,* 122(6), 3725–3731.

Parsons, E. C. M., Dolman, S. J., Wright, A. J., Rose, N. A., & Burns, W. C. G. (2008). Navy sonar and cetaceans: Just how much does the gun need to smoke before we act? *Marine Pollution Bulletin,* 56(7), 1248–1257.

Parsons, E. C. M., Dolman, S. J., Jasny, M., Rose, N. A., Simmonds, M. P., & Wright, A. J. (2009). A critique of the UK's JNCC seismic survey guidelines for minimising acoustic disturbance to marine mammals: Best practise? *Marine Pollution Bulletin,* 58(5), 643–651.

Payne, R., & Webb, D. (1971). Orientation by means of long range acoustic signaling in baleen whales. *Annals of the New York Academy of Sciences,* 188, 110–141.

Payne, R. S., & McVay, S. (1971). Songs of humpback whales. *Science,* 173(3997), 585–597.

Rendell, L. E., & Gordon, J. C. D. (1999). Vocal response of long-finned pilot whales (*Globicephala melas*) to military sonar in the Ligurian Sea. *Marine Mammal Science,* 15(1), 198–204.

Richardson, W. J., Greene, C. R., Malme, C. I., & Thomson, D. H. (1995). *Marine Mammals and Noise*. San Diego, CA: Academic Press.

Rolland, R. M., Parks, S. E., Hunt, K. E., Castellote, M., Corkeron, P. J., Nowacek, D. P., Wasser, S. K., & Kraus, S. D. (2012). Evidence that ship noise increases stress in right whales. *Proceedings of the Royal Society B: Biological Sciences,* 279(1737), 2363–2368.

Roman, J., Altman, I., Dunphy-Daly, M. M., Campbell, C., Jasny, M., & Read, A. J. (2013). The Marine Mammal Protection Act at 40: Status, recovery, and future of US marine mammals. *Annals of the New York Academy of Sciences,* 1286, 29–49.

Romano, T. A., Keogh, M. J., Kelly, C., Feng, P., Berk, L., Schlundt, C. E., Carder, D. A., & Finneran, J. J. (2004). Anthropogenic sound and marine mammal health: measures of the nervous and immune systems before and after intense sound exposure. *Canadian Journal of Fisheries and Aquatic Sciences,* 61(7), 1124–1134.

Romero, M. L., & Butler, L. K. (2007). Endocrinology of stress. *International Journal of Comparative Psychology,* 20(2), 89–95.

Rommel, S. A., Costidid, A. M., Fernández, A., Jepson, P. D., Pabst, D. A., McLellan, W. W., Houser, D. S., Cranford, T. W., Van Helden, A. L., Allen, D. M., & Barros, N. B. (2006). Elements of beaked whale anatomy and diving physiology and some hypothetical causes of sonar-related stranding. *Journal of Cetacean Research and Management,* 7, 189–209.

Salter, R. E. (1979). Site utilisation, activity budgets, and disturbance responses of Atlantic walruses during terrestrial haul-out. *Canadian Journal of Zoology,* 57(6), 1169–1180.

Scheifele, P. M., Andrew, S., Cooper, R. A., Darre, M., Musiek, F. E., & Max, L. (2005). Indication of a Lombard vocal response in the St. Lawrence River beluga. *The Journal of the Acoustical Society of America,* 117(3), 1486–1492.

Schorr, G. S., Falcone, E. A., Moretti, D. J., & Andrews, R. D. (2014). First long-term behavioral records from Cuvier's beaked whales (*Ziphius cavirostris*) reveal record-breaking dives. *PLoS ONE,* 9(3), e92633.

Scott, K., & Dolman, S. J. (2006). *Current Noise Pollution Issues*. Paper presented at the 25th Meeting of the Convention for the Conservation of Antarctic and Marine Living Resources (CCAMLR), Hobart, Australia, October 23 to November 3, 2006.

Simmonds, M. P., Dolman, S. J., Jasny, M., Parsons, E. C. M., Weilgart, L., Wright, A. J., & Leaper, R. (2014). Marine noise pollution-Increasing recognition but need for more practical action. *Journal of Ocean Technology,* 9, 71–90.

Southall, B. L., Bowles, A. E., Ellison, W. T., Finneran, J. J., Gentry, R. L., Greene, C. R., Jr., Kastak, D., Ketten, D. R., Miller, J. H., Nachtigall, P. E., Richardson, W. J., Thomas, J. A., & Tyack, P. L. (2007). Marine mammal noise exposure criteria: Initial scientific recommendations. *Aquatic Mammals,* 33(4), 411–521.

Suzuki, M., Uchida, S., Ueda, K., Tobayama, T., Katsumata, E., Yoshioka, M., & Aida, K. (2003). Diurnal and annual changes in serum cortisol concentrations in Indo-Pacific bottlenose dolphins *Tursiops aduncus* and killer whales *Orcinus orca*. *General and Comparative Endocrinology, 132*(3), 427–433.

Thompson, P. M., Brookes, K. L., Graham, I. M., Barton, T. R., Needham, K., Bradbury, G., & Merchant, N. D. (2013). Short-term disturbance by a commercial two-dimensional seismic survey does not lead to long-term displacement of harbour porpoises. *Proceedings of the Royal Society B: Biological Sciences, 280*(1771), 20132001. https://doi.org/10.1098/rspb.2013.2001.

Tougaard, J., Wright, A. J., & Madsen, P. T. (2015). Cetacean noise criteria revisited in the light of proposed exposure limits for harbour porpoises. *Marine Pollution Bulletin, 90*(1–2), 196–208.

Trana, M. R., Roth, J. D., Tomy, G. T., Anderson, W. G., & Ferguson, S. H. (2015). Influence of sample degradation and tissue depth on blubber cortisol in beluga whales. *Journal of Experimental Marine Biology and Ecology, 462*, 8–13.

Turnbull, S. D. (1994). Changes in masked thresholds of a harbor seal *Phoca vitulina* associated with angular separation of signal and noise sources. *Canadian Journal of Zoology, 72*, 1863–1866.

Tyack, P. L. (2000). Functional aspects of cetacean communication. In J. Mann, R. C. Connor, P. L. Tyack, & H. Whitehead (Eds.), *Cetacean Societies: Field Studies of Dolphins and Whales* (pp. 270–307). Chicago, IL: The University of Chicago Press.

Tyack, P. L., Johnson, M., Soto, N. A., Sturlese, A., & Madsen, P. T. (2006). Extreme diving of beaked whales. *Journal of Experimental Biology, 209*(21), 4238–4253.

Tyack, P. L., Zimmer, W. M. X., Moretti, D., Southall, B. L., Claridge, D. E., Durban, J. W., Clark, C. W., D'Amico, A., DiMarzio, N., Jarvis, S., McCarthy, E., Morrissey, R., Ward, J., & Boyd, I. L. (2011). Beaked whales respond to simulated and actual navy sonar. *PLoS ONE, 6*(3), e17009.

US Department of the Navy. (2008). *Atlantic Fleet Active Sonar Training Environmental Impact Statement/Overseas Environmental Impact Statement*. Norfolk, VA: Naval Facilities Engineering Command Atlantic.

US Department of the Navy. (2009). *Gulf of Alaska Navy Training Activities Draft Environmental Impact Statement/Overseas Environmental Impact Statement*. Silverdale, WA: Naval Facilities Engineering Command Northwest.

van Bree, P. J. H., & Kristensen, I. (1974). On the intriguing stranding of four Cuvier's beaked whales, *Ziphius cavirostris* G. Cuvier, 1823, on the Lesser Antillean island of Bonaire. *Bijdragen tot de Dierkunde, 44*, 235–238.

van der Graaf, A. J., Ainslie, M. A., Andre, M., Brensing, K., Dalen, J., Dekeling, R. P. A., Robinson, S. M., Tasker, M. L., Thomsen, F., & Werner, S. (2012). *European Marine Strategy Framework Directive--Good Environmental Status (MSFD GES)*. Report of the Technical Subgroup on Underwater Noise and other forms of energy, TSG Noise & Milieu Ltd., Brussels.

Weir, C. R., & Dolman, S. J. (2007). Comparative review of the Regional Marine Mammal Mitigation Guidelines implemented during industrial seismic surveys, and guidance towards a worldwide standard. *Journal of International Wildlife Law and Policy, 10*, 1–27.

Wenz, G. M. (1962). Acoustic ambient noise in the ocean: spectra and sources. *The Journal of the Acoustical Society of America, 34*(12), 1936–1956.

Wenz, G. M. (1969). Low-frequency deep-water ambient noise along the Pacific Coast of the United States. *US Navy Journal of Underwater Acoustics, 19*, 423–444.

Williams, R., Wright, A. J., Ashe, E., Blight, L. K., Bruintjes, R., Canessa, R., Clark, C. W., Cullis-Suzuki, S., Dakin, D. T., Erbe, C., Hammond, P. S., Merchant, N. D., O'Hara, P. D., Purser, J., Radford, A. N., Simpson, S. D., Thomas, L., & Wale, M. A. (2015). Impacts of anthropogenic noise on marine life: Publication patterns, new discoveries, and future directions in research and management. *Ocean & Coastal Management, 115*, 17–24.

Wright, A. J. (2014). Reducing impacts of human ocean noise on cetaceans: Knowledge gap analysis and recommendations. Gland, Switzerland: WWF International.

Wright, A. J., Soto, N. A., Baldwin, A. L., Bateson, M., Beale, C. M., Clark, C., Deak, T., Edwards, E. F., Fernandez, A., Godinho, A., Hatch, L. T., Kakuschke, A., Lusseau, D., Martineau, D., Romero, L. M., Weilgart, L. S., Wintle, B. A., Notarbartolo-di-Sciara, G., & Vidal, M. (2007). Do marine mammals experience stress related to anthropogenic noise? *International Journal of Comparative Psychology,* 20, 274–316.

Wright, A. J., Deak, T., & Parsons, E. C. M. (2011). Size matters: Management of stress responses and chronic stress in beaked whales and other marine mammals may require larger exclusion zones. *Marine Pollution Bulletin,* 63(1–4), 5–9.

Wyatt, R. (2008). *Review of Existing Data on Underwater Sounds Produced by the Oil and Gas Industry*. Report Submitted to the Joint Industry Programme on Sound and Marine Life, London.

Zirbel, K., Balint, P., & Parsons, E. C. M. (2011). Navy sonar, cetaceans and the US Supreme Court: A review of cetacean mitigation and litigation in the US. *Marine Pollution Bulletin,* 63(1–4), 40–48.

The manufacturer's authorised representative in the EU is Springer Nature Customer Service Centre GmbH, Europaplatz 3, 69115 Heidelberg, Germany. If you have any concerns regarding our products, please contact ProductSafety@springernature.com

Printed and bound by CPI Group (UK) Ltd, Croydon, CR0 4YY
10/10/2026
02231220-0003